P9-CFW-654

DNA Topoisomerase Protocols
Part II: Enzymology and Drugs

METHODS IN MOLECULAR BIOLOGY™

John M. Walker, SERIES EDITOR

171. **Proteoglycan Protocols,** edited by *Renato V. Iozzo, 2001*
170. **DNA Arrays:** *Methods and Protocols,* edited by *Jang B. Rampal, 2001*
169. **Neurotrophin Protocols,** edited by *Robert A. Rush, 2001*
168. **Protein Structure, Stability, and Folding,** edited by *Kenneth P. Murphy, 2001*
167. **DNA Sequencing Protocols,** *Second Edition,* edited by *Colin A. Graham and Alison J. M. Hill, 2001*
166. **Immunotoxin Methods and Protocols**, edited by *Walter A. Hall, 2001*
165. **SV40 Protocols,** edited by *Leda Raptis, 2001*
164. **Kinesin Protocols,** edited by *Isabelle Vernos, 2001*
163. **Capillary Electrophoresis of Nucleic Acids, Volume 2:** *Practical Applications of Capillary Electrophoresis,* edited by *Keith R. Mitchelson and Jing Cheng, 2001*
162. **Capillary Electrophoresis of Nucleic Acids, Volume 1:** *The Capillary Electrophoresis System as an Analytical Tool,* edited by *Keith R. Mitchelson and Jing Cheng, 2001*
161. **Cytoskeleton Methods and Protocols**, edited by *Ray H. Gavin, 2001*
160. **Nuclease Methods and Protocols,** edited by *Catherine H. Schein, 2000*
159. **Amino Acid Analysis Protocols,** edited by *Catherine Cooper, Nicole Packer, and Keith Williams, 2000*
158. **Gene Knockoout Protocols,** edited by *Martin J. Tymms and Ismail Kola, 2000*
157. **Mycotoxin Protocols,** edited by *Mary W. Trucksess and Albert E. Pohland, 2000*
156. **Antigen Processing and Presentation Protocols,** edited by *Joyce C. Solheim, 2000*
155. **Adipose Tissue Protocols,** edited by *Gérard Ailhaud, 2000*
154. **Connexin Methods and Protocols,** edited by *Roberto Bruzzone and Christian Giaume, 2000*
153. **Neuropeptide Y Protocols**, edited by *Ambikaipakan Balasubramaniam, 2000*
152. **DNA Repair Protocols:** *Prokaryotic Systems*, edited by *Patrick Vaughan, 2000*
151. **Matrix Metalloproteinase Protocols,** edited by *Ian M. Clark, 2000*
150. **Complement Methods and Protocols,** edited by *B. Paul Morgan, 2000*
149. **The ELISA Guidebook**, edited by *John R. Crowther, 2000*
148. **DNA–Protein Interactions:** *Principles and Protocols* **(2nd ed.)**, edited by *Tom Moss, 2000*
147. **Affinity Chromatography:** *Methods and Protocols*, edited by *Pascal Bailon, George K. Ehrlich, Wen-Jian Fung, and Wolfgang Berthold, 2000*
146. **Mass Spectrometry of Proteins and Peptides,** edited by *John R. Chapman, 2000*
145. **Bacterial Toxins:** ***Methods and Protocols***, edited by *Otto Holst, 2000*
144. **Calpain Methods and Protocols**, edited by *John S. Elce, 2000*
143. **Protein Structure Prediction:** *Methods and Protocols*, edited by *David Webster, 2000*
142. **Transforming Growth Factor-Beta Protocols**, edited by *Philip H. Howe, 2000*
141. **Plant Hormone Protocols**, edited by *Gregory A. Tucker and Jeremy A. Roberts, 2000*
140. **Chaperonin Protocols**, edited by *Christine Schneider, 2000*
139. **Extracellular Matrix Protocols**, edited by *Charles Streuli and Michael Grant, 2000*
138. **Chemokine Protocols**, edited by *Amanda E. I. Proudfoot, Timothy N. C. Wells, and Christine Power, 2000*
137. **Developmental Biology Protocols, Volume III**, edited by *Rocky S. Tuan and Cecilia W. Lo, 2000*
136. **Developmental Biology Protocols, Volume II,** edited by *Rocky S. Tuan and Cecilia W. Lo, 2000*
135. **Developmental Biology Protocols, Volume I,** edited by *Rocky S. Tuan and Cecilia W. Lo, 2000*
134. **T Cell Protocols:** *Development and Activation*, edited by *Kelly P. Kearse, 2000*
133. **Gene Targeting Protocols,** edited by *Eric B. Kmiec, 2000*
132. **Bioinformatics Methods and Protocols,** edited by *Stephen Misener and Stephen A. Krawetz, 2000*
131. **Flavoprotein Protocols,** edited by *S. K. Chapman and G. A. Reid, 1999*
130. **Transcription Factor Protocols,** edited by *Martin J. Tymms, 2000*
129. **Integrin Protocols,** edited by *Anthony Howlett, 1999*
128. **NMDA Protocols,** edited by *Min Li, 1999*
127. **Molecular Methods in Developmental Biology:** Xenopus *and Zebrafish*, edited by *Matthew Guille, 1999*
126. **Adrenergic Receptor Protocols,** edited by *Curtis A. Machida, 2000*
125. **Glycoprotein Methods and Protocols:** *The Mucins,* edited by *Anthony P. Corfield, 2000*
124. **Protein Kinase Protocols,** edited by *Alastair D. Reith, 2000*
123. ***In Situ*** **Hybridization Protocols (2nd ed.),** edited by *Ian A. Darby, 2000*
122. **Confocal Microscopy Methods and Protocols,** edited by *Stephen W. Paddock, 1999*
121. **Natural Killer Cell Protocols:** *Cellular and Molecular Methods,* edited by *Kerry S. Campbell and Marco Colonna, 2000*
120. **Eicosanoid Protocols,** edited by *Elias A. Lianos, 1999*
119. **Chromatin Protocols,** edited by *Peter B. Becker, 1999*
118. **RNA–Protein Interaction Protocols,** edited by *Susan R. Haynes, 1999*
117. **Electron Microscopy Methods and Protocols,** edited by *M. A. Nasser Hajibagheri, 1999*
116. **Protein Lipidation Protocols,** edited by *Michael H. Gelb, 1999*
115. **Immunocytochemical Methods and Protocols (2nd ed.),** edited by *Lorette C. Javois, 1999*
114. **Calcium Signaling Protocols,** edited by *David G. Lambert, 1999*
113. **DNA Repair Protocols:** *Eukaryotic Systems,* edited by *Daryl S. Henderson, 1999*
112. **2-D Proteome Analysis Protocols,** edited by *Andrew J. Link, 1999*
111. **Plant Cell Culture Protocols,** edited by *Robert D. Hall, 1999*
110. **Lipoprotein Protocols,** edited by *Jose M. Ordovas, 1998*
109. **Lipase and Phospholipase Protocols,** edited by *Mark H. Doolittle and Karen Reue, 1999*
108. **Free Radical and Antioxidant Protocols,** edited by *Donald Armstrong, 1998*

METHODS IN MOLECULAR BIOLOGY™

DNA Topoisomerase Protocols

Part II: Enzymology and Drugs

Edited by

Neil Osheroff

Vanderbilt University School of Medicine, Nashville, TN

and

Mary-Ann Bjornsti

St. Jude Children's Research Hospital, Memphis, TN

Humana Press Totowa, New Jersey

999 Riverview Drive, Suite 208
Totowa, New Jersey 07512

This publication is printed on acid-free paper. ∞
ANSI Z39.48-1984 (American Standards Institute) Permanence of Paper for Printed Library Materials.

Cover design by Patricia F. Cleary.

For additional copies, pricing for bulk purchases, and/or information about other Humana titles, contact Humana at the above address or at any of the following numbers: Tel.: 973-256-1699; Fax: 973-256-8341; E-mail: humana@humanapr.com; Website: http://humanapress.com

Printed in the United States of America. 10 9 8 7 6 5 4 3 2 1

Library of Congress Cataloging in Publication Data

Main entry under title:

Methods in molecular biology™.

DNA Topoisomerase Protocols / edited by Mary-Ann Bjornsti and Neil Osheroff.
v. <2 >; cm.—(Methods in molecular biology™ ; vols. 94, 95)
Includes bibliographical references and index.
Contents: pt. 2. Enzymology and Drugs.
ISBN 0-89603-444-5 (alk. paper); ISBN 0-89603-512-3 (alk. paper)
1. DNA—Structure—Laboratory manuals. 2. DNA—Conformation—Laboratory manuals. 3. DNA topoisomerases. I. Bjornsti, Mary-Ann. II. Osheroff, Neil. III. Series: Methods in molecular biology™ (Totowa, N.J.); 95.
QP624.5.S78P76 2001 vol. 2 99-13390
572.8'663—dc21 CIP

Preface

Beginning with the *Escherichia coli* ω protein, or bacterial DNA topoisomerase I, an ever-increasing number of enzymes have been identified that catalyze changes in the linkage of DNA strands. DNA topoisomerases are ubiquitous in nature and have been shown to play critical roles in most processes involving DNA, including DNA replication, transcription, and recombination. These enzymes further constitute the cellular targets of a number of clinically important antibacterial and anticancer agents. Thus, further studies of DNA topology and DNA topoisomerases are critical to advance our understanding of the basic biological processes required for cell cycle progression, cell division, genomic stability, and development. In addition, these studies will continue to provide critical insights into the cytotoxic action of drugs that target DNA topoisomerases. Such mechanistic studies have already played an important role in the development and clinical application of antimicrobial and chemotherapeutic agents.

The two volumes of *DNA Topoisomerase Protocols* are designed to help new and established researchers investigate all aspects of DNA topology and the function of these enzymes. The chapters are written by prominent investigators in the field and provide detailed background information and step-by-step experimental protocols. The topics covered in *Part I: DNA Topology and Enzymes* range from detailed methods to analyze various aspects of DNA structure, from linking number, knotting/unknotting, site-specific recombination, and decatenation to the overexpression and purification of bacterial and eukaryotic DNA topoisomerases from a variety of cell systems and tissues. Additional chapters deal with such specialized topics as phosphopeptide mapping, band depletion assays, and visualizing DNA topoisomerases by electron microscopy. The last chapter contains a compendium of DNA topoisomerase type I and II gene sequences.

Part II: Enzymology and Drugs contains detailed methodologies for the analysis of enzyme activities, including the relaxation of supercoiled DNA, DNA unknotting and decatenation, ATP hydrolysis, and the formation of drug-stabilized, covalent enzyme–DNA intermediates. Additional chapters describe methods for analyzing topoisomerase-targeted drug action. A variety of experimental systems are presented to assess the cytotoxic action of putative topoisomerase-directed agents, as well as methods for selecting and analyzing drug-resistant DNA topoisomerase mutants.

We thank the authors for their fine contributions and patience, as well as John Walker for his editorial skills. We also wish to thank our colleagues and members of our laboratories for making this all worthwhile.

Neil Osheroff
Mary-Ann Bjornsti

Contributors

ANNI H. ANDERSON • *Department of Molecular and Structural Biology, Aarhus University, Aarhus, Denmark*

PIERO BENEDETTI • *Department of Pharmaceutical Science, St. Jude Children's Research Hospital, Memphis, TN*

MARY-ANN BJORNSTI • *Department of Pharmaceutical Science, St. Jude Children's Research Hospital, Memphis, TN*

CLAIRE BOUTHIER DE LA TOUR • *Laboratoire d'Enzymologie des Acides Nucleiques, Institute Genetique et Microbiologie, Universite Paris-Sud, Orsay, France*

D. ANDREW BURDEN • *Department of Biochemistry, Vanderbilt University School of Medicine, Nashville, TN*

ALEX B. BURGIN, JR. • *Enzymology Division, Ribozyme Pharmaceuticals, Boulder, CO*

THOMAS BURKE • *Division of Medicinal Chemistry and Phamaceutics, College of Pharmacy, University of Kentucky, Lexington , KY*

JAMES J. CHAMPOUX • *Department of Microbiology, University of Washington, Seattle, WA*

KENT CHRISTIANSEN • *Department of Molecular and Structural Biology, Aarhus University, Aarhus, Denmark*

FABRICE CONFALONIERI • *Laboratoire d'Enzymologie des Acides Nucleiques, Institute Genetique et Microbiologie, Universite Paris-Sud, Orsay, France*

MARY K. DANKS • *Department of Molecular Pharmacology, St. Jude Children's Research Hospital, Memphis, TN*

ZBIGNIEW DARZYNKIEWICZ • *The Cancer Research Institute, New York Medical College, Elmsford, NY*

ANNE-CÉCILE DÉCLAIS • *Laboratoire d'Enzymologie des Acides Nucleiques, Institute Genetique et Microbiologie, Universite Paris-Sud, Orsay, France*

MICHEL DUGUET • *Laboratoire d'Enzymologie des Acides Nucleiques, Institute Genetique et Microbiologie, Universite Paris-Sud, Orsay, France*

PAOLA FIORANI • *Department of Pharmaceutical Science, St. Jude Children's Research Hospital, Memphis, TN*

L. Mark Fisher • *Division of Biochemistry, St. George's Hospital Medical School, University of London, London, UK*
John M. Fortune • *Department of Biochemistry, Vanderbilt University School of Medicine, Nashville, TN*
Catherine H. Freudenreich • *Department of Microbiology, Duke University Medical Center, Durham, NC*
Stacie J. Froelich-Ammon • *Department of Biochemistry, Vanderbilt University School of Medicine, Nashville, TN*
Christine Sommer Furbee • *Department of Molecular Genetics, Ohio State University, Columbus, OH*
Ram Ganapathi • *Department of Pharmacology and Therapeutics, University of Florida College of Medicine, Gainesville, FL*
Martin L. Goble • *Division of Biochemistry, St. George's Hospital Medical School, University of London, London, UK*
Thomas D. Gootz • *Department of Immunology and Infectious Diseases, Central Research Division, Pfizer, Groton, CT*
Dale Grabowski • *Department of Pharmacology and Therapeutics, University of Florida College of Medicine, Gainesville, FL*
David E. Graves • *Department of Chemistry, University of Mississippi, Oxford, MS*
Andrea Haldane • *Division of Bone Marrow Transplant, Moffitt Cancer Center and Research Institute, University of South Florida, Tampa, FL*
Christine L. Hann • *Department of Pharmaceutical Science, St. Jude Children's Research Hospital, Memphis, TN*
Cynthia E. Herzog • *Division of Pediatrics, University of Texas, M. D. Anderson Cancer Center, Houston, TX*
Wai Mun Huang • *Department of Oncological Sciences, Division of Molecular Biology and Genetics, Univeristy of Utah Medical Center, Salt Lake City, UT*
Christine Jaxel • *Laboratoire d'Enzymologie des Acides Nucleiques, Institute Genetique et Microbiologie, Universite Paris-Sud, Orsay, France*
Gloria Juan • *The Cancer Research Institute, New York Medical College, Elmsford, NY*
Paul S. Kingma • *Department of Biochemistry, Vanderbilt University School of Medicine, Nashville, TN*
Kenneth N. Kreuzer • *Department of Microbiology, Duke University Medical Center, Durham, NC*
Janet E. Lindsley • *Department of Biochemistry, University of Utah, School of Medicine, Salt Lake City, UT*

1

Assaying DNA Topoisomerase I Relaxation Activity

Lance Stewart and James J. Champoux

1. Introduction

Type I topoisomerases catalyze topological changes in duplex DNA by reversibly nicking one strand, whereas type II enzymes catalyze the transient breakage of both strands simultaneously. The type I enzymes alter the linking number of covalently closed circular DNA in steps of one, presumably by allowing an unbroken segment of one strand of the DNA to move through the transient single-strand break in the other strand. The type II enzymes alter the linking number in steps of two by allowing an unbroken segment of duplex DNA to pass through the transient double-strand break.

All topoisomerases conserve phosphodiester bond energy during catalysis by transiently forming a phosphotyrosine bond between the active-site tyrosine residue and the phosphate at one end of the broken strand(s) ***(1)***. The type I topoisomerases fall into two categories depending on the polarity of their covalent attachment. The cellular and viral eukaryotic topoisomerase I enzymes ***(2)*** and the archebacterial topoisomerase V ***(3)*** are classified as type I-3', since they become linked to the 3'-end of the broken strand ***(1)***. These enzymes also have the distinctive characteristic that they can relax both positively and negatively supercoiled DNA in the absence of an energy cofactor or divalent metal cation ***(2)***. The other type I enzymes—topoisomerases I and III of prokaryotes ***(1,4,5)***, reverse gyrase and topoisomerase III of archebacteria ***(6–9)***, and eukaryotic topoisomerase III ***(10)***—become linked to the 5'-end of the broken strand and are classified as type I-5'. These enzymes require Mg^{2+} for activity and can only relax negatively supercoiled DNA. The type II topoisomerases (*see* Chapters **2** and **3**)—whether virally encoded or isolated from archebacteria, eubacteria, or eukaryotes—invariably require both ATP and Mg^{2+} for activity and become linked to the 5'-end of the broken strands ***(1,11,12)***.

From: *Methods in Molecular Biology, Vol. 95: DNA Topoisomerase Protocols, Part II: Enzymology and Drugs*
Edited by N. Osheroff and M.A. Bjornsti © Humana Press Inc., Totowa, NJ

A plasmid relaxation assay for topoisomerase I activity was initially described by Wang together with the identification of the *Escherichia coli* topoisomerase I ***(13)***. In this assay, sedimentation through a CsCl gradient was used to examine the topological state of the plasmid DNA following reaction with topoisomerase I. Other early assays employed equilibrium centrifugation in the presence of propidium diiodide ***(14)*** or fluorometric analysis of the changes in ethidium binding that accompany relaxation of the DNA ***(15,16)***. Subsequently, Keller described the method of agarose-gel electrophoresis to separate individual topological isomers of covalently closed SV40 DNA circles ***(17)***. With this technique, the compact nature of supercoiled topoisomers enables them to migrate through the porous gel matrix with less resistance than relaxed topoisomers whose migration is impeded owing to their more open configuration. Agarose-gel electrophoresis is now the method of choice for visualizing the products of topoisomerase relaxation assays.

This chapter describes the methodology for assaying topoisomerase I activity, in either crude cell extracts or purified preparations, by following the relaxation of negatively supercoiled DNA by agarose-gel electrophoresis. By taking into account the different requirements for Mg^{2+} or ATP by the various topoisomerases (described above), the assay can discriminate between the type I-3' or I-5' enzymes, and eliminates type II topoisomerase activity altogether. Methods for measuring type II topoisomerase relaxation activity are the subject of Chapter 2.

2. Materials

2.1. Enzymes and Closed Circular DNA

1. Topoisomerase I from either eubacterial, archebacterial, or eukaryotic cells may be present in crude cell extracts or purified according to protocols outlined in Volume 94 of this series. Alternatively, at least some of the enzymes can be purchased from commercial sources, such as Gibco BRL (calf thymus topo I) and Promega (wheat germ topo I).
2. The substrate for the relaxation assay is a bacterial plasmid DNA (3–7 kbp) that has been purified by CsCl centrifugation in the presence of ethidium bromide (EtBr), or by column chromatographic methods of Quiagen (cat. no. 12143). These purification methods yield plasmid DNA that is free of contaminating protein and is primarily composed of negatively supercoiled covalently closed circular DNA molecules (form I DNA), with nicked circles (form II DNA) representing no more than 20% of the total DNA. A CsCl-purified 2.6-kbp plasmid (pKSII+, Stratagene) was employed in the assays shown in **Subheading 2.2.**

2.2. Buffers

1. 10X TBE: 0.89 *M* Tris-borate, 20 m*M* EDTA, pH 8.0. The final 1X TBE is a 10-fold dilution of the concentrated stock.

2. 1X TBE-EtBr: 89 m*M* Tris-borate, 2 m*M* EDTA, pH 8.0, 0.25 µg/mL EtBr.
3. Standard buffer: 150 m*M* KCl, 10 m*M* Tris-HCl, pH 7.5, 1 m*M* DTT, 1 m*M* EDTA, and 0.1 mg/mL BSA (New England Biolabs).
4. Universal type I assay buffer: 150 m*M* KCl, 10 m*M* Tris-HCl, pH 7.5, 15 m*M* $MgCl_2$, 1 m*M* DTT, 1 m*M* EDTA, 0.1 mg/mL BSA, 25 ng/µL plasmid DNA.
5. Type I-3' assay buffer: 150 m*M* KCl, 10 m*M* Tris-HCl, pH 7.5, 1 m*M* DTT, 1 m*M* EDTA, 0.1 mg/mL BSA, 25 ng/µL plasmid DNA.
6. 5X Stop buffer: 2.5% SDS, 15% Ficoll-400, 0.05% bromphenol blue, 0.05% xylene cyanol, 25 m*M* EDTA.

3. Methods

In the relaxation assays described below, a universal assay buffer containing Mg^{2+} (**Subheading 2.2.**) is used to detect the activity of either type I-5' or I-3' enzymes. Since the buffer contains no ATP, any potential type II activity is eliminated. A type I-3' assay buffer is used to assay only the type I-3' enzymes, and is prepared by excluding Mg^{2+} from the universal buffer. The Tris buffer, 150 m*M* KCl, and DTT components of the assay buffers have been chosen in order to approximate the physiological environment in terms of its pH, ionic strength, and reducing nature. Bovine serum albumin (BSA) is included to eliminate loss of activity owing to binding of low concentrations of enzyme to the walls of microtubes. The 1 m*M* EDTA is included to chelate low concentrations (<1 m*M*) of divalent metal cations.

The level of topoisomerase I activity present in crude cell extracts containing overexpressed or mutant forms of topoisomerase I may be unknown and could vary over three orders of magnitude ***(18)***. In such cases, a "serial dilution" assay is used to provide an initial estimate of the level of activity (**Subheading 3.1.**). Subsequently, a more accurate "time-course" assay (**Subheading 3.2.**) is used to define the exact level of activity to within 10% error. Together, the "serial dilution" and "time-course" assays produce visually quantifiable results that are linear for essentially any possible enzyme concentration or level of activity.

3.1. Topoisomerase I Relaxation Assay by Serial Dilution

With the serial dilution assay, protein samples are sequentially diluted twofold, and each dilution is incubated with plasmid DNA under the reaction conditions. After analyzing the reaction products by agarose-gel electrophoresis (**Subheading 3.3.**), relative levels of activity between samples and standards are quantified to within a factor of 4 by visually determining which enzyme dilution is just sufficient to relax fully all of the plasmid substrate (**Fig. 1**). Step-by-step details of the assay are given below.

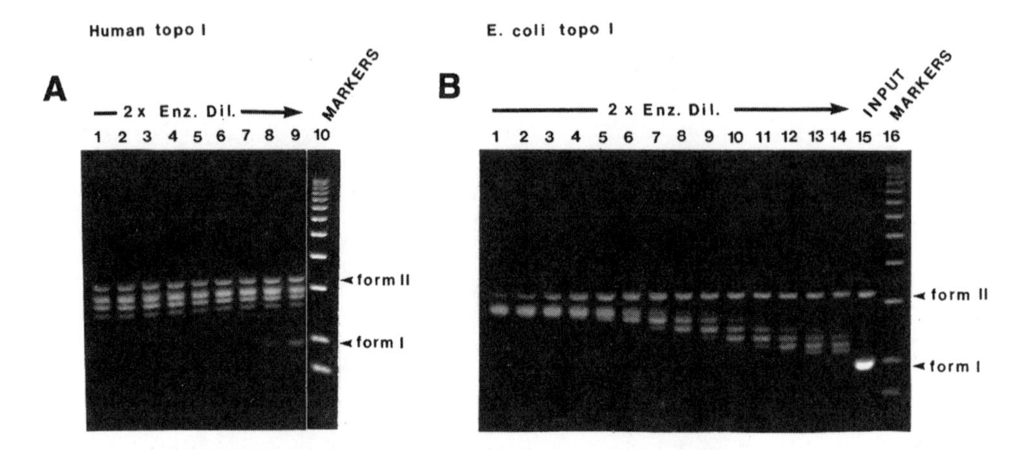

Fig. 1. Serial dilution assay of topoisomerase I. **(A)** A 4.0 ng/µL stock (diluted from concentrated stock in standard buffer) of purified recombinant human topoisomerase I was subjected to a serial twofold dilution assay as outlined in **Subheading 3.1.** Lanes 1–9 are the assays for the first nine serial twofold dilutions. Lane 10, 1-kbp ladder (Gibco BRL). **(B)** Lanes 1–14, a 125 ng/µL stock (diluted from concentrated stock in standard buffer) of purified *E. coli* topoisomerase I (a gift from K. Marians), was subjected to a serial twofold dilution assay. Lane 15, untreated substrate plasmid DNA.

1. Prepare 14 twofold serial dilutions of protein sample by mixing 50 µL of one sample with 50 µL of standard buffer to make the next sample, and so on (*see* **Note 1**). Store the diluted samples on ice for as short a time as is practical.
2. Initiate the reactions in sequence, from dilution #1 to 14, at 1-min intervals by adding 10 µL of each serial dilution to 20 µL of assay buffer in individual microtubes that have been prewarmed to 37°C. The final reaction conditions will be: 150 m*M* KCl, 10 m*M* Tris-HCl, pH 7.5, 1 m*M* DTT, 1 m*M* EDTA, 0.1 mg/mL BSA, 0.017 µg/µL plasmid DNA, with or without 10 m*M* $MgCl_2$.
3. Incubate at 37°C for 10 min.
4. Terminate each reaction in sequence, from dilution #1 to 14, at 1-min intervals by adding 7.5 µL of 5X stop buffer and mixing rapidly.
5. Analyze the products by agarose-gel electrophoresis (**Subheading 3.3.**).
6. The highest serial dilution sample that produces a fully relaxed topoisomer distribution (form Ir) is said to be the amount of enzyme that is just sufficient to fully relax 0.5 µg of plasmid DNA in 10 min at 37°C (**Fig. 1**, lane 7 of panel A, and lane 2 of panel B).

3.2. Topoisomerase I Time-Course Relaxation Assay

Once an approximate level of activity has been established by the serial dilution assay (**Subheading 3.1.** *above*), a "time-course" assay is performed to

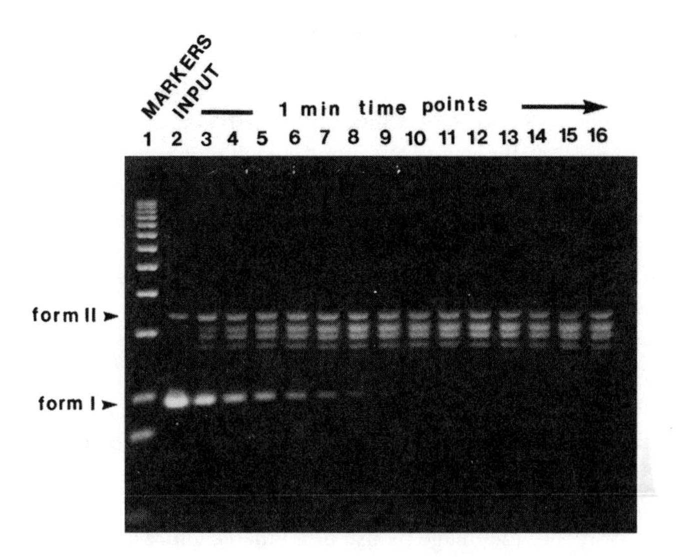

Fig. 2. A time-course assay of human topoisomerase I. A 25 ng/µL stock of purified recombinant human topoisomerase I was diluted to a concentration of 0.03 ng/µL and used in a time-course relaxation experiment as outlined in **Subheading 3.2.** Lane 1, 1-kbp ladder (Gibco BRL). Lane 2, unreacted plasmid DNA. Lanes 3–16, samples terminated at 1-min time-points, up to 14 min after initiation of the reaction.

obtain a more accurate measure of topo I activity. The time-course assay is initiated by adding the appropriate amount of enzyme, determined from the preliminary serial dilution assay, to a single large-scale reaction. At the appropriate time-points, aliquots of the reaction are terminated and analyzed by gel electrophoresis (**Fig. 2**). Step-by-step details of the assay are given below.

1. Dilute the topoisomerase I enzyme in standard buffer to a level that is just sufficient to relax the plasmid DNA fully in the 10-min serial dilution assay (**Subheading 3.1.**) (*see* **Note 3**).
2. Add 100 µL of the diluted enzyme to 200 µL of assay buffer that has been prewarmed to 37°C.
3. At 1-min time intervals (or longer time intervals as appropriate; *see* **Notes 2–4**), terminate a 20-µL aliquot of the reaction by mixing rapidly with 5 µL of 5X stop buffer.
4. Analyze the products by agarose-gel electrophoresis (**Subheading 3.3.**).
5. The end point of relaxation is taken as the earliest time where no differences are observed in the relaxed topoisomer (form Ir) distribution from that time-point to the next (*see* **Fig. 2**).

3.3. Agarose-Gel Electrophoretic Analysis of Relaxation Assay Products

1. Cast a horizontal 0.8% agarose minigel in 1X TBE (10 cm wide × 5 cm long × 7 mm thick), with a 16-well comb (4 × 1 mm for each tooth, with 2-mm space between teeth). Low concentrations of chloroquine diphosphate may be included in the gel, allowing for improved resolution of topological isomers (*see* **Note 5**).
2. Prepare samples by adding 1/4 vol of 5X stop buffer.
3. Load 15 µL of each sample, as well as 0.2 µg of 1-kbp DNA ladder (Gibco BRL, cat. no. 15615-016), which serves as size standards.
4. Electrophorese at ~1 V/cm for 16 h or 5 V/cm for 4 h in 1X TBE.
5. Stain the gel for 30 min in 1X TBE with 0.25 µg/mL EtBr.
6. To resolve the relaxed topoisomers (form Ir) and nicked circles (form II) better, which often display similar mobilities, electrophorese the gel for an additional hour at ~2 V/cm after staining with EtBr (*see also* Chapter 9).
7. Transfer the gel to a UV transilluminator, and photograph the image with Polaroid X-5 film, or record the image by use of a video capture system.

3.4. Type I-5' vs I-3' Topoisomerase I

A type I-3' activity can be unambiguously defined by assaying activity in the absence of Mg^{2+}. In order to detect a type I-5' activity, the assay buffer must include Mg^{2+}, which could stimulate the type I-3' enzymes by as much as 25-fold ***(18,19)***. Therefore, even a small contaminant of a type I-3' activity could mask the presence of a much larger quantity of type I-5' activity. For example, this condition would be approximated by extracts of eukaryotic cells in which the *E. coli* topo I has been overexpressed. In such cases, it is advisable to carry out assays with both the universal and type I-3' buffers. However, it should be realized that since Mg^{2+} can stimulate the type I-3' enzymes, the difference in activity measured in two buffer systems may not reflect type I-5' activity.

An alternative method to distinguish a type I-3' from a type I-5' activity is to examine the topological distribution of the fully relaxed topoisomers. At thermodynamic equilibrium, the type I-3' enzymes generate a Poisson distribution of fully relaxed (form Ir) topoisomers differing in their number of superhelical turns about some mean value (**Fig. 3**, lane 4) ***(17)***. This distribution results from the fact that the type I-3' enzymes can relax both negatively and positively supercoiled DNA. Therefore, when a relaxation assay is terminated, the resulting distribution of topoisomers will be a function of the probability that a given plasmid will have a certain level of internal energy as described by a Boltzmann distribution ***(20–23)***. Complete relaxation by a eukaryotic topoisomerase leads to a distribution of topoisomers that resembles that which would be obtained following ligation of a linear plasmid molecule under identical conditions ***(22,23)***. In contrast, since the type I-5' enzymes incom-

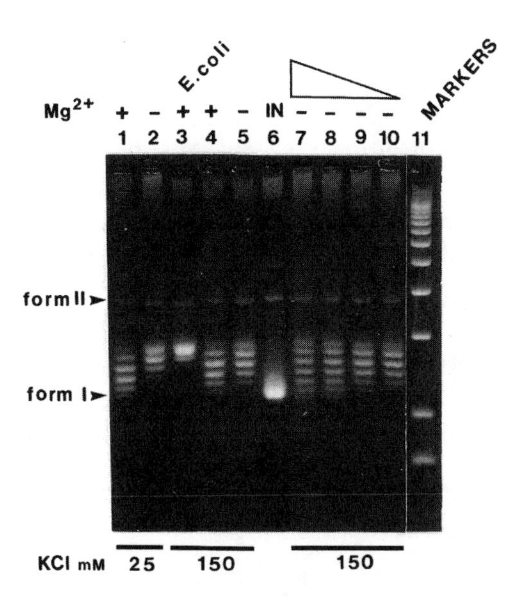

Fig. 3. A comparison of relaxed topoisomer distributions produced by type I-3' and I-5' enzymes under various conditions. Purified type I-3′ human topoisomerase (20 ng in lanes 1, 2, and 4–6) and type I-5' *E. coli* topoisomerase I (350 ng, lane 3) were allowed to relax fully 0.5 μg of plasmid DNA in a 30-μL vol for 10 min at 37°C containing the following buffers. Lane 1, universal buffer modified to contain 25 m*M* KCl. Lane 2, type I-3' buffer modified to contain 25 m*M* KCl. Lanes 3 and 4, universal buffer. Lane 5, type I-3' buffer. Lane 6, contained unreacted plasmid DNA. Lanes 7–10, respectively, contained plasmid DNA that was reacted with 2.5, 1.25, 0.6, or 0.3 μg human topoisomerase I under conditions described for lane 6. Lane 11 contained 1-kbp ladder (Gibco BRL). Samples were electrophoresed for 16 h at 1 V/cm in the presence of 1.5 μg/mL of chloroquine, stained with 0.25 μg/mL of EtBr for 1 h, electrophoresed for 2 h at 2 V/cm, and then photographed.

pletely relax only negatively supercoiled DNA ***(1,13)***, the end product of relaxation will be a population of plasmid molecules with a unique linking number (**Fig. 3**, lane 3), which is often approximately three turns fewer than the mean linking number of the same plasmid relaxed by a type I-3' enzyme under identical conditions (**Fig. 3**, lane 4).

3.5. Distributive vs Processive Activity

An important qualitative aspect of topoisomerase activity relates to the number of superhelical turns released per substrate binding event. A topoisomerase is said to be highly processive if, after binding to plasmid substrate, it catalyzes the complete relaxation of the substrate without dissociating from it. In contrast, a topoisomerase is said to be highly distributive if it readily dissociates from the substrate following the release of only one or a few superhelical turns. Processive activity manifests itself in a plasmid relaxation assay by the distinctive absence of topological isomers with intermediate superhelicities between fully supercoiled form I and fully relaxed form Ir. For example, human topoisomerase I displays highly processive activity under the type I-3' assay conditions (**Figs. 1A** and **2**). In contrast, the *E. coli* topoisomerase I acts in a highly distributive manner as evidenced by the fact that the plasmid molecules are relaxed together as a population (**Fig. 1B**), and at moderate enzyme concentrations after 10 min, all of the covalently closed molecules exist as a population with superhelicities intermediate between form I and Ir.

3.6. Effects of Mg^{2+} and High Salt on Relaxed Topoisomer Distribution

Divalent cations, such as Mg^{2+}, are known to effectively shield the negative charge of the phosphate backbone of duplex DNA. This allows the two strands to wind *(24)*. Therefore, in the presence of Mg^{2+}, fully relaxed covalently closed circles will have fewer basepairs per turn and consequently have a higher linking number than in the absence of Mg^{2+} (**Fig. 3**, compare lanes 1 and 2). Salt can also shield the phosphodiester backbone, although much less effectively than Mg^{2+}. Therefore, the linking number of fully relaxed circles will increase with the concentration of salt (**Fig. 3**, compare lanes 2 and 5).

It should also be noted that since gel-electrophoretic separation of topoisomers is carried out in 89 m*M* Tris-borate and 10 m*M* EDTA, fully relaxed topoisomers that were formed in the presence of Mg^{2+} (**Fig. 3**, lanes 1 and 4) or higher salt (150 m*M* KCl) (**Fig. 3**, lanes 4 and 5) will become somewhat overwound or positively supercoiled during electrophoresis. Consequently, these topoisomers run faster in the agarose gel than the relaxed topoisomers formed in the absence of Mg^{2+} (**Fig. 3**, lanes 2 and 5) or in low salt (25 m*M* KCl) (**Fig. 3**, lanes 1 and 2).

3.7. Large Quantities of the Node Binding Eukaryotic Topoisomerase I Expand the Poisson Distribution of Relaxed Topoisomers

The eukaryotic topoisomerase I enzyme has been shown to bind preferentially to supercoiled DNA *(25)*. This preference for supercoiled DNA appears to be mediated by high-affinity binding to DNA nodes, the points at which two duplexes cross *(26,27)*. Node binding is expected to stabilize the supercoiled nature of a covalently closed molecule as has been shown for the eukaryotic type II enzyme *(27)*, a condition that would manifest itself as an expansion of the Poisson distribution of relaxed topoisomers. Indeed, large quantities of human topoisomerase I will generate an expanded distribution of relaxed topoisomers (**Fig. 3**, compare lanes 7 and 10). The larger the quantity of topoisomerase I, the greater the expansion of the relaxed topoisomer distribution (**Fig. 3**, lanes 7–10). Since the distribution is expanded in both directions about a mean topoisomer that does not change position with increasing enzyme, it can be concluded that the human enzyme shows no preference for the handedness of nodes.

4. Notes

1. When generating enzyme dilutions, pipet at least 10 µL to ensure that small pipeting errors (which can be as much as 0.2 µL) do not affect the final pipeted volume by more than 2%.
2. Since it is difficult to terminate reactions accurately at time intervals shorter than 30 s, it is recommended that time-points be taken at least 1 min apart.
3. The amount of enzyme used in the time-course assay should be adjusted to ensure that all of the plasmid substrate will be fully relaxed in approx 8–12 min. By sampling the reaction every 1 min for 15 min, and visually determining the point at which the plasmid substrate has become fully relaxed, the total activity can be measured with an error of ±10%.
4. Time-course assays are linear for enzyme concentrations that lead to complete relaxation of the substrate within a range of 5 min to 2 h. The lower time limit merely reflects the inability to sample accurately a reaction at time intervals <30 s. At times longer than 2 h, the assay not only becomes time-consuming, but also looses linearity, presumably owing to low-level, time-dependent enzyme inactivation.
5. Relaxed topoisomer populations are often poorly resolved in the conventional 1X TBE electrophoretic run buffer. To resolve topoisomer populations better, low concentrations of chloroquine diphosphate (typically 0.5–10.0 µg/mL) can be added to the gel and the 1X TBE electrophoretic run buffer. Like EtBr, chloroquine will intercalate into DNA, causing it to unwind at the site of binding. With the appropriate concentration of chloroquine, topological isomers with relatively high negative superhelicities can be electrophoretically resolved *(27)*.

Acknowledgments

We thank the following past and present members of the Champoux lab for their support, helpful comments, and valuable discussions: Gregory C. Ireton, Leon H. Parker, Knut R. Madden, Sam Whiting, and Sharon Schultz. We thank Kenneth Marians for supplying the purified *E. coli* topo I. This work was supported by Grant GM49156 to J. J. C. from the National Institutes of Health. L. S. was supported by an American Cancer Society Grant PF-3905.

References

1. Roca, J. (1995) The mechanisms of DNA topoisomerases. *Trends Biochem. Sci.* **20,** 156–160.
2. Champoux, J. J. (1990) Mechanistic aspects of type-I topoisomerases, in *DNA Topology and Its Biological Effects* (Cozzarelli, N. R. and Wang, J. C., eds.), Cold Spring Harbor Laboratory Press, Cold Spring Harbor, NY, p. 217–242.
3. Slesarev, A. I., Stetter, K. O., Lake, J. A., Gellert, M., Krah, R., and Kozyavkin, S. A. (1993) DNA topoisomerase V is a relative of eukaryotic topoisomerase I from a hyperthermophilic prokaryote. *Nature* **364,** 735–737.
4. Zhang, H. L., Malpure, S., and DiGate, R. J. (1995) *Escherichia coli* DNA topoisomerase III is a site-specific DNA binding protein that binds asymmetrically to its cleavage site. *J. Biol. Chem.* **270,** 23,700–23,705.
5. Srivenugopal, K. S., Lockshon, D., and Morris, D. R. (1984) *Escherichia coli* DNA topoisomerase III: purification and characterization of a new type I enzyme. *Biochemistry* **23,** 1899–1906.
6. Kovalsky, O. I., Kozyavkin, S. A., and Slesarev, A. I. (1990) Archaebacterial reverse gyrase cleavage-site specificity is similar to that of eubacterial DNA topoisomerases I. *Nucleic Acids Res.* **18,** 2801–2805.
7. Confalonieri, F., Elie, C., Nadal, M., de La Tour, C., Forterre, P., and Duguet, M. (1993) Reverse gyrase: a helicase-like domain and a type I topoisomerase in the same polypeptide. *Proc. Natl. Acad. Sci. USA* **90,** 4753–4757. Published erratum appears in *Proc. Natl. Acad. Sci. USA* **8,** 3478.
8. Jaxel, C., Nadal, M., Mirambeau, G., Forterre, P., Takahashi, M., and Duguet, M. (1989) Reverse gyrase binding to DNA alters the double helix structure and produces single-strand cleavage in the absence of ATP. *EMBO J.* **8,** 3135–3139.
9. Slesarev, A. I., Zaitzev, D. A., Kopylov, V. M., Stetter, K. O., and Kozyavkin, S. A. (1991) DNA topoisomerase III from extremely thermophilic archaebacteria. ATP-independent type I topoisomerase from *Desulfurococcus amylolyticus* drives extensive unwinding of closed circular DNA at high temperature. *J. Biol. Chem.* **266,** 12,321–12,328.
10. Kim, R. A. and Wang, J. C. (1992) Identification of the yeast TOP3 gene product as a single strand-specific DNA topoisomerase. *J. Biol. Chem.* **267,** 17,178–17,185.
11. Wigley, D. B. (1995) Structure and mechanism of DNA topoisomerases. *Annu. Rev. Biophys. Biomol. Struct.* **24,** 185–208.
12. Liu, L. F. (1990) Anticancer drugs that convert DNA topoisomerases into DNA damaging agents, in *DNA Topology and Its Biological Effects* (Cozzarelli, N. R. and Wang, J. C., eds.), Cold Spring Harbor Laboratory Press, Cold Spring Harbor, NY, pp. 371–389.
13. Wang, J. C. (1971) Interaction between DNA and an *Escherichia coli* protein ω. *J. Mol. Biol.* **55,** 523–533.
14. Champoux, J. J., and Dulbecco, R. (1972) An activity from mammalian cells that untwists superhelical DNA—a possible swivel for DNA replication (polyoma-

ethidium bromide-mouse-embryo cells-dye binding assay). *Proc. Natl. Acad. Sci. USA* **69,** 143–146.

15. McConaughy, B. L., Young, L. S., and Champoux, J. J. (1981) The effect of salt on the binding of the eucaryotic DNA nicking-closing enzyme to DNA and chromatin. *Biochim. Biophys. Acta* **655,** 1–8.
16. Morgan, A. R. and Pulleyblank, D. E. (1974) Native and denatured DNA, cross-linked and plaindromic DNA and circular covalently-closed DNA anlaysed by a sensitive fluorometric procedure. *Biochem. Biophys. Res. Commun.* **61,** 396–403.
17. Keller, W. (1975) Determination of the number of superhelical turns in simian virus 40 DNA by gel electrophoresis. *Proc. Natl. Acad. Sci. USA* **72,** 4876–4880.
18. Stewart, L. Ireton, G. C., Parker, L. H., Madden, K. R., and Champoux, J. J. (1996) Biochemical and biophysical analyses of recombinant forms of human topoisomerase I. *J. Biol. Chem.* **271,** 7593–7601.
19. Wang, J. C. and Becherer, K. (1983) Cloning of the gene topA encoding for DNA topoisomerase I and the physical mapping of the cysB-topA-trp region of *Escherichia coli. Nucleic Acids Res.* **11,** 1773–1790.
20. Shure, M., Pulleyblank, D. E., and Vinograd, J. (1977) The problems of eukaryotic and prokaryotic DNA packaging and in vivo conformation posed by superhelix density heterogeneity. *Nucleic Acids Res.* **4,** 1183–1205.
21. Liu, L. F. and Miller, K. G. (1981) Eukaryotic DNA topoisomerases: two forms of type I DNA topoisomerases from HeLa cell nuclei. *Proc. Natl. Acad. Sci. USA* **78,** 3487–3491.
22. Depew, D. E. and Wang, J. C. (1975) Conformational fluctuations of DNA helix. *Proc. Natl. Acad. Sci. USA* **72,** 4275–4279.
23. Pulleyblank, D. E., Shure, M., Tang, D., Vinograd, J., and Vosberg, H. P. (1975) Action of nicking-closing enzyme on supercoiled and nonsupercoiled closed circular DNA: formation of a Boltzmann distribution of topological isomers. *Proc. Natl. Acad. Sci. USA* **72,** 4280–4284.
24. Wang, J. C. (1969) Degree of superhelicity of covalently closed cyclic DNA's from *Escherichia coli. J. Mol. Biol.* **43,** 263–272.
25. Madden, K. R., Stewart, L., and Champoux, J. J. (1995) Preferential binding of human topoisomerase I to superhelical DNA. *EMBO J.* **14,** 5399–5409.
26. Zechiedrich, E. L. and Osheroff, N. (1990) Eukaryotic topoisomerases recognize nucleic acid topology by preferentially interacting with DNA crossovers. *EMBO J.* **9,** 4555–4562.
27. Roca, J., Berger, J. M., and Wang, J. C. (1993) On the simultaneous binding of eukaryotic DNA topoisomerase II to a pair of double-stranded DNA helices. *J. Biol. Chem.* **268,** 14,250–14,255.

2

DNA Topoisomerase II-Catalyzed DNA Decatenation

Andrea Haldane and Daniel M. Sullivan

1. Introduction

A major role of DNA topoisomerase II in vivo is to catalyze the double-stranded cleavage of DNA, allowing passage of a second DNA duplex through the break. This activity requires adenosine triphosphate (ATP) and is necessary for separating catenated DNA duplexes found at the end of replication. The decatenation of DNA molecules is a topoisomerase II-specific reaction, and is a convenient assay for measuring topoisomerase II activity in vitro *(1)*.

Kinetoplast DNA (kDNA), which is the DNA substrate used in the in vitro decatenation assay, is found in the mitochondria of trypanosomes and related protozoa, and consists mainly of a large network of interlocked or catenated covalently closed DNA minicircles. Each network of form I *Crithidia fasciculata* kDNA contains about 5000 minicircles (2.5 kb each) and about 25 maxicircles (37 kb each). For further details regarding the replication of kDNA networks in *C. fasciculata*, *see* refs. *(2)* and *(3)*. DNA topoisomerase II is able to decatenate kDNA networks isolated from various trypanosomes into their respective minicircles and maxicircles.

The minicircles released from kDNA networks by topoisomerase II are most easily visualized by agarose-gel electrophoresis. In its catenated form, kDNA cannot enter a 1% agarose gel, whereas minicircles released by topoisomerase II will migrate into the gel. If a more quantitative assessment of catalytic activity at initial velocities is required, kDNA labeled with [^{3}H]thymidine can be used as a substrate *(4)*. The isolation techniques for both labeled and unlabeled kDNA described below are those we routinely use in our laboratory and are based on methods previously described by others *(2,5,6)*. The decatenation assay utilizing this kDNA is outlined, as are several reaction conditions that can optimize this assay (*see* **Notes 1–5**).

From: *Methods in Molecular Biology, Vol. 95: DNA Topoisomerase Protocols, Part II: Enzymology and Drugs*
Edited by N. Osheroff and M.A. Bjornsti © Humana Press Inc., Totowa, NJ

2. Materials

2.1. Isolation of Unlabeled kDNA

1. Buffer 1: 100 m*M* NaCl, 10 m*M* Tris-HCl, 200 m*M* Na_2EDTA, pH 8.0.
2. 6% (w/v) *N*-lauroylsarcosine (Sigma, catalog no. L-5125) dissolved in buffer 1.
3. Proteinase K: dissolve 20 mg/mL in buffer 1 (make immediately before use; DNase-free Sigma type XI, cat. no. P-0390).
4. 0.5 *M* Na_2EDTA (adjust to pH 8.0 to dissolve).
5. CsCl (enough for ten Beckman 25 × 89 mm centrifuge tubes): For solution A, combine 151.5 g of CsCl (Sigma optical-grade, cat. no. C-3139) and 6.0 mL of 0.5 *M* Na_2EDTA. Bring to a total volume of 300 mL with H_2O (n = 1.3702). Solution B is made by adding 48 mL of H_2O to 69.6 g CsCl and 1.8 mL of 0.5 *M* Na_2EDTA (n = 1.4040).
6. TE buffer: 10 m*M* Tris-HCl, 1 m*M* Na_2EDTA, pH 8.0.
7. Hemin (Sigma, cat. no. H-2250): dissolve at 2 mg/mL in 50 m*M* NaOH. Filter-sterilize and store in aliquots at –20°C.
8. BHI media (Fisher Scientific, cat. no. 0037-17-8): make according to the instructions, and then add 100 IU/mL penicillin and 100 µg/mL streptomycin. Immediately before use, add 20 µg/mL sterile hemin.
9. *C. fasciculata* (*see* Chapter 9 for routine growth conditions).
10. Ethidium bromide solution: dissolve 5 mg/mL in TE buffer.

2.2. Isolation of [^{3}H]kDNA

1. [Methyl-^{3}H]thymidine (20.0 Ci/mmol, New England Nuclear): for the general isolation procedure, a total of 6 mCi is required (10 µCi/mL of *Crithidia* culture).
2. All of the materials listed in **Subheading 2.1.**

2.3. kDNA Decatenation: Assayed by Agarose-Gel Electrophoresis (Unlabeled)

1. Decatenation buffer (final concentration): 50 m*M* Tris-HCl, 85 m*M* KCl, 10 m*M* $MgCl_2$, 0.5 m*M* DTT, 0.5 m*M* Na_2EDTA, 1 m*M* ATP, 30 µg/mL BSA (Sigma, cat. no. A-6793), pH 7.5. This buffer is made at 10 times the final concentration and stored in 50- to 100-µL aliquots at –20°C. Buffers without KCl and/or ATP may also be needed.
2. kDNA (refer to **Subheading 3.1.**).
3. TBE buffer: 89 m*M* Tris-HCl, 89 m*M* boric acid, 2 m*M* Na_2EDTA, pH 8.0. This buffer can be made at 10 times this concentration.
4. Stop solution: 2% SDS, 0.05% bromophenol blue, 50% glycerol (v/v).
5. 1% Agarose gel made in TBE buffer.
6. Ethidium bromide solution: Dilute 1.0 mL of 5 mg/mL ethidium bromide with 500 mL TBE (10 µg/mL). Store in a dark bottle at room temperature.

2.4. kDNA Decatenation: Assayed by Centrifugation (Labeled)

1. Decatenation buffer (*see* **Subheading 2.3.**).
2. [^{3}H]kDNA (*see* **Subheading 2.2.**).
3. Stop solution: 2.25% SDS.

3. Methods

3.1. Isolation of kDNA from C. fasciculata

C. fasciculata double every 4–6 hours and should be grown in a shaking incubator at 27°C. The following procedure should yield 500–1000 µg of kDNA if the *Crithidia* are set up at the suggested concentrations.

1. Day 1: Set up *Crithidia* at 2.5 × 10^6/mL in 200 mL BHI media containing hemin. To count on a hemacytometer, add 100 µL of 4% formalin to 100 µL *Crithidia* broth and 800 µL of PBS.
2. Day 2: Dilute *Crithidia* to 5 × 10^6/mL with BHI media. The volume should now be approx 600 mL in a 4-L Erlenmeyer flask, and fresh hemin should be added at 20 µg/mL to the total volume. Shake at 27°C overnight.
3. Day 3: To begin the kDNA network isolation, the *Crithidia* should be at 1.5–2.0 × 10^8/mL. Pellet the trypanosomes at 5000*g* (Sorvall GSA rotor at 6000 rpm) for 10 min at 4°C.
4. Wash the pellet once with 100 mL cold PBS, and centrifuge as above.
5. The trypanosomes should be lysed at approx 1 × 10^9/mL. Resuspend the pellet in buffer 1 on ice (gently triturate, and do not vortex). This volume should be half of the final lysis volume (generally 40–50 mL). Add an equal volume of 6% *N*-lauroylsarcosine, and swirl to mix. (For example, 600 mL at 1.5 × 10^8/mL is equal to 9 × 10^{10} total trypanosomes. This will require a total combined volume of buffer 1 and 6% *N*-lauroylsarcosine of 90 mL to lyse at 1 × 10^9/mL; 45 mL of each is added).
6. Add proteinase K solution so that the final concentration is 1 mg/mL (e.g., to 90 mL from above, add 4.5 mL of 20 mg/mL proteinase K). Gently mix and place in a 37°C water bath for 30 min.
7. CsCl gradients: In a thin-walled disposable plastic ultracentrifuge tube (e.g., Beckman 25 × 89 mm tubes, cat. no. 344058) place 24 mL of solution A. Carefully underlay this with 4.0 mL of solution B to which 10 µL of 5 mg/mL ethidium bromide have been added (we use a blunted spinal needle attached to a syringe). Alternately, solution A can be very carefully layered over solution B. Carefully layer the lysate from above on top of solution A in several tubes and balance them. Centrifuge at 53,000*g* for 10 min at 4°C (20,000 rpm in a Beckman SW 28 rotor).
8. Examine the centrifuge tubes carefully with a handheld UV light (254 nm); we secure the full centrifuge tube with a clamp on a ring stand. The kDNA should be

in a discrete band at the interface of solutions A and B. Carefully penetrate the centrifuge tube above the band with a #20 gage needle attached to a 5- to 10-mL syringe, and aspirate the kDNA band (remove 3–4 mL). Repeat this procedure on all tubes, and measure the total volume aspirated. **Note:** when working with the UV lamp, it is important to wear protective eye glasses to preclude corneal damage from UV exposure.

9. Extract the ethidium bromide from the above solution containing kDNA with an equal volume of *n*-butanol saturated with TE. Repeat this procedure for a total of three extractions (*n*-butanol is the top pink layer and the kDNA is in the bottom aqueous layer).
10. Dialyze against 200 vol of TE for 2 h at room temperature or overnight at 4°C (change TE once).
11. Concentrate the kDNA networks by centrifugation at 100,000*g* for 1 h at 4°C (27,000 rpm in a Beckman SW 28 rotor). The same ultracentrifuge tubes as above can be used (filled and balanced with TE buffer). Remove the supernatant and resuspend the pellet (very small) in a total of 0.5–1.0 ml TE buffer (vortex gently).
12. Concentration of kDNA: To 990 µL water add 10 µL of the kDNA solution. Read the absorbance at 260 and 280 nm. DNA at a concentration of 1.0 mg/mL has an $A_{260} = 20$. An A_{260}/A_{280} ratio of 1.86 denotes minimal protein contamination. The expected yield of kDNA is 500–1000 µg.
13. To determine if the kDNA is intact, a 1% agarose gel loaded with 1, 2, and 3 µg of kDNA can be run. The procedure for this is detailed below in **Subheading 3.3.**

3.2. Isolation of [³H]kDNA from Crithidia fasciculata

1. Grow *Crithidia* as described above for unlabeled kDNA isolation. When they reach a density of 0.5–1 × 10^7/mL, add [³H]thymidine at 10 µCi/mL culture. This is at d 2 of the procedure described in **Subheading 3.1.** Grow until a density of 1.5–2 × 10^8 is reached. **Note:** appropriate precautions should be used when handling radioactive material.
2. Continue isolation as described above for unlabeled kDNA. This procedure should yield 500–1000 µg of [³H]kDNA with a specific activity of 10,000–15,000 cpm/µg kDNA.

3.3. kDNA Decatenation: Assayed by Agarose-Gel Electrophoresis (Unlabeled)

Each component in this assay has an optimal concentration (*see* **Note 1**). This may have to be established for a given sample, depending on the extraction method and origin of the sample. In general, it is very important to have the combined NaCl and KCl concentrations within 75–125 m*M* (*see* **Note 2**). ATP is essential, but can also be in excess (*see* **Note 3**). Similarly, too much protein sample can be used (*see* **Note 4**), and the presence of >1.0% DMSO will inhibit the reaction (*see* **Note 5**). With this assay, 1 U of enzyme activity is defined as the amount of protein required to decatenate 1 µg of kDNA fully in 30 min at 30°C.

1. Combine 2 µl of 10X decatenation buffer (±KCl depending on the total salt concentration), 0.5–1 µg kDNA, topoisomerase II inhibitors if required, and various concentrations of topoisomerase II-containing extracts or purified topoisomerase II. The optimal amount of 0.35 *M* NaCl nuclear extract protein for decatenation is generally 0.5–1.0 µg (usually in a volume of 1–3 µL) or 5–30 ng of purified topoisomerase II. The reaction is brought to a total volume of 20 µL with H_2O. The enzyme is added last to start the reaction at 30-s intervals.
2. Incubate in a 30°C water bath for 30 min. Stop the reaction at 30-s intervals by adding 5 µL of stop solution (**Subheading 2.3.**) and vortexing.
3. The samples are then electrophoresed in a 1% agarose gel (20 × 25 cm made in TBE) in TBE buffer at 100 V (4 V/cm) for 2–3 h. The voltage can be increased to 150 V (6 V/cm) for 1 h if the amount of buffer in the electrophoresis chamber is increased (1.5–2 cm above the upper surface of the gel); this is to dissipate the heat from the increased voltage. A 1-h electrophoresis time is often convenient when assaying column fractions during the purification of topoisomerase II.
4. Submerge the gel in the ethidium bromide solution (**Subheading 2.3.**) for 2–3 min and then wash it several times with H_2O. Alternatively, ethidium bromide can be included in both the agarose gel and in the TBE buffer at a concentration of 0.5 µg/mL. However, we do not follow this procedure, because too much waste solution containing ethidium bromide is generated. In addition, ethidium bromide is a DNA intercalator and can thus change the mobility of different DNA species when they are electrophoresed in an agarose gel. **Note:** remember to wear gloves when working with ethidium bromide and dispose of it properly. The ethidium bromide staining solution can be reused several times if stored in the dark; we have used the same 500-mL bottle for 3–4 mo.
5. The gel is then photographed under UV illumination. If the background is too high (i.e., too bright because of Etbr such that the minicircle band is not discrete or poorly visualized), the gel can be soaked in H_2O for several hours (or overnight) and a second photograph taken.
6. The large kDNA networks remain in the well, and the released minicircles migrate into the gel, usually as a discrete band. Intermediate complexes at various stages of decatenation can be seen when there is less than optimal topoisomerase II decatenating activity. To obtain semiquantitative data with this assay, the negative of the Polaroid photograph can be scanned with a densitometer to determine the amount of kDNA which has been decatenated.

3.4. kDNA Decatenation: Assayed by Centrifugation (Labeled)

1. Combine 4 µL of 10X decatenation buffer (±KCl depending on the total salt concentration), 5000–10,000 cpm [^{3}H] kDNA (usually 0.5–1 µg kDNA), topoisomerase II inhibitor if required, and topoisomerase II (e.g., 0.5–1.0 µg

nuclear extract in 1–3 µl or 5–30 ng purified topoisomerase II). Bring to a total volume of 40 µL with H_2O. Remember that topoisomerase II should be added last to start the reaction. To measure rates of decatenation at initial velocities use time-points of 15 s to 10 min. If the specific activity of the [^{3}H]kDNA is too high, it can be diluted with cold kDNA.

This reaction can be performed where either the concentration of topoisomerase II or the time of reaction is the variable. If time is the variable, then an amount of topoisomerase II that totally decatenates 0.5–1 µg kDNA in 10 min needs to be used in the assay, and should have been predetermined in an assay where topoisomerase II concentration was the variable.

2. Incubate at 30°C for 15 s to 10 min. Stop the reaction by adding 5 µL of 2.25% SDS and vortexing.
3. Centrifuge at room temperature in a microfuge at 13,000*g* for 10 min.
4. Carefully remove 30 µL of the supernatant (two-thirds of the total volume of the supernatant), which contains the released minicircles, place it in 2 mL of H_2O, and add 2 mL aqueous scintillation fluid. Count [^{3}H] in a scintillation counter.
5. Decatenation is quantified as the percentage decatenation of available cpm (available cpm are determined by counting an equivalent amount of [^{3}H]kDNA as was used in each decatenation reaction). This should be adjusted to 100% cpm of the supernatant (i.e., to avoid disturbing the kDNA network pellet, one-third of the supernatant was not counted and needs to be accounted for) minus the solvent control in absence of enzyme. These data are plotted as time vs µg [^{3}H]kDNA decatenated/µg topoisomerase II (or µg nuclear extract protein). The rate of decatenation can be determined from the linear (first-order) part of the curve as µg kDNA decatenated/µg topoisomerase II/min (*see* **Fig. 1**).

4. Notes

1. The decatenation assays described above are a relatively easy way to determine the catalytic activity of topoisomerase II in a nuclear extract, whole-cell extract, or in a pure preparation of enzyme. Decatenation of interlocked closed-circular double-strand DNA minicircles is specific for topoisomerase II. A control reaction that lacks ATP alone should demonstrate no migration of DNA into the agarose gel. The agarose gel assay is semiquantitative and useful when information regarding topoisomerase II catalytic activity is required rapidly, e.g., when assaying column fractions during FPLC purification of the enzyme. The [^{3}H]kDNA assay is a more quantitative means by which to study enzyme kinetics when comparing different isoforms, examining potentially mutated enzymes with altered activity, or determining the effects of topoisomerase II inhibitors on catalysis. However, these assays involve small volumes of each component, all of which need to be kept within a narrow concentration range to optimize the reaction. We have illustrated below the concentration dependence of a number of these components with purified topoisomerase II and/or nuclear extracts from wild-type Chinese hamster ovary cells. It is important to optimize these conditions for the samples you are investigating.

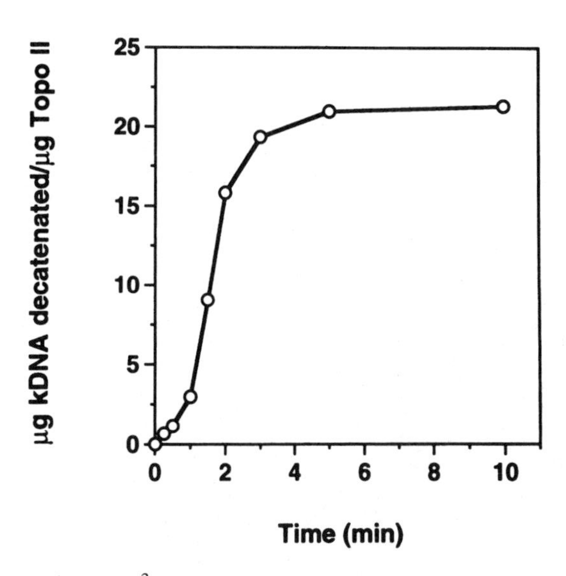

Fig. 1. Decatenation of [^{3}H]kDNA by purified topoisomerase IIα isolated from human small-cell lung cancer H209 cells. In this experiment, the decatenation of 0.7 µg [^{3}H]kDNA (20,000 cpm) by 31 ng purified topoisomerase IIα at 1 m*M* ATP and 100 m*M* NaCl was assayed over a time range of 15 s to 10 min. The linear part of this curve shows that the rate of decatenation for this enzyme preparation is about 6 µg kDNA decatenated/µg topoisomerase II/min.

2. The total concentration of salt (NaCl and/or KCl) present in the decatenation reaction is critical when measuring topoisomerase II activity. This becomes important when measuring the activity present in nuclear extracts, where the NaCl concentration is generally 0.35–1.0 *M*. A 4-µL aliquot of a 1.0 *M* NaCl nuclear extract will give a salt concentration of 200 m*M* when diluted to 20 µL; topoisomerase II is essentially inactive at this NaCl concentration. The salt dependency of the decatenation reaction should also be kept in mind when assaying column fractions during enzyme purification, since the topoisomerase II isoforms are generally eluted at 300–400 m*M* NaCl or KCl. **Figure 2** demonstrates the salt dependency of decatenation by a homogeneous preparation of topoisomerase IIα purified from wild-type Chinese hamster ovary cells. This figure demonstrates that decatenation by topoisomerase IIα is maximal over the range of 75–125 m*M* NaCl (the same results are also obtained with KCl).
3. The binding of ATP by the topoisomerase II/DNA complex is required for passing a DNA strand through the break site, whereas ATP hydrolysis is necessary for enzyme turnover. Again, there is a narrow range of ATP concentrations that are optimal for topoisomerase II decatenation activity. **Figure 3** demonstrates the ATP concentration dependence of decatenation by purified Chinese hamster ovary topoisomerase IIα. These reactions were performed in the presence of 100 m*M* NaCl and show that the total decatenation of 1 µg of kDNA occurs at

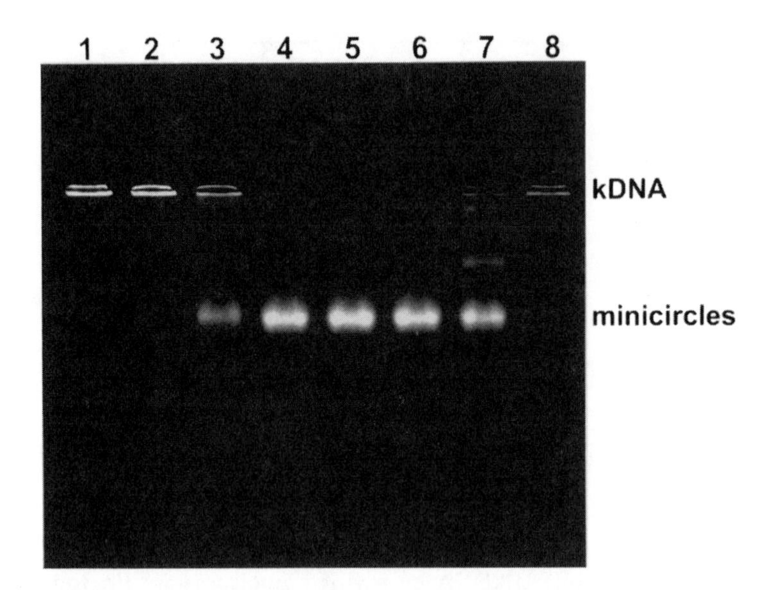

Fig. 2. Decatenation of kDNA by topoisomerase IIα purified from wild-type Chinese hamster ovary cells in the presence of increasing concentrations of NaCl. Equal concentrations of purified topoisomerase II (80 ng) were incubated with 1 μg kDNA in the absence (lane 1) or presence (lanes 2–8) of NaCl. Each sample has a constant ATP concentration of 1 m*M*. Lanes 1–8 represent 0, 25, 50, 75, 100, 125, 150, and 200 m*M* NaCl. The reaction was for 30 min at 30°C, and a 1% agarose gel was used for electrophoresis in TBE at 100 V for 2 h.

0.75–1.25 m*M* ATP. One millimolar ATP is generally used for decatenation assays.

4. The decatenation of kDNA by topoisomerase II-containing nuclear extracts is also dependent on the amount of nuclear extract protein used in the decatenation assay, independent of the salt concentration. Nuclear extracts prepared from mammalian cell lines that contain topoisomerase II are obtained with 0.35–1.0 *M* NaCl and generally have protein concentrations of 2–5 mg/mL. Optimal decatenation activity is found with 0.5–1.0 μg nuclear extract protein. Greater amounts of protein inhibit the decatenation reaction (perhaps because of histones). This inhibition of decatenation by increased protein concentrations is not seen with purified preparations of topoisomerase II. **Figure 4** is an example of this phenomenon. In this experiment, a 0.35 *M* NaCl nuclear extract was obtained from wild-type Chinese hamster ovary cells, and a decatenation assay was performed in the presence of increasing amounts of extract protein at a constant concentration of both ATP (1 m*M*) and NaCl (100 m*M*). Assaying high protein concentrations from eluted column fractions for topoisomerase II activity during the early stages of enzyme purification can mask the true elution profile of topoisomerase II isoforms.

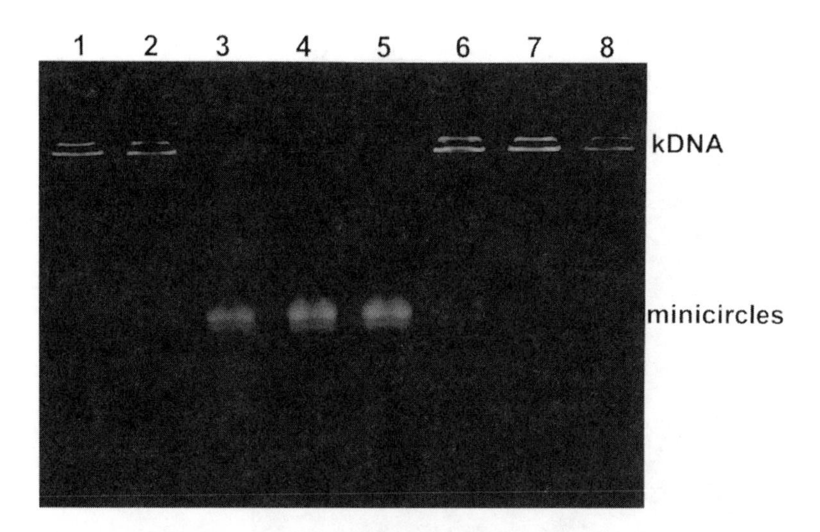

Fig. 3. Decatenation of kDNA by purified Chinese hamster ovary topoisomerase IIα at various concentrations of ATP. Equal concentrations of purified topoisomerase II (80 ng) were incubated with 1 µg kDNA in the absence (lane 1) or presence (lanes 2–8) of ATP. Each sample has a constant NaCl concentration of 100 m*M*. Lanes 1–7 represent 0, 0.5, 0.75, 1.0, 1.25, 1.5, and 2.0 m*M* ATP. Lane 8 is the decatenation with 1.0 m*M* ATP in the absence of topoisomerase II. The time of reaction and gel electrophoresis conditions are the same as **Fig. 2**.

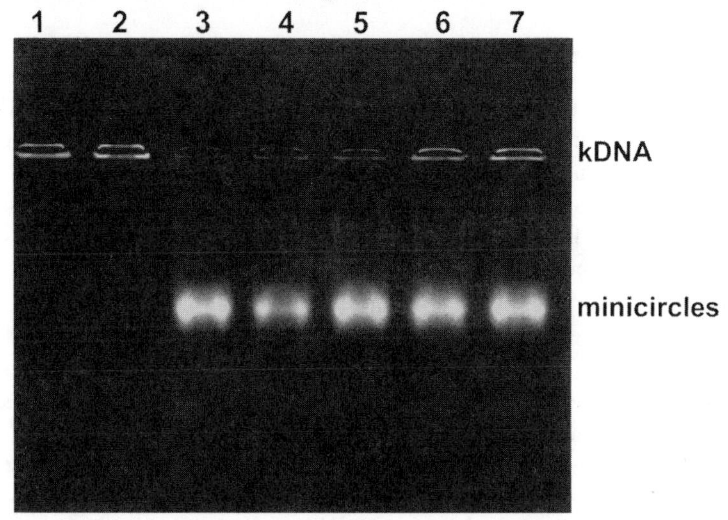

Fig. 4. Decatenation of kDNA by topoisomerase II-containing 0.35 *M* NaCl nuclear extracts obtained from wild-type Chinese hamster ovary cells. Increasing amounts of a nuclear extract were incubated with 1 µg kDNA in the presence of 1 m*M* ATP and 100 m*M* NaCl. Lanes 1–7 represent 0, 0.25, 0.5, 0.75, 1.0, 2.0, and 3.0 µg of nuclear extract protein. The time of reaction and gel electrophoresis conditions are the same as **Fig. 2**.

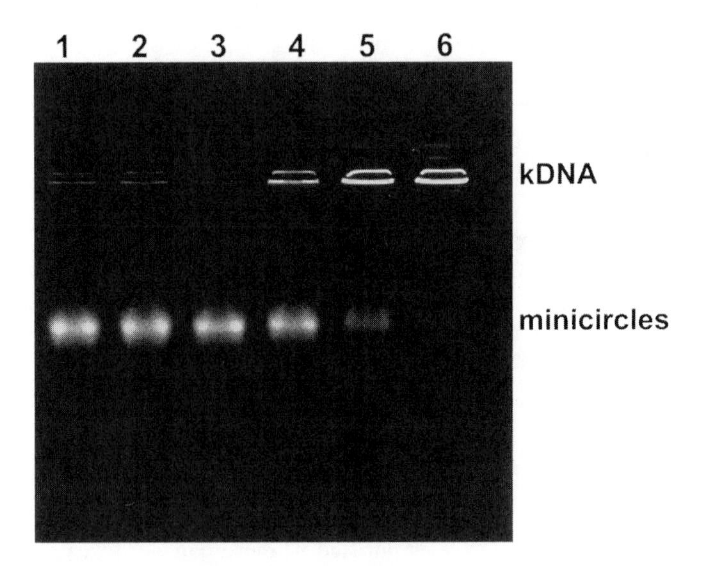

Fig. 5. Decatenation of kDNA by topoisomerase II-containing 0.35 *M* NaCl nuclear extracts from Chinese hamster ovary cells in the presence of DMSO. Equal amounts of nuclear extract (0.5 μg) were incubated with 1 μg kDNA in the presence of 1 m*M* ATP and 100 m*M* NaCl in the absence (lane 1) or presence (lanes 2–6) of DMSO. Lanes 1–6 represent 0, 0.5, 1, 2, 3, and 5 % (v/v) DMSO. The time of reaction and gel electrophoresis conditions are the same as **Fig. 2**.

5. Certain solvents used to dissolve topoisomerase II inhibitors can also inhibit the decatenation reaction, independent of the activity of the topoisomerase II poison. This is seen with both nuclear extracts and homogeneous preparations of the enzyme. A common example of this is found with VP-16 (etoposide), which is dissolved in DMSO. **Figure 5** shows the decatenation of kDNA by 0.5 μg protein of a 0.35 *M* NaCl nuclear extract in the presence of fixed amounts of ATP and NaCl in the presence of increasing concentrations of DMSO. Concentrations of DMSO that are >1% (v/v) generally inhibit the decatenation of kDNA.

References

1. Marini, J. C., Miller, K. G., and Englund, P. T. (1980) Decatenation of kinetoplast DNA by topoisomerases. *J. Biol. Chem.* **255,** 4976–4979.
2. Englund, P. T. (1978) The replication of kinetoplast DNA networks in *Crithidia fasciculata*. *Cell* **14,** 157–168.
3. Chen, J., Englund, P. T., and Cozzarelli, N. R. (1995) Changes in network topology during the replication of kinetoplast DNA. *EMBO J.* **14,** 6339–6347.
4. Sullivan, D. M., Latham, M. D., Rowe, T. C., and Ross, W. E. (1989) Purification and characterization of an altered topoisomerase II from a drug-resistant Chinese hamster ovary cell line. *Biochemistry* **28,** 5680–5687.

5. Simpson, A. M., and Simpson, L. (1974) Isolation and characterization of kinetoplast DNA networks and minicircles from *Crithidia fasciculata*. *J. Protozool.* **21,** 774–781.
6. Simpson, A. M., and Simpson, L. (1974) Labeling of *Crithidia fasciculata* DNA with [^{3}H]Thymidine. *J. Protozool.* **21,** 379–382.

3

Plasmid DNA Supercoiling by DNA Gyrase

Penny J. Sayer, Martin L. Goble, Mark Oram, and L. Mark Fisher

1. Introduction

DNA gyrase catalyzes DNA supercoiling in a reaction coupled to ATP hydrolysis *(1)*. The enzyme has been found in many eubacterial species and is unique among the topoisomerases in promoting negative supercoiling of DNA *(2,3)*. A variety of studies have shown that gyrase is essential for bacterial growth with roles in DNA replication, transcription, and recombination. Moreover, in combination with the relaxing activity of DNA topoisomerase I, gyrase is responsible for the homeostatic regulation of DNA supercoiling in bacteria (reviewed in **refs.** ***3,4***). The enzyme from *Escherichia coli* has been the most extensively characterized, and is a tetramer made up of two GyrA and two GyrB subunits encoded by the *gyrA* and *gyrB* genes, respectively. DNA supercoiling takes place by the directional crossing of a DNA duplex through a transient enzyme-bridged double-strand break in a 120–150 bp segment of DNA wrapped on the enzyme *(5–7)*. This process changes the linking number of DNA in steps of two, and together with an ability to form and resolve DNA knots and catenanes, establishes gyrase as a type II topoisomerase.

Aside from its essential biological role, gyrase is also of interest as a target for the quinolone and coumarin families of antibacterial agents. Quinolones, and particularly fluoroquinolones, such as ciprofloxacin and norfloxacin, inhibit gyrase by interfering with DNA breakage-reunion mediated by the GyrA subunits, whereas coumarins, such as novobiocin and coumermycin, act as competitive inhibitors for ATP binding to the GyrB subunits *(2,3)*. Many studies have shown that bacterial resistance to coumarins arises from mutational alterations in GyrB *(3,8)*: resistance to quinolones involves mutations in GyrA and Gyr B (reviewed in **ref.** ***9***). Clearly, purification of gyrase from bacterial

From: *Methods in Molecular Biology, Vol. 95: DNA Topoisomerase Protocols, Part II: Enzymology and Drugs*
Edited by N. Osheroff and M.A. Bjornsti

pathogens and resistant mutants, and study of its interactions with various inhibitors will be a continuing area of interest.

Gyrase activity is conveniently assayed by following the supercoiling of plasmid DNA. Gyrase may be in the form of purified GyrA and GyrB subunits (obtained from overexpressing strains), which can be mixed to reconstitute activity ***(10)***. Alternatively, the activity present in bacterial extracts may be used ***(1)***. The assay involves incubating gyrase with a relaxed plasmid DNA substrate in the presence of Mg^{2+} and ATP, and then analyzing the DNA products by agarose-gel electrophoresis. Supercoiled DNA, being more compact than its relaxed counterpart, migrates more rapidly in the gel allowing separation and quantitation. The prerequisite for the assay is a relaxed DNA substrate free of nicked DNA circles. Although, in principle, almost any relaxed plasmid DNA could be employed as a substrate, it is convenient and now customary to use the small (4.3-kb) ampicillin resistance plasmid pBR322, thus maximizing the electrophoretic separation of relaxed and supercoiled DNA species. pBR322 is commercially available, is readily propagated in *E. coli*, and is a substrate for gyrase from a variety of sources. The following sections describe the isolation of supercoiled pBR322, preparation of relaxed pBR322 substrate, and its use in the supercoiling assay.

2. Materials

2.1. Isolation of Supercoiled pBR322 from E. coli

1. LB medium: 10 g/L bacto-tryptone, 5 g/L yeast extract, 5 g/L NaCl made up in deionized water and adjusted to pH 7.5 with 5 *M* NaOH. For solid media, add 15 g/L of agar. Autoclave for 20 min at 15 lb/sq. in. Sodium ampicillin (Sigma) (10 mg/mL) dissolved in water, sterilized by passage through a 0.2-μ filter (Nalgene) is added where appropriate to 50 μg/mL.
2. Solution 1: 25 m*M* Tris-HCl, pH 8.0, containing 50 m*M* glucose and 10 m*M* NaEDTA.
3. Solution 2: 0.2 *M* NaOH containing 1% sodium dodecyl sulfate (SDS) (*see* **Note 1**).
4. Solution 3: 5 *M* potassium acetate, pH 5.2. (Prepare 600 mL of 5 *M* potassium acetate and add 115 mL glacial acetic acid. Make up to 1 L with water).
5. TE buffer: 10 m*M* Tris-HCl, pH 7.5, 1 m*M* EDTA.
6. 10 mg/mL ethidium bromide (EtBr) (*see* **Note 2**).
7. CsCl: Boehringer-Mannheim, crystallized for molecular biology uses.
8. UV lamp: Mineralight Model UVSL-58 (Ultraviolet Products, Cambridge, UK).
9. Propan-2-ol, 1-butanol.

2.2. Preparation of Relaxed pBR322 DNA

1. 4X relaxation buffer: 80 m*M* Tris-HCl, pH 8.0, 1 m*M* EDTA, 20% w/v glycerol, 60 μg/mL bovine plasma albumin, 0.48 *M* KCl, 2 m*M* DTT.

2. Agarose gel: 0.8% agarose in TBE (90 m*M* Tris base, 90 m*M* boric acid, 2.5 m*M* EDTA).
3. 5X dye mix: 5% SDS, 25% glycerol, 0.25 mg/mL bromophenol blue.
4. Topoisomerase I: Human enzyme from Topogen (*see* **Notes 5** and **6**).

2.3. DNA Supercoiling Assay

1. 3X Gyrase assay mix: 105 m*M* Tris-HCl, pH 7.5, 18 m*M* $MgCl_2$, 5.4 m*M* spermidine, 72 m*M* KCl, 15 m*M* DTT, 1.08 mg/mL BSA, 19.5% glycerol (w/v), and (optional) 90 µg/mL *E. coli* tRNA (Calbiochem).
2. Gyrase dilution buffer: 50 m*M* Tris-HCl, pH 7.5, 0.2 *M* KCl, 5 m*M* DTT, 1 m*M* EDTA, 20 mg/mL BSA, 50% glycerol. (Both assay mix and dilution buffer can be made up and stored in aliquots at –20°C.)
3. 50 m*M* ATP solution: 27.5 mg ATP (disodium salt) dissolved in 1 mL 100 m*M* NaOH.
4. Relaxed pBR322 DNA.
5. 5X Dye mix, as in previous section.
6. Gyrase preparation or the individual purified GyrA and GyrB subunits.

3. Methods

3.1. Isolation of Supercoiled pBR322 DNA

Plasmid pBR322 should be maintained in a recA- strain of *E. coli*, e.g., MG1182 or DH5, thereby minimizing the intracellular formation of concatenated plasmid dimers and trimers that arise through recombination. Supercoiled pBR322 dimers migrate in agarose gels with a similar mobility to relaxed monomeric circles and, when present in large amounts, can confuse the outcome of the supercoiling assay. Plasmid dimers can be avoided by retransforming the *E. coli recA* strain with plasmid DNA and selecting ampicillin-resistant colonies, which by the plasmid miniprep procedure are shown to contain only monomeric plasmid DNA. Although a variety of methods are now available for isolating plasmid DNA from bacteria, including various column procedures, e.g., Qiagen columns, it is preferable to use a method that employs cesium chloride-ethidium bromide (CsCl-EtBr) isopycnic density ultracentrifugation. This technique separates the DNA into two bands: one contains supercoiled plasmid, the other chromosomal DNA and nicked (open circular) plasmid, which are not substrates for gyrase and are undesirable contaminants in the supercoiling assay. A purification protocol, modified from that described in **ref. *11*** follows.

1. Streak out the *E. coli* strain transformed with pBR322 on LB-ampicillin plates, and incubate overnight at 37°C.
2. Use a sterile wire loop to pick three to four bacterial colonies, and use to inoculate 20 mL of LB broth containing ampicillin. Incubate for about 12 h at 37°C on an orbital shaker set at 250 rpm.

3. Use 5 mL of this culture to inoculate each of four 2-L conical flasks each containing 500 mL of LB broth plus ampicillin, and grow overnight at 37°C with shaking.
4. Pellet the cells at 4200*g* for 10 min at 4°C. Process the pellet from each 500-mL culture as described below.
5. Discard the supernatant, and resuspend each pellet in 20 mL of solution 1 (*see* **Note 1**). Leave on ice for 10 min while preparing solution 2.
6. Add 40 mL of freshly prepared solution 2, swirl gently, and leave on ice for 10 min.
7. To the lysed cells, add 40 mL of solution 3, and leave on ice for a further 10 min.
8. Spin the sample at 4200*g* for 30 min at 4°C in a Sorvall GS3 to pellet the cell debris.
9. Pour the supernatant through muslim to remove any contamination by loose pellet, add 0.6 vol of propan-2-ol, mix, and leave at room temperature for 10 min.
10. Pellet the DNA by spinning at 4200*g* for 30 min at room temperature. Pour off the supernatant, invert the tubes, and allow to drain for 10–30 min.
11. Resuspend the pellet in 4.5 mL of TE. Add 6.2 g CsCl, and allow to dissolve by inversion. Then in the dark, add 100 µL of 10 mg/mL EtBr (*see* **Note 2**) and mix.
12. Transfer the solutions to polyallomer tubes suitable for the Beckman VTi80 or VTi70 ultracentrifuge rotor, and balance the tubes carefully to within 0.02 g (*see* **Note 3**). Heat-seal the tubes, and centrifuge either for 14 h at 190,000*g* or for 4 h at 320,000*g*.
13. In a darkened room, remove the tubes from the rotor and pierce at the top with a syringe to make an air hole. Two bands of DNA should be visible under normal light: the upper (usually smaller) band contains chromosomal and nicked plasmid DNA, whereas the lower band contains closed circular supercoiled plasmid DNA. If the bands are faint, they may be visualized in a darkened room using a handheld UV lamp on the long-wavelength (366-nm) setting. In this case, a face visor must be worn to protect the eyes from UV light. Puncture the tube by inserting a syringe needle just below the lower band and draw off the solution containing the closed circular DNA into the syringe (*see* **Note 4**).
14. In the dark, transfer the retrieved solution into a 15-mL Falcon tube, add an equal volume of water-saturated 1-butanol, shake, and allow the phases to separate. Remove and discard the upper butanol phase containing EtBr. Repeat this process (usually four to five times) until the pink color of the EtBr is no longer visible in the aqueous phase.
15. Dialyze the DNA solution against 1 L TE buffer overnight at 4°C and then against 1 L TE over the next day to remove CsCl.
16. Precipitate the DNA with ethanol, resuspend in 0.2–1.0 mL of TE, and determine the DNA concentration by measuring the OD_{260} on a suitably diluted aliquot. The yield of supercoiled plasmid pBR322 DNA varies depending on the preparation, but is approx 0.25–0.5 mg/500 mL bacterial culture. The DNA should be checked by electrophoresis on a 0.8% agarose gel. It should be >90% supercoiled and should be stored at 4°C. The DNA is stable, but over a period of several months, may slowly undergo conversion to the nicked form.

3.2. Preparation of Relaxed pBR322 DNA

The most convenient way of obtaining relaxed pBR322 is to relax the supercoiled form using a preparation of mammalian DNA topoisomerase I. Previous work has used topoisomerase I activity released by a 0.35 *M* NaCl wash of cell nuclei, e.g., from calf thymus, chicken erythrocytes, or from the human HeLa cell line. If such extracts are used, it is essential that they be free of nuclease activity to avoid nicking of the plasmid DNA. We have successfully used such extracts. However, recently, human topoisomerase I has become available commercially from Topogen, Columbus, OH, and we prefer to use this purified preparation (*see* **Notes 5** and **6**). It may be mentioned that bacterial topoisomerases I, e.g., the commercially available enzymes from *E. coli* or *Micrococcus luteus*, do not produce a completely relaxed product and should be avoided.

1. Place approx 200 µg of supercoiled pBR322 (in TE buffer) in an Eppendorf centrifuge tube on ice.
2. To the DNA, add an appropriate volume of 4X relaxation buffer to give a final concentration of 1X when all the topoisomerase I is added (final volume typically 0.5–1.0 mL).
3. Calculate the required volume of topoisomerase I (1 U relaxes 0.4 µg of pBR322 DNA in 1 h at 30°C). Add half the required units to the relaxation reaction. Incubate in a circulating water bath at 30°C for 30 min.
4. Add the remaining topoisomerase I, and continue the incubation at 30°C for a further 30 min.
5. Remove an aliquot of the reaction mixture (2–3 µL), add 2 µL 5X dye mix, and make up to 10 µL with distilled water. Check that the DNA has all been relaxed by running the sample on a 0.8% agarose gel at 100 V (5–10 V/cm) for 1–2 h using a sample of supercoiled DNA run alongside as a control. Leave the remaining reaction at 30°C while the gel is run, stained with EtBr, and the DNA examined under UV transillumination (*see* **Note 7**).
6. If the DNA has all relaxed, terminate the reaction by adding SDS to the tube still incubating at 30°C to give a final concentration of 1% Vortex (*see* **Note 8**).
7. Add an equal volume of phenol (equilibrated with 50 m*M* Tris-HCl, pH 8.0), vortex, and spin at 1200*g* in an MSE Centaur centrifuge (any comparable centrifuge will suffice). Remove the upper aqueous layer to a fresh tube, and back-extract the phenol layer with an equal volume of TE. Vortex and spin as above, and combine the aqueous phases.
8. Vortex sample with an equal volume of 1-butanol to remove phenol. Repeat two to three times to reduce the volume of the aqueous phase, and concentrate the DNA to ~200 µg/mL.
9. Dialyze the DNA at 4°C against 800 mL TE for 2 h and again for 4 h.
10. Check the quality and quantity of DNA by electrophoresis on a 0.8% agarose gel.

3.3. DNA Supercoiling Assay

The main uses of the assay are 1) to monitor purification of gyrase from bacterial extracts, and 2) to test the action of inhibitors on the supercoiling activity of purified enzyme. Crude extracts from bacteria normally contain substantial nuclease activity that can nick plasmid DNA and obfuscate the supercoiling assay. Therefore, bacterial extracts are usually subjected to Polymin P or ammonium sulfate fractionation, or column chromatography prior to assay. The effects of nucleases are suppressed by inclusion of tRNA in the assay buffer. Highly purified DNA gyrases from *E. coli* and *M. luteus* are available from Lucent, University of Leicester, UK and from Gibco BRL. Alternatively, *E. coli* GyrA and GyrB proteins can be individually purified either from bacterial extracts by affinity chromatography on novobiocin-Sepharose ***(12)*** or from overproducing strains engineered with inducible *gyrA* and *gyrB* plasmids. Neither subunit alone is active and must be mixed to reconstitute enzyme activity. Each subunit is assayed in the presence of a 10-fold or greater molar excess of the complementing protein allowing the specific activity of each subunit to be assessed independently. The latter approach based on **ref. *10*** is described below.

1. Gyrase assays are conducted in 35 m*M* Tris-HCl, pH 7.5, 24 m*M* KCl, 6 m*M* $MgCl_2$, 1.8 m*M* spermidine, 0.36 mg/mL BSA containing 1.4 m*M* ATP, 0.4 µg relaxed pBR322, and various dilutions of the gyrase activity to be assayed (final volume 35 µL). In the case of purified GyrA and GyrB subunits, various dilutions of the subunit to be assayed are made in gyrase dilution buffer (**Note 9**) (on ice) and added to reaction mix on ice containing an excess (>10 U) of the complementing subunit.
2. Make up a cocktail on ice consisting of 11.7 µL 3X gyrase assay mix, 2 µL relaxed DNA (i.e., 0.4 µg), ATP (**Note 10**), complementing subunit (where appropriate), and sterile water to 33 µL. (The quantities can be increased accordingly depending on the number of assays to be run.) The mix is distributed to 1.5-mL Eppendorf tubes on ice, and the gyrase (2 µL) is added. Mix by gently tapping the tube; do not vortex. To initiate the reaction, transfer the tubes to a circulating water bath at 25°C, and incubate for 1 h.
3. Terminate the reaction by adding 8 µL of 5X dye mix, and analyze the products by electrophoresis on a 0.8% agarose gel run overnight at 2 V/cm (**Note 11**). Gels should be stained in TBE containing 0.5 mg/mL EtBr for 30 min to 1 h, destained for 1 h in TBE, and photographed on a transilluminator using a red Wratton filter. An adjustable Land camera and Polaroid 665 film give good photographic results.
4. One unit of gyrase is defined as the amount of enzyme that supercoils 50% of the input pBR322 DNA under these reaction conditions.
5. **Figure 1** shows a typical assay for gyrase activity in which a suitably diluted *E. coli* GyrA subunit is assayed in the presence of excess GyrB (**Notes 12** and **13**).

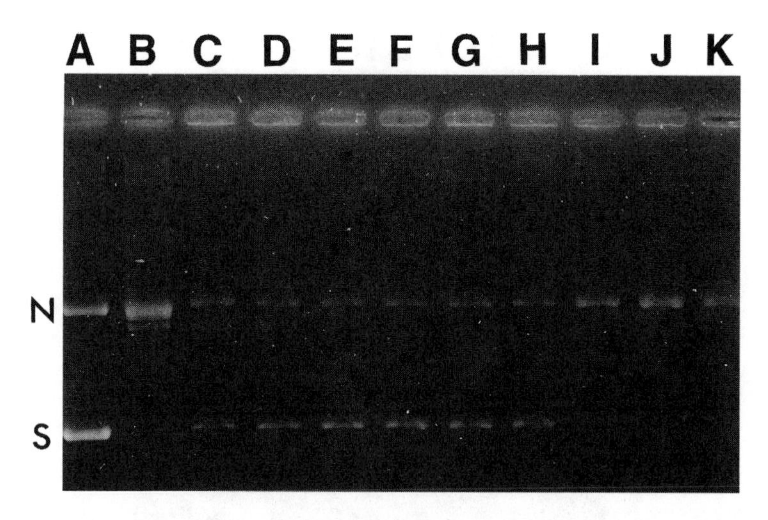

Fig. 1. DNA supercoiling by *E. coli* DNA gyrase. Relaxed pBR322 DNA (lanes B–K) was incubated with 100 U GyrB and fourfold serial dilutions of GyrA subunit (C–K) in the presence of ATP. Products were resolved by electrophoresis on a 0.8% agarose gel. Lane A is a sample of supercoiled pBR322 DNA (S), which contained a small amount of the nicked species (N). Bands running between N and S are partially supercoiled pBR322 circles.

6. **Figure 2** shows the use of the assay to examine the effect of a gyrase inhibitor on supercoiling activity (**Note 14**). For inhibitor studies, it is usual to employ 2 U of gyrase activity so that the inhibitory effect can be detected in the region of greatest sensitivity of the assay.

4. Notes

1. Solutions 1 and 3 should be ice-cold: Solution 2 should be made up just before use.
2. **Care:** EtBr is a powerful mutagen. Gloves should be worn when handling EtBr solutions, and they should be disposed of using the proper procedures.
3. It is essential that the ultracentrifuge tubes be correctly balanced.
4. The top band can be withdrawn first using a syringe, thereby minimizing any contamination of the supercoiled plasmid DNA (*see* **ref. *11*** for further explanation).
5. Eukaryotic DNA topoisomerase II also relaxes DNA in a reaction requiring ATP and Mg^{2+} ions. However, inclusion of magnesium ions alters the pitch of the DNA helix such that DNA relaxed in the presence of Mg^{2+} is not relaxed in its absence, e.g., when analyzed under the conditions of agarose-gel electrophoresis. Mammalian topoisomerase I is functional in the absence of Mg^{2+} and is preferable.
6. Topoisomerase I should be stored in aliquots at –80°C. It is not advisable to use enzyme that has been thawed and refrozen.
7. EtBr must not be included in the agarose gel: by intercalating into relaxed DNA, it alters its mobility on gels to that of the supercoiled form.

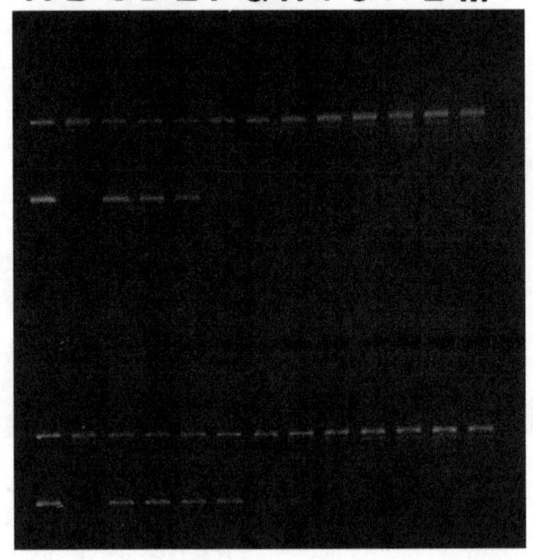

Fig. 2. Inhibition of DNA gyrase activity by the cyclothialidine GR122222X. Relaxed pBR322 (B–M) was incubated in the absence (B, C) or presence of the cyclothialidine inhibitor at 5, 10, 25, 50, 75, 100, 150, 200, 250, and 500 ng/mL (D–M). Top: 4 U GyrB, 20 U GyrA, 0.2 m*M* ATP. Bottom: 2 U GyrB, 20 U GyrA and 1.4 m*M* ATP. Lane A contained supercoiled pBR322.

8. The distribution of relaxed DNA topoisomers generated by topoisomerase I is affected by temperature.
9. Do not add more than 2 μL of gyrase diluent to the assay, since it contains high salt concentrations and can be inhibitory.
10. ATP solutions are best stored in aliquots at –20°C and discarded after use.
11. Gels are best run slowly overnight to obtain the optimum separation of supercoiled and relaxed DNA.
12. Relaxed DNA consists of a Gaussian distribution of topoisomers of differing linking numbers, which are resolved by the gel into several discrete bands (*see* **Fig. 1**, lane B) and **Fig. 2**. Nucleases present in a gyrase preparation cause DNA nicking, which is revealed by the absence of observable supercoiling and the conversion of the relaxed DNA ladder to a single band, i.e., nicked DNA. This can be problematical in assaying crude bacterial extracts.
13. Purified *E. coli* GyrA subunit has an SA of ~10^6 U/mg, whereas that of purified GyrB subunit is some 10-fold less. This difference can be minimized by preincubating GyrA and GyrB subunits on ice for 30 min prior to assay.
14. Cyclothialidines are competitive inhibitors of ATP binding to the GyrB subunit ***(13)***.

Acknowledgments

M. L. G. and P. S. were supported by a European Union Concerted Action grant PL931318 and by an MRC Collaborative Studentship Award (jointly with Glaxo-Wellcome), respectively.

References

1. Gellert, M., Mizuuchi, K., O'Dea, M. H., and Nash, H. A. (1976) DNA gyrase: an enzyme that introduces superhelical turns into DNA. *Proc. Natl. Acad. Sci. USA* **73,** 3872–3876.
2. Reece, R. J. and Maxwell, A. (1991) DNA gyrase: structure and function. *Crit. Rev. Biochem. Mol. Biol.* **26,** 335–371.
3. Wang, J. C. (1996) DNA topoisomerases. *Ann. Rev. Biochem.* **65,** 635–692.
4. Fisher, L. M. (1986) DNA supercoiling and gene expression. *Nature* **307,** 686–687.
5. Brown, P. O. and Cozzarelli, N. R. (1979) A sign inversion model for enzymatic supercoiling of DNA. *Science* **206,** 1081–1083.
6. Mizuuchi, K., Fisher, L. M., O'Dea, M. H., and Gellert, M. (1980) DNA gyrase action involves the introduction of transient double strand breaks into DNA. *Proc. Natl. Acad. Sci. USA* **77,** 1847–1851.
7. Fisher, L. M., Mizuuchi, K., O'Dea, M. H., Ohmori, H., and Gellert, M. (1981) Site-specific interaction of DNA gyrase with DNA. *Proc. Natl. Acad. Sci. USA* **78,** 4165–4169.
8. Maxwell, A. (1993) The interaction between coumarin drugs and DNA gyrase. *Mol. Microbiol.* **9,** 681–686.
9. Fisher, L. M., Oram, M., and Sreedharan, S. (1992) DNA gyrase: mechanism and resistance to 4-quinolone antibacterial agents, in *Molecular Biology of DNA Topoisomerases and Its Application to Chemotherapy* (Andoh, T., ed.), CRC, Boca Raton, pp. 145–155.
10. Mizuuchi, K., Mizuuchi, M., O'Dea, M. H., and Gellert, M. (1984) Cloning and simplified purification of *Escherichia coli* DNA gyrase A and B proteins. *J. Biol. Chem.* **259,** 9199–9201.
11. Sambrook, J., Fritsch, E. F., and Maniatis, T. (1989) Molecular Cloning, 2nd ed. Cold Spring Harbor Laboratory Press, Cold Spring Harbor, NY.
12. Staudenbauer, W. L. and Orr, E. (1982) DNA gyrase: affinity chromatography on novobiocin-sepharose and catalytic properties. *Nucleic Acids Res.* **9,** 3589–3603.
13. Oram, M., Dosanjh, B., Gormley, N. A., Smith, C. V., Fisher, L. M., Maxwell, A., and Duncan, K. (1996) The mode of action of GR122222X, a novel inhibitor of DNA gyrase. *Antimicrob. Agents Chemother.* **40,** 473–476.

4

Analyzing Reverse Gyrase Activity

Michel Duguet, Christine Jaxel, Anne-Cécile Déclais, Fabrice Confalonieri, Janine Marsault, Claire Bouthier de la Tour, Marc Nadal, and Christiane Portemer

1. Introduction

In "inventing" reverse gyrase, the living world has built a unique and remarkably sophisticated enzyme that is more than a simple topoisomerase. Reverse gyrase possesses the unique ability to catalyze the production of positive supercoils in a closed-circular DNA at the expense of ATP *(1)*; *see* **ref. 2** for review.

The enzyme is widely distributed, both in Archaea and Bacteria, but restricted to thermophilic species *(3,4)*. However, recent data suggest that a reverse gyrase activity could be present in eukaryotes *(5)*.

The biological function of reverse gyrase is still a matter of speculation. It could be required to stabilize DNA at high temperatures in thermophilic organisms or to control the level of recombination in eukaryotes.

The positive supercoiling reaction catalyzed by reverse gyrase is apparently symmetrical to that of the "classical" gyrase. However, the possible mechanism of reverse gyrase, suggested by enzymatic and biochemical experiments conjugated to sequence analysis, is of a completely new kind: reverse gyrase is a single polypeptide apparently made of two domains of almost the same size. The C-terminal half is a type I topoisomerase, clearly related to top A gene product, the 5' topoisomerase I in *Escherichia coli*, whereas the N-terminal half presents all of the seven boxes found in DNA and RNA helicases *(6,7)*.

Thus, it seems that in the course of evolution, reverse gyrase was built by the fusion of a "helicase-like" domain to an ATP-independent type I topoisomerase. Remarkably, the unique ATP binding site resides within the

From: *Methods in Molecular Biology, Vol. 95: DNA Topoisomerase Protocols, Part II: Enzymology and Drugs*
Edited by N. Osheroff and M.A. Bjornsti © Humana Press Inc., Totowa, NJ

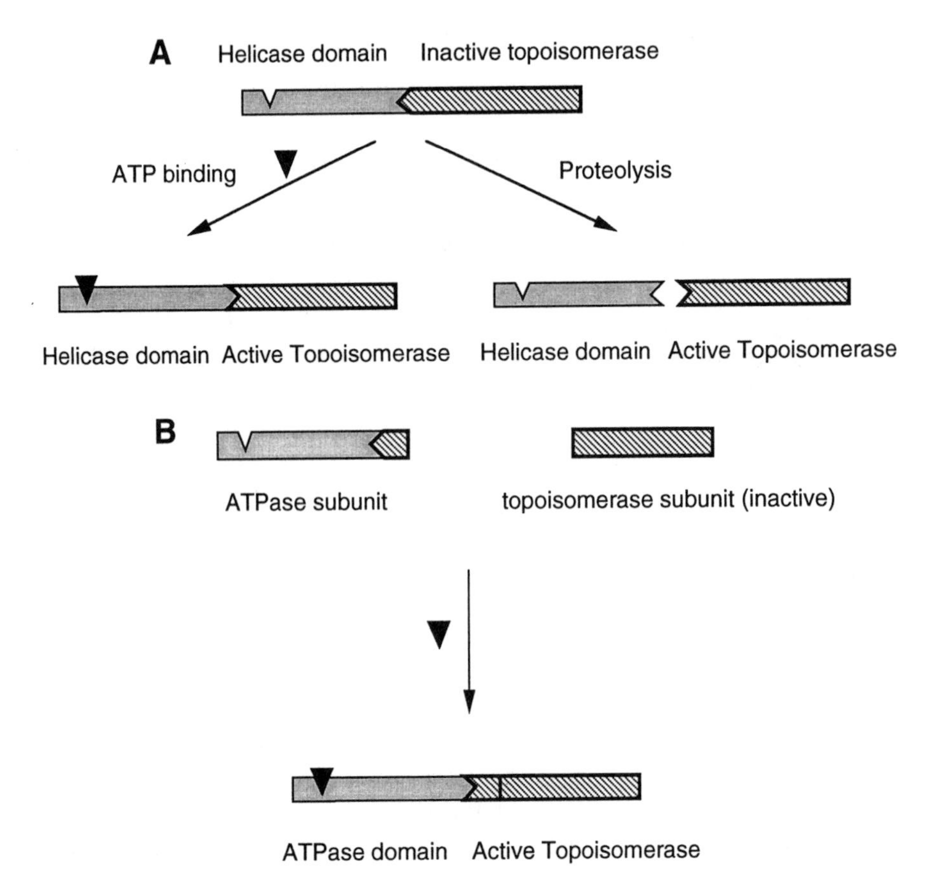

Fig. 1. Schematic structures of one-subunit **(A)** or two-subunit **(B)** reverse gyrase. The ATP molecule is represented by the black triangle. Conformational changes in the helicase (gray) and topoisomerase (hatched) domains from inactive to active are figurated.

"helicase" domain: when this site is not occupied, the activity of the topoisomerase domain is repressed (**Fig. 1A**). This view was confirmed by the appearence of an ATP-independent topoisomerase I activity on limited proteolytic cleavage of reverse gyrase *(8)*. Recently, a two-subunit reverse gyrase was described in which the topoisomerase subunit is truncated and, therefore, only active in the presence of the other subunit and ATP *(9)* (**Fig. 1B**).

The structure in two domains also provides clues to the mechanism of reverse gyration. One possibility is that the N-terminal part, acting as a helicase, produces a partition between two independent topological domains, one positively supercoiled and the other negatively supercoiled (or with local strand separation). Specific relaxation of this last domain by the topoisomerase I

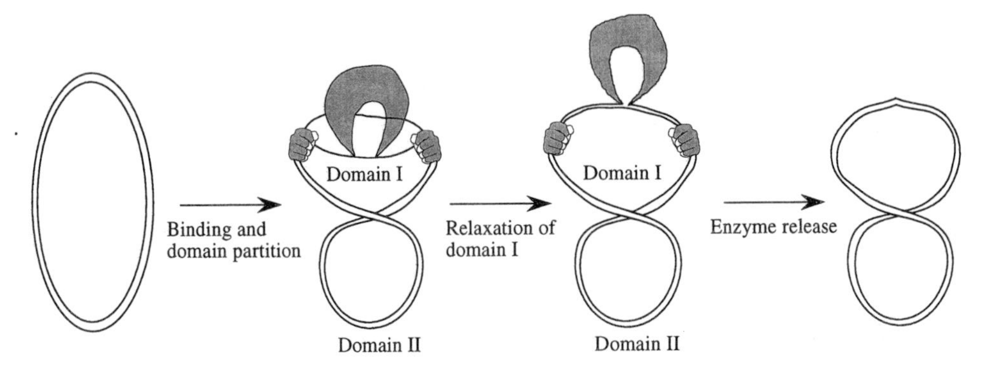

Fig. 2. Schematic model for positive supercoiling. The "hands" represent anchorage points of reverse gyrase to DNA required to maintain the separation between the two topological domains.

activity contained in the C-terminal part would yield net positive supercoils (**Fig. 2**).

Evidently, the composite structure of reverse gyrase that is described above results in multiple enzymatic activities and makes the analysis of its properties more difficult. Moreover, some of the putative activities of reverse gyrase (i.e., helicase) have not so far been demonstrated. This chapter will only describe the best-characterized activities of the enzyme.

Defining an accurate assay for measuring topoisomerase activity is always difficult because the structure of the substrate DNA and consequently its affinity for the enzyme change after each topoisomerization cycle. Reverse gyrase catalyzes the increase of the linking number between the two strands of a duplex: starting from a negatively supercoiled substrate, it is able, in the same reaction, first to relax the substrate and then to introduce positive supercoils. In this reaction, as the Lk increases, the affinity of the enzyme regularly decreases (**Fig. 3**). Therefore, the first part of the reaction, ATP-dependent relaxation of negative supercoils, is extremely efficient and is used as a standard assay.

2. Materials

2.1. DNA

The closed-circular substrate can either be a viral DNA or a bacterial plasmid DNA. The various conditions to obtain enzymatic activities can be adapted as a function of the size of the DNA. As for all topoisomerase reactions, it is important that the starting material contains a low level of nicked circles. It is therefore necessary to purify the DNA further by CsCl equilibrium centrifugation in the presence of ethidium bromide.

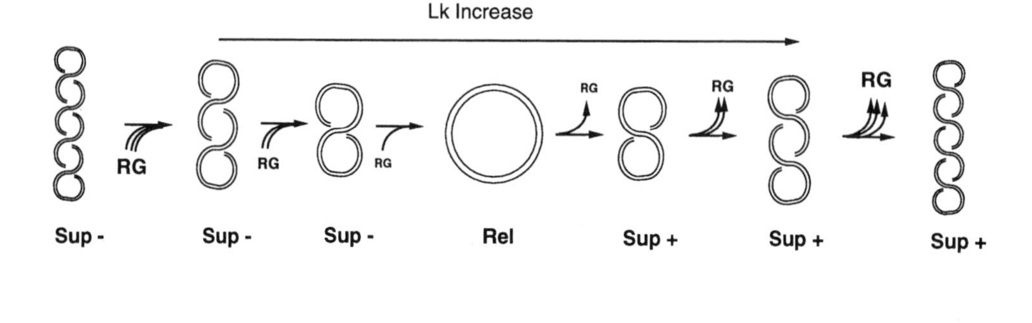

Fig. 3. Topological changes catalyzed by reverse gyrase starting from negatively supercoiled substrate. In the first part of the reaction (relaxation), the relative affinity of reverse gyrase for its substrate decreases, as represented by the decreasing number of arrows originating from RG. In the second part, the increasing number of arrows represents the increasing tendency of reverse gyrase to dissociate from the substrate as it is more and more positively supercoiled. Negative supercoils (Sup^-) are represented as a right-handed superhelix, whereas positive supercoils (Sup^+) are left-handed.

1. Preparation of DNA relaxed at 37°C: plasmid DNA (about 100 µg) is incubated with eukaryotic topoisomerase I (e.g., calf thymus topoisomerase I) for 1 h at 37°C in the following buffer: 50 m*M* Tris-HCl, pH 8.0, 0.5 m*M* DTT, 0.5 m*M* EDTA, 30 µg/mL bovine serum albumin, 170 m*M* KCl. In some cases, 4 m*M* $MgCl_2$ are added to the reaction mixture.
2. Preparation of DNA relaxed at 75°C: plasmid DNA is incubated with a thermophilic ATP-independent topoisomerase I (e.g., *Fervidobacterium islandicum* topoisomerase I *[10]*) for 1 h at 75°C in the following buffer: 50 m*M* Tris-HCl, pH 8.0, 0.5 m*M* DTT, 0.1 m*M* EDTA, 30 µg/mL bovine serum albumin, 10 m*M* $MgCl_2$, 30 m*M* NaCl, and 6% ethylene glycol.

   For both preparations, after incubation with the enzyme, the sample is treated as follows: an incubation with 0.5 mg/mL proteinase K at 65°C in the presence of 1% SDS for 30 min, and two chloroform/isoamyl alcohol extractions. The aqueous phase is precipitated by ethanol in the presence of ammonium acetate, and the pellet is washed twice with ethanol 70%. The pellet is dried and resuspended in 10 m*M* Tris-HCl, pH 8.0, 1 m*M* EDTA (TE) buffer.
3. Preparation of form II DNA: plasmid DNA (about 2 µg) is incubated with 1.2 U of DNase I ***(11)*** in the presence of 80 µg of ethidium bromide for 30 min at 31°C in the following buffer: 50 m*M* Tris-HCl, pH 7.4, 10 m*M* $MgCl_2$, 20 m*M* NaCl. The total incubation volume is 100 µL, and the incubation is performed in the dark. After proteolysis by proteinase K, the sample is extracted once with phenol, once with chloroform/isoamyl alcohol, and ethidium bromide (EtBr) is removed

by several extraction/concentrations with pure butanol. The obtained sample is extracted once again with chloroform/isoamyl alcohol and ethanol-precipitated in the presence of ammonium acetate. After several washes with 70% ethanol, the DNA is dried and resuspended in 20 µL of TE buffer.

For these three preparations, the DNA concentration is determined spectrophotometrically.

4. Preparation of one end-labeled DNA: 0.4 pmol of a linear DNA fragment is incubated with 3 U of Klenow polymerase for 30 min at 37°C in the following buffer: 50 m*M* Tris-HCl, pH 8.0, 0.5 mM DTT, 0.5 m*M* EDTA, 30 µg/mL bovine serum albumin, 10 m*M* $MgCl_2$ and 16 µmol of the corresponding (α-^{32}P) dNTP (3000 Ci/mmol) (i.e., the first desoxynucleotide after the cleavage point is added at the 3′-OH terminus generated by a 5'-overhanging ends restriction endonuclease). The total incubation volume is 30 µL. After incubation with proteinase K, the sample is extracted with chloroform/isoamyl alcohol, ethanol-precipitated in the presence of ammonium acetate, and the pellet is washed with ethanol 70% (to eliminate unincorporated desoxynucleotide) until the radioactive measurement of the supernatant is about 0.5% that of the pellet.

2.2 Reverse Gyrase

Reverse gyrase can be purified in a relatively high amount from an archaebacterial cell type ***(8,12,13)***. In order to avoid DNA degradation owing to contaminating endonuclease activity, it is important to use a highly pure enzyme. The enzyme can be stored at 4°C for several years in a convenient buffer containing 25 m*M* NaH_2PO_4/Na_2HPO_4, pH 7.0, 0.5 m*M* DTT, 0.5 m*M* EDTA, 100 m*M* NaCl, and 0.05% Triton X-100 (purchased from Pierce) as a detergent (*see* **Note 1**).

2.3. Agarose Gels

1. 1X TEP buffer: 36 m*M* Tris, 30 m*M* NaH_2PO_4, 1 m*M* EDTA, pH 7.8.
2. Gel-loading buffer: 0.01% bromophenol blue and 15% sucrose or 10% glycerol (final concentrations).
3. EtBr staining: gel soaked for 20–30 min in a bath of 2 µg/mL of ethidium bromide.
4. Magnesium sulfate destaining: gel soaked for at least 30 min in a bath of 1 m*M* $MgSO_4$.

3. Methods

3.1. Catalytic Activities of Reverse Gyrase

3.1.1. Standard ATP-Dependent Relaxation Assay in Distributive Conditions

1. The reaction mixture (usually 20 µL final volume) contains 50 m*M* Tris-HCl, pH 8.0, at 20°C, 1 m*M* ATP, 0.5 m*M* DTT, 0.5 m*M* EDTA, 30 µg/mL bovine serum

albumin, 10 m*M* $MgCl_2$, 120 m*M* NaCl, 300 ng of negatively supercoiled plasmid DNA (e.g., pTZ 18), and 2 µL of the fraction to be assayed *(8)*. Finally, parafin oil (4 µL) may be added on top of the reaction mixture (*see* **Note 2**).
2. After 30 min of incubation at 75°C (*see* **Note 3**), the reaction is stopped by addition of 1% SDS (final concentration).
3. The reaction products are analyzed at room temperature by 1.2% agarose-gel electrophoresis for 4 h at 3.5 V/cm.
4. The gel is stained with ethidium bromide, followed by destaining with 1 m*M* $MgSO_4$, and the image is recorded on a CCD camera. **Figure 4A** shows the disappearence of form I DNA substrate and the appearence of positively supercoiled DNA for a series of reverse gyrase dilutions (*see* **Note 4**). The reaction is very efficient, 100% of the substrate being converted even with high enzyme dilutions (**Fig. 4A**, lane 5).
5. Quantitation of the activity is usually made by measuring the disappearence of the input DNA band. One unit of enzyme is defined as the amount of protein required to relax 50% of form I substrate in the presence of ATP under the conditions of this assay (at 120 m*M* NaCl).

3.1.2. ATP-Dependent Positive Supercoiling Assay in Processive Conditions

1. The reaction mixture is as in the **Subheading 3.1.1.**, except for the NaCl concentration, which is 10–30 m*M*. At this low salt concentration, the enzymatic reaction is processive. Even if only a fraction of the substrate is transformed, the extent of positive supercoiling is higher (**Fig. 4B**).
2. A way to distinguish the positive and the negative DNA topoisomers is to analyze the products in a two-dimensional gel instead of a mono-dimensional gel (*see* **Note 5**). The first dimension is a standard migration (4 h at 3.5 V/cm) 1% agarose-gel electrophoresis, whereas the second dimension in a perpendicular direction, contains 3–4 µg/mL chloroquine and is run at 0.9 V/cm for 10 h ***(14)***. After chloroquine elimination (several washes in distilled water) and staining with ethidium bromide, the distribution of positive topoisomers is revealed (**Fig. 5B**).
3. A better way to monitor the production of positive supercoiling is to perform a two-dimensional gel in the presence of netropsin (at the concentration of about 13 µ*M*) instead of chloroquine in the second dimension ***(13)*** (*see* **Note 6**).
4. Another difficulty in quantitating positive DNA supercoiling comes from the temperature difference between the incubation mixture (75°C) and the electrophoresis (25°C) that tends to reduce the apparent positive supercoiling in the gels ***(15)*** (*see* **Note 7**).

3.1.3. Stimulation of Reverse Gyrase Reaction by DNA Condensing Agents

Since in the course of positive supercoiling reverse gyrase has less and less affinity for its substrate (*see* **Fig. 3**), it can be stimulated by agents that increase the contacts between macromolecules.

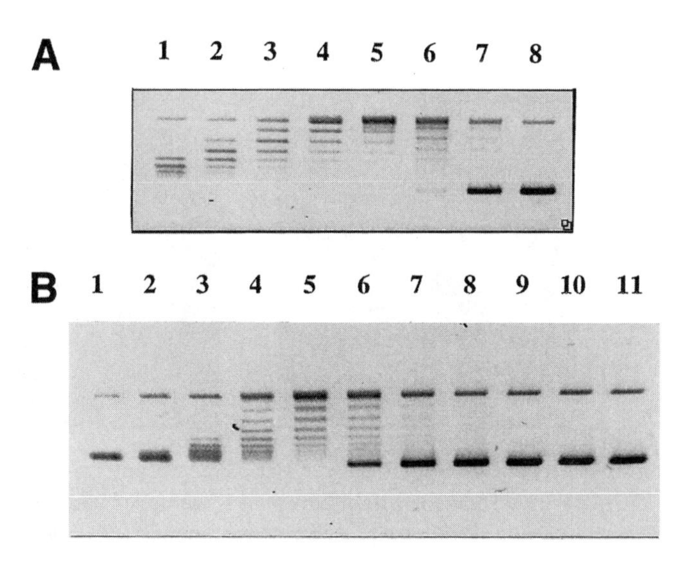

Fig. 4. ATP-dependent relaxation and positive supercoiling assays for reverse gyrase. **(A)** Standard ATP-dependent relaxation assay. The assay takes place in distributive conditions (120 m*M* NaCl) with the protocol described in the text. The fast migrating band (lane 8) is negatively supercoiled (form I) pTZ 18 DNA used as a substrate. Lanes 1–8 are incubations with increasing dilutions of pure reverse gyrase fraction: 2048 U (lane 1); 512 U (lane 2); 128 U (lane 3); 32 U (lane 4); 8 U (lane 5); 2 U (lane 6); 0.5 U (lane 7), and 0.125 U (lane 8). The individual topoisomers visible in lanes 1–5 are positively supercoiled. Those visible in lanes 6 and 7 are negatively supercoiled and do not migrate exactly at the same location. **(B)** ATP-dependent positive supercoiling assay. The assay takes place in processive conditions (10 m*M* NaCl). The other conditions are identical to A. Lane 11 is a control without enzyme. Lanes 1–10 are incubations with increasing dilutions of reverse gyrase: 8192 U (lane 1); 2048 U (lane 2); 512 U (lane 3); 128 U (lane 4); 32 U (lane 5); 8 U (lane 6); 2 U (lane 7); 0.5 U (lane 8); 0.125 U (lane 9), and 0.031 U (lane 10). Note that in processive conditions, the form I substrate (lower band in lanes 6–11) is still present for high amounts of reverse gyrase (lane 6), but individual positive topoisomers are already present in lane 6 and a high degree of positive supercoiling is obtained (lower bands in lanes 1–4).

1. The more commonly used condensing agent is polyethylene glycol (*see* **Note 8**). Addition of 7–9% PEG 6000 (purchased from Merck) to the reaction mixture in low salt (10–20 m*M* NaCl) yields highly positive DNA supercoils ***(16)*** (**Fig. 6**). For this, the following conditions may be used: 50 m*M* Tris-HCl, pH 8.0, 1.25 m*M* ATP, 0.5 m*M* DTT, 0.5 m*M* EDTA, 30 μg/mL bovine serum albumin, 10 m*M* $MgCl_2$, 10 m*M* NaCl, 7% PEG 6000, 300 ng plasmid DNA, and 2 μL of

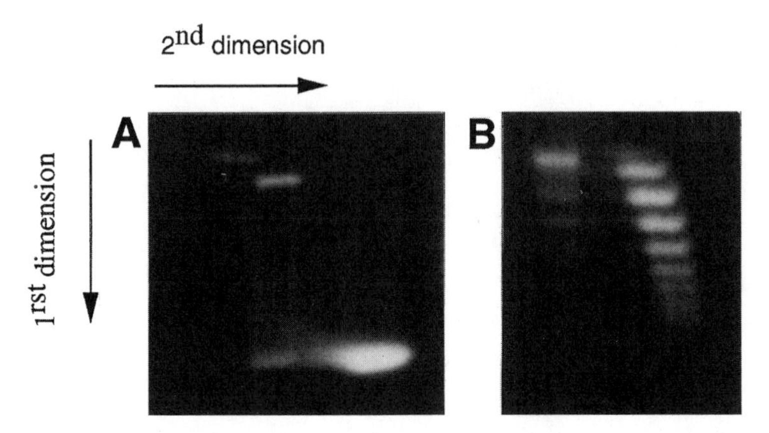

Fig. 5. Analysis of positive supercoiling in two-dimensional gels. **(A)** Negatively supercoiled DNA substrate (control without enzyme) The upper band is nicked (form II) DNA. **(B)** Distribution of positive topoisomers after incubation with reverse gyrase (dilution 1/16).

a dilution of pure reverse gyrase (equivalent to 32 U). The incubation is in 20 μL reaction volume for 30 min at 75°C.

2. Spermidine (1 m*M*) has been widely used (*see* **Note 9**) both in purification and in assay buffers ***(12)***.

3.1.4. Positive Supercoiling from Relaxed DNA

This reaction corresponds to the second part of the topoisomerization scheme of **Fig. 3**. It is clearly the best, unambiguous assay for positive supercoiling activity.

1. The assay of positive supercoiling is performed in the reaction mixture of **Subheading 3.1.1.** at low salt concentration (10 m*M* NaCl) and with 300 ng of DNA relaxed as a substrate. The results are shown in **Fig. 7**. The affinity of reverse gyrase for relaxed DNA being low, higher concentrations of the enzyme are required for this reaction (*see* **Note 10**).
2. Relaxed circular DNA may be prepared at 75°C by a thermophilic topoisomerase I ***(10)***. As a control, this DNA may be compared to a form II closed by a thermophilic ligase at 75°C ***(17)*** (*see* **Note 11**).
3. Relaxed circular DNA may be prepared at 37°C by a topoisomerase. This DNA is in fact naturally positively supercoiled at 75°C owing to the change in the twist with temperature and salt concentrations (**Fig. 7**, lane 1). This positively supercoiled DNA is transformed in a more positively supercoiled DNA by reverse gyrase. Consequently, this enzymatic reaction may be easily visualized by using a classical one-dimensional gel.

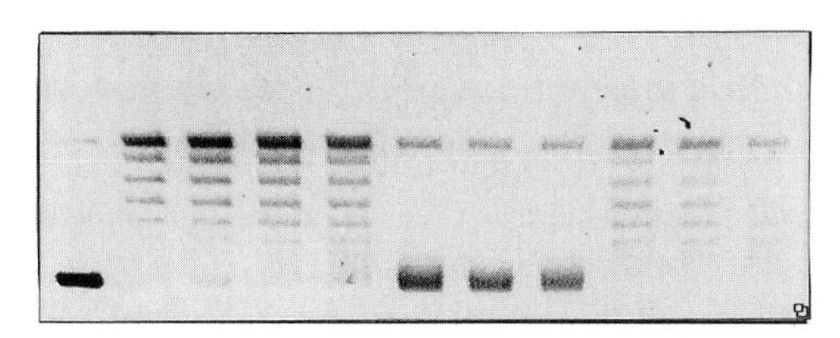

Fig. 6. Stimulation of positive supercoiling by DNA condensing agents. Experimental conditions are described in the text. Lane 1 is plasmid DNA control without enzyme. Lanes 2–11 are incubations with increasing concentrations of polyethylene glycol 6000: 0% PEG (lane 2); 2% (lane 3); 4% (lane 4); 6% (lane 5); 7% (lane 6); 8% (lane 7); 9% (lane 8); 10% (lane 9); 11% (lane 10); and 12% (lane 11). Highly positive supercoils are obtained for 7–9% PEG in the incubation.

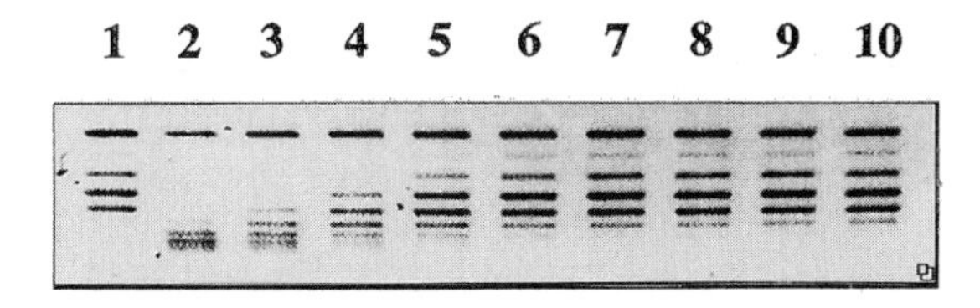

Fig. 7. Positive supercoiling of relaxed DNA. Lane 1 is a control of pTZ 18 DNA relaxed at 37°C. Lanes 2–10 are incubations with increasing dilutions of reverse gyrase: 512 U (lane 2); 256 U (lane 3); 128 U (lane 4); 64 U (lane 5); 32 U (lane 6); 16 U (lane 7); 8 U (lane 8); 4 U (lane 9); 2 U (lane 10).

3.1.5. ATPase Activity of Reverse Gyrase

Reverse gyrase hydrolyzes ATP to ADP like other ATP-dependent topoisomerases. Other ribonucleoside triphosphates are not hydrolyzed ***(18)***, and the reaction is strictly dependent on the presence of DNA. However, this ATPase activity is atypical for a topoisomerase.

1. Reverse gyrase only requires 10 µ*M* ATP, when classical ATP-dependent topoisomerases usually require 1 m*M* ATP to function.
2. ATPase activity is considerably stimulated by the addition of single-stranded DNA, which is otherwise a competitor of reverse gyrase topoisomerization activity (*see* **Note 12**).
3. Several classical assays for ATPase activity, including HPLC and TLC are appli-

cable to reverse gyrase, with the incubation mixture of **Subheading 3.1.2.** and labeled ATP *(18)*.

3.2. Stoichiometric Interactions of Reverse Gyrase with DNA

3.2.1. Efficient DNA Cleavage Without Drug

Since it is a type I topoisomerase, reverse gyrase transiently cleaves a single DNA strand. A cleavage reaction can be observed in the presence of stoichiometric amounts of the enzyme bound to DNA ***(19,20)*** (*see* **Note 13**).

1. The reaction takes place in a 20-µL reaction mixture, containing 50 m*M* Tris-HCl, pH 8.0, 0.5 m*M* DTT, 0.5 m*M* EDTA, 30 µg/mL bovine serum albumin, 10 m*M* $MgCl_2$, 10 m*M* NaCl, and **no ATP**. A linear DNA fragment (4.6 fmol), labeled either at the 3'- or at the 5'-end on only one strand is incubated at 75°C with pure reverse gyrase (0.1 pmol corresponding also to 160 enzymatic units or 14.6 ng of pure enzyme).
2. After 5 min of incubation at 75°C, the reaction is stopped by 1% SDS (final concentration) or by NaOH (12.5 m*M*) and 0.5 mg/mL proteinase K. Incubation with the protease is performed at 55°C during 30 min.
3. The reaction products are chloroform/isoamyl alcohol-extracted twice.
4. After DNA denaturation by heating at 85°C, the samples are analyzed by electrophoresis in agarose or in polyacrylamide gels. The results are shown in **Fig. 8**.

3.2.2. DNA "Unwinding" by Reverse Gyrase

Stoichiometric binding of proteins often modifies the helical repeat of DNA. The classical test consists of the covalent closure of a form II DNA in the presence of the protein of interest. After deproteinization, the DNA is analyzed by electrophoresis and compared to the substrate DNA closed in the absence of protein. The principle of the test is given in **Fig. 9A**.

1. The assay for reverse gyrase ***(19)***, in the absence of ATP, contains 50 m*M* Tris-HCl, pH 8.0, 0.05 m*M* EDTA, 6 m*M* DTT, 30 µg/mL bovine serum albumin, 6 m*M* $MgCl_2$, 50 m*M* NaCl, 1 m*M* NAD (cofactor for ligase), 600 ng form II plasmid DNA with one single-strand break per circle, and various amounts of pure reverse gyrase (0–24 mol/DNA circle).
2. Two microliters of a dilution of the thermophilic DNA ligase (i.e., HB8 ligase from *Thermus thermophilus*) ***(17)*** are introduced prior to incubation, as a drop on the inner wall of the siliconized reaction tube placed horizontally, but not mixed with the 22 µL reaction medium. The total incubation volume is 24 µL.
3. After preincubation for 10 min at 75°C to form the complex, the tube is quickly vortexed to mix the ligase drop with the other components, and the incubation is continued for 5 min at 75°C.
4. The reaction is stopped by quick cooling, centrifugation for 20 s, and addition of EDTA (50 m*M*) and SDS (1%) as final concentrations.

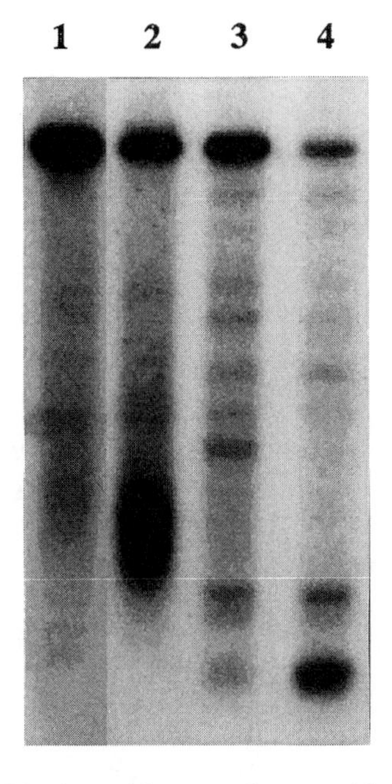

Fig. 8. Formation of DNA cleavable complexes with reverse gyrase. Lane 1 is a control showing the migration of the uncleaved linear DNA fragment. Lanes 2–4 are incubations with stoichiometric amounts of reverse gyrase, the molar ratio reverse gyrase/DNA being about 20 (*see* **Subheading 3.2.**). Lane 2: the reaction is stopped by SDS and the DNA directly analyzed by 1.4% agarose-gel electrophoresis, revealing covalent complexes between reverse gyrase and DNA. Lane 3: the reaction is stopped by SDS and the complex is treated by proteinase K before electrophoresis, revealing multiple fragments. Lane 4: as in lane 3, but the reaction was stopped by NaOH treatment.

5. After phenol extraction, the DNA is precipitated and analyzed by agarose gels. **Figure 9B** shows an increasing electrophoretic mobility as the ratio of reverse gyrase to DNA is increased during the initial incubation. More precise analysis indicates that the DNA becomes more and more negatively supercoiled, suggesting that stoichiometric binding of reverse gyrase induces DNA unwinding.

4. Notes

1. This buffer allows storage of reverse gyrase with the same specific activity for several years.
2. Taking into account the high temperature of the enzymatic incubation, care is taken to avoid evaporation.
3. In some cases, the temperature of the incubation is increased up to 80 or 90°C, depending on the source of reverse gyrase.

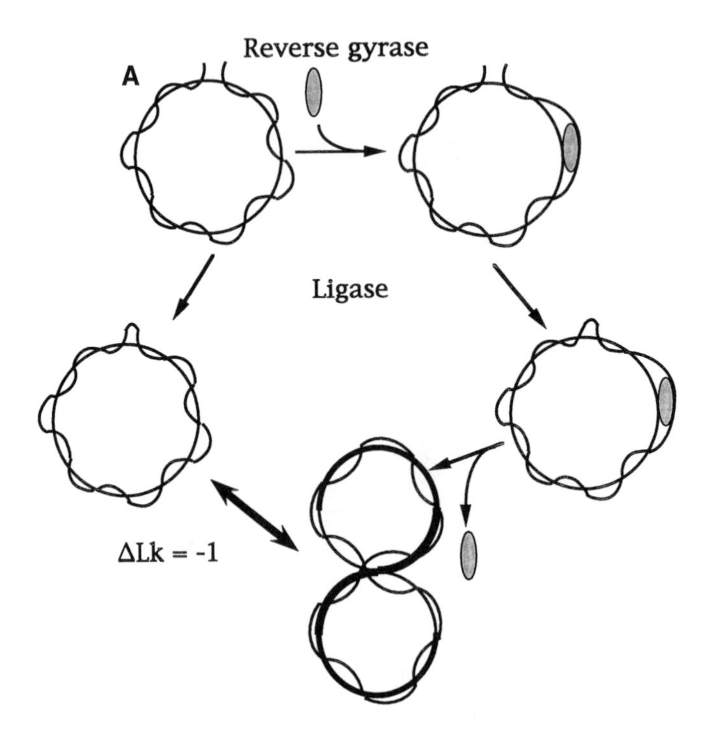

Fig. 9. Structural changes in DNA on stoichiometric binding of reverse gyrase. **(A)** Principle of the "DNA unwinding" assay *(21)*. **(B)** Incubation of form II DNA with increasing concentrations of reverse gyrase (the molar ratio reverse gyrase/DNA are 0 [lane 1]; 3 [lane 2]; 6 [lane 3]; 12 [lane 4]; 18 [lane 5], and 24 [lane 6]) ligation, and electrophoretic analysis.

4. The assay at 120 m*M* NaCl can also be used to look for the appearence of positively supercoiled DNA. However, in these salt conditions, the reaction is distributive, i.e., the negatively supercoiled substrate is rapidly relaxed to various extents, but positive supercoiling is limited (*see* **Fig. 4A**). As reverse gyrase concentration increases, the negative topoisomers are converted to positive topoisomers that migrate with a slight different mobilities (compare lane 5 with lane 6 in **Fig. 4A**).
5. Starting from negatively supercoiled DNA, it is difficult to identify positively supercoiled topoisomers appearing in the course of the reaction, unless the resolution of the gel is high.
6. Netropsin is known to bind to the minor groove of the DNA and increase the path of the double helix of DNA. As a consequence, the mobility of the positive topoisomers is reduced, and these topoisomers are better resolved than the negative ones in the second dimension of the gel.
7. The actual extent of supercoiling is higher than estimated from the gel. Indeed, with a 3-kbp plasmid, the change in the twist is about three to four turns *(15)*: that

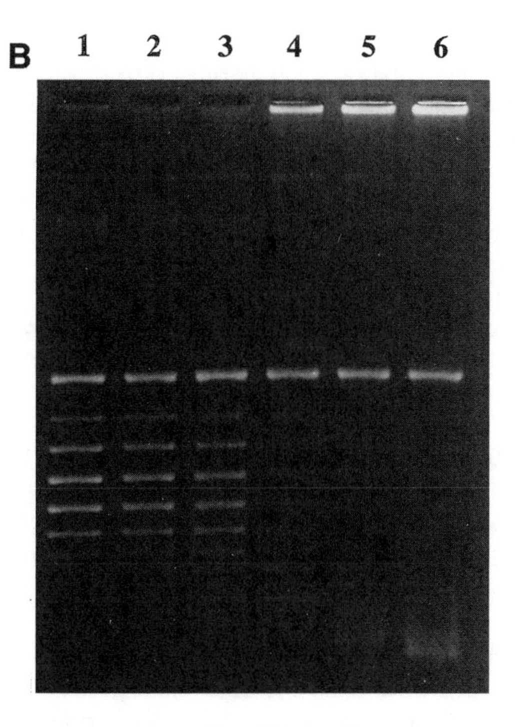

Fig. 9B.

means three to four extra positive supercoils are present in the reaction products at 75°C compared to the agarose gel at 25°C.

8. The use of PEG is applicable to the preparation of highly positively supercoiled DNA. Nevertheless, it is difficult is to keep good control over DNA condensation by PEG in order to avoid DNA aggregation and precipitation.
9. The spermidine stimulatory effect has not been studied in detail. However, spermidine inhibits nuclease activities, which is an advantage in a reverse gyrase assay, but may mask a possible nuclease contaminant in the course of the enzyme purification.
10. The amount of reverse gyrase is 50-fold higher than that used to detect relaxation activity, but is consistent with the amount necessary for positive DNA supercoiling assay.
11. Even if relaxed DNA prepared at 75°C may be considered a good substrate from an enzymatic point of view, this kind of DNA substrate is not used in the assays, because a two-dimensional gel is needed to distinguish between substrate and product.
12. These properties are reminiscent of ATP hydrolysis by helicases; indeed, the putative ATP binding site resides in the helicase-like domain of reverse gyrase. At this time, the hypothetical helicase activity of reverse gyrase has not been described.

13. None of the drugs classically used to stabilize topoisomerase cleavable complexes are efficient on reverse gyrase reaction. However, cleavage is considerably more efficient in the absence of ATP (or in the presence of a nonhydrolyzable analog), presumably because of the longer time of reverse gyrase residence on DNA. Since reverse gyrase is transiently bound at the 5'-end of the DNA, the labeling of the DNA fragment at the 3'-terminus permits distinguishing the cleaved fragment linked to the enzyme (in the absence of proteinase K) and the free cleaved fragment (in the presence of proteinase K) (*see* **Fig. 8**). However, the labeling of the DNA fragment at the 5'-terminus allows a more precise determination of the size of the completely free cleaved fragment and, consequently, the localization (at the nucleotide level) of the cleavage sites in a particular DNA.

Acknowledgments

We would like to thank E. Couderc and D. Thevenet for technical assistance and V. Borde for critical reading of this chapter. The Laboratoire d'Enzymologie des Acides Nucléiques is supported by funds from CNRS, Université Paris-Sud, and Association pour la Recherche contre le Cancer (ARC).

References

1. Kikuchi, A. and Asai, K. (1984) Reverse gyrase—a topoisomerase which introduces positive superhelical turns into DNA. *Nature* **309,** 677–681.
2. Duguet, M. (1995) Reverse gyrase, in *Nucleic Acids and Molecular Biology*, vol. 9 (Eckstein, F. and Lilley, D. M. J., eds.), Springer Verlag, Berlin, pp. 85–114.
3. Collin, R. G., Morgan, H. W., Musgrave, D. R., and Daniel, R. M. (1988) Distribution of reverse grease in representative species of eubacteria and archaeabacteria. *FEMS Microbiol. Lett.* **55,** 235–240.
4. Bouthier de la Tour, C., Portemer, C., Huber, R., Forterre, P., and Duguet, M. (1991) Reverse gyrase in thermophilic eubacteria. *J. Bacteriol.* **173,** 3921–3923.
5. Gangloff, S., McDonald, J. P., Bendixen, C., Arthur, L., and Rothstein, R. (1994) The yeast type I topoisomerase, Top 3, interacts with Sgs1, a DNA helicase homologue: a potential eukaryotic reverse gyrase. *Mol. Cell. Biol.* **14,** 8391–8398.
6. Confalonieri, F., Elie, C., Nadal, M., Bouthier de la Tour, C., Forterre, P., and Duguet, M. (1993) Reverse gyrase: a helicase-like domain and a type I topoisomerase in the same polypeptide. *Proc. Natl. Acad. Sci. USA* **90,** 4753–4757.
7. Schmid, S. R. and Linder, P. (1992) D-E-A-D protein family of putative RNA helicases. *Mol. Microbiol.* **6,** 283–292.
8. Nadal, M., Couderc, E., Duguet, M., and Jaxel, C. (1994) Purification and characterization of reverse gyrase from *Sulfolobus shibatae*. Its proteolytic product appears as an ATP-independent topoisomerase. *J. Biol. Chem.* **269,** 5255–5263.
9. Krah, R., Kozyavkin, S. A., Slesarev, A. I., and Gellert, M. (1996) A two-subunit type I DNA topoisomerase (reverse gyrase) from an extreme hyperthermophile. *Proc. Natl. Acad. Sci. USA* **93,** 106–110.

10. Bouthier de la Tour, C., Portemer, C., Forterre, P., Huber, R., and, Duguet, M. (1993) ATP-independent DNA topoisomerase from *Fervidobacterium islandicum. Biochim. Biophys. Acta* **1216,** 213–220.
11. Barzilai, R. (1973) SV40 DNA: quantitative conversion of closed circular to open circular form by an ethidium bromide-restricted endonuclease. *J. Mol. Biol.* **74,** 739–742.
12. Nakasu, S. and Kikuchi, A. (1985) Reverse gyrase; ATP-dependent type I topoisomerase from *Sulfolobus. EMBO J.* **4,** 2705–2710.
13. Nadal, M., Jaxel, C., Portemer, C., Forterre, P., Mirambeau, G., and Duguet, M. (1988) Reverse gyrase of *Sulfolobus*: purification to homogeneity and characterization. *Biochemistry* **27,** 9102–9108.
14. Peck, L. J. and Wang, J. C. (1983) Energetics of B-to-Z transition in DNA. *Proc. Natl. Acad. Sci. USA* **80,** 6206–6210.
15. Duguet, M. (1993) The helical repeat of DNA at high temperature. *Nucleic Acids Res.* **21,** 463–468.
16. Forterre, P., Mirambeau, G., Jaxel, C., Nadal, M., and Duguet, M. (1985) High positive supercoiling *in vitro* catalyzed by an ATP and polyethylene glycol-stimulated topoisomerase from *Sulfolobus acidocaldarius. EMBO J.* **4,** 2123–2128.
17. Takahashi, M., Yamaguchi, E., and Uchida, T. (1984) Thermophilic DNA ligase. *J. Biol. Chem.* **259,** 10,041–10,047.
18. Shibata, T., Nakasu, S., Yasui, K., and Kikuchi, A. (1987) Intrinsic DNA-dependent ATPase activity of reverse gyrase. *J. Biol. Chem.* **262,** 10,419–10,421.
19. Jaxel, C., Nadal, M., Mirambeau, G., Forterre, P., Takahashi, M., and Duguet, M. (1989) Reverse gyrase binding to DNA alters the double helix structure and produces single-strand cleavage in the absence of ATP. *EMBO J.* **8,** 3135–3139.
20. Kovalsky, O. I., Kozyavkin, S. A., and Slesarev, A. I. (1990) Archaebacterial reverse gyrase cleavage-site specificity is similar to that of eubacterial DNA topoisomerases I. *Nucleic Acids Res.* **18,** 2801–2805.
21. Nadal, M. (1990) The reverse gyrase of *Sulfolobus*: DNA positive supercoiling *in vitro* and *in vivo*. Ph.D. thesis, Paris.

5

Topoisomerase II-Catalyzed ATP Hydrolysis as Monitored by Thin-Layer Chromatography

Paul S. Kingma, John M. Fortune, and Neil Osheroff

1. Introduction

Although type I topoisomerases do not require a high-energy cofactor, type II topoisomerases require ATP in order to carry out their essential catalytic functions *(1–4)*. ATP binding is necessary to close the protein clamp *(5,6)* and trigger DNA strand passage *(7,8)*, whereas hydrolysis is necessary for topoisomerase II recycling *(7,8)*. Beyond the the critical roles played by ATP in the catalytic cycle of the enzyme, interactions between type II topoisomerases and ATP are also affected by compounds with antimicrobial and anticancer properties. For example, coumarin-based antimicrobials block DNA gyrase function specifically by inhibiting ATP hydrolysis *(7–11)*. In addition, several anticancer drugs that act by enhancing topoisomerase II-mediated DNA cleavage also impair interactions between the enzyme and ATP *(12)*.

Currently, two methods are used for monitoring the ATP hydrolysis activity of type II topoisomerases. The first, which is described in this chapter, uses thin-layer chromatography (TLC) to monitor the release of phosphate from [γ-^{32}P]ATP *(8,13,14)*. The second, which is described in Chapter 6, couples topoisomerase II ATPase activity with an ATP regenerating system *(15,16)*. This latter assay monitors ATP hydrolysis indirectly by quantitating decreases in the optical absorbance of the NADH pool that is consumed during regeneration.

The TLC method is simple, requires relatively small amounts of enzyme, and is extremely sensitive. However, it is not optimal for kinetic analyses, since the number of samples required to generate multiple time-courses may exceed the capacity of several TLC plates. Alternatively, the optical method is ideal

From: *Methods in Molecular Biology, Vol. 95: DNA Topoisomerase Protocols, Part II: Enzymology and Drugs*
Edited by N. Osheroff and M.A. Bjornsti © Humana Press Inc., Totowa, NJ

for kinetic experiments, since it generates continuous time curves. However, this technique requires large amounts of enzyme and is inherently less sensitive than the TLC method. Furthermore, it cannot be used to analyze the effects of compounds that potentially compete with ATP (such as nucleotide analogs or drugs), since any resulting change may be owing to alterations in the rate of ATP regeneration rather than topoisomerase II activity.

The assay described below is based on the method of Osheroff et al. *(8)*. Although it has been optimized for *Drosophila* topoisomerase II, it can easily be adapted for use with virtually any type II topoisomerase.

2. Materials

1. Topoisomerase II.
2. 5X Assay buffer stock: 50 m*M* Tris-HCl, pH 7.9, 250 m*M* NaCl, 250 m*M* KCl, 0.5 m*M* EDTA, 25 m*M* $MgCl_2$, 12.5% glycerol (stored at 4°C).
3. Nonradioactive 20 m*M* ATP in H_2O or 5 m*M* Tris-HCl, pH 7.9, 0.5 m*M* EDTA (stored at –20°C).
4. [γ-^{32}P]ATP (3000 Ci/mmol, Amersham or equivalent) (stored at –80°C) (*see* **Note 1**).
5. DNA.
6. 20 × 20 cm polyethyleneimine-impregnated cellulose TLC (0.1-mm) plates (Baker-flex precoated flexible TLC sheets or equivalent) (stored at 4°C).
7. Freshly prepared 400 m*M* NH_4HCO_3 (~250 mL/chromatography tank).
8. Chromatography tank with lid.
9. Hair dryer (optional).
10. Scintillation counter.
11. Aqueous scintillation fluid (Ecolume or equivalent).
12. Autoradiography film (DuPont, Kodak, or equivalent).
13. Visible light box (optional).
14. Phosphorimager (optional; if used, **items 10–13** are unnecessary).

3. Methods

1. Prepare the TLC plate by drawing a horizontal pencil line ~2 cm from the bottom of the plate (*see* **Notes 2** and **3**). Make pencil marks along this origin line (no closer than ~1.5 cm from each other) to serve as points of sample application. It is important to draw lightly to ensure that the chromatographic resin is not scraped from the TLC plate. A representation of a TLC plate is shown in **Fig. 1**.
2. Prepare the assay buffer at room temperature (*see* **Note 4**) as follows (for a 20-µL reaction): mix 4 µL of 5X assay buffer stock, 1 µL of nonradioactive ATP (1 m*M* final concentration), appropriate volume of [γ-^{32}P]ATP to contain ~10 µCi, appropriate volume of DNA to contain 200-µ*M* base pairs (50 n*M* plasmid of pBR322) (*see* **Note 5**), and H_2O so that the volume will be 20 µL following the addition of topoisomerase II (*see* **Note 6**).
3. Start the reaction at room temperature by the addition of topoisomerase II (~5 n*M* final concentration). The ATP hydrolysis reaction of topoisomerase II terminates

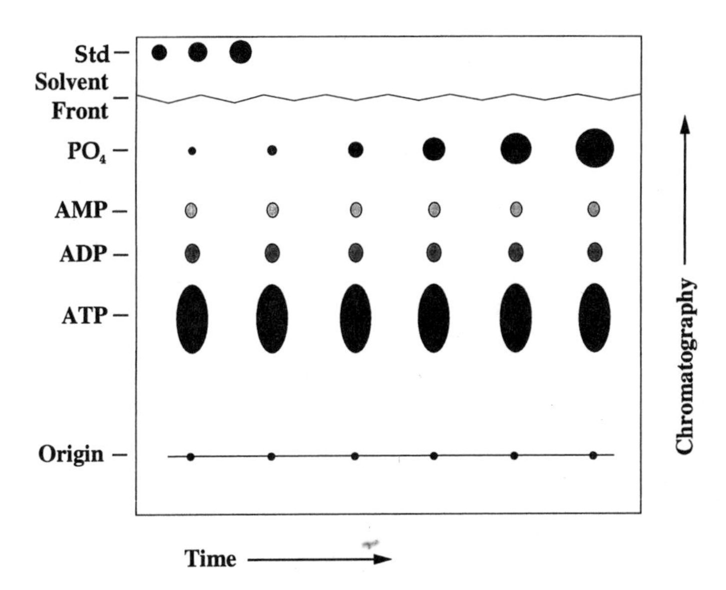

Fig. 1. Representation of a TLC plate from a topoisomerase II ATP hydrolysis assay. The predominant species at time zero should be ATP; however, slight contaminants of radioactive ADP, AMP, and phosphate are often found in commercial preparations of [γ-^{32}P]ATP. The free phosphate region of the autoradiogram should become noticeably darker and larger with increasing reaction time. Representative reaction standards of 1, 2, and 3 µL are spotted at the top of the plate so that they remain above the solvent front.

immediately on adsorption of the assay mixture to the TLC plate. Therefore, the reaction can be followed at specific time-points simply by spotting samples onto the TLC plate. No additional processing is required to stop reactions.

4. Remove a 2.5-µL sample as rapidly as possible following the addition of topoisomerase II and spot it onto the TLC plate at a point of sample application (take care not to dislodge the chromatographic resin from the point of sample application). This sample represents the time = 0 point and will provide the background-free phosphate present in the [γ-^{32}P]ATP preparation. (Alternatively, a zero time-point sample can be removed prior to the addition of topoisomerase II; however, the sample volume must be adjusted to reflect the actual reaction volume before the enzyme is added.).
5. Under the reaction conditions described, the ATPase reaction of topoisomerase II is linear up to ~20 min. Remove 2.5-µL samples at desired time intervals, and spot them at points of sample application on the TLC plate. Reserve a portion of the reaction mixture for the reaction standards described below.
6. In order to convert the cpm of released phosphate into moles of ATP hydrolyzed, it is necessary to determine the precise specific activity of the ATP used in the reaction. The most straightforward way to accomplish this is to spot samples (for

example, 1, 2, and 3 µL) of the reserved reaction mixture ~1 cm from the top of the TLC plate. Since these standards are located above the eventual solvent front and will not be resolved by chromatography, they will provide a linear scale that can be extrapolated to the total cpm in the initial reaction mix (*see* **Note 7**).

7. Allow the samples on the TLC plate to dry. This can be accomplished by air-drying for a few minutes or by gentle treatment with a hair dryer.
8. While the samples are drying, rinse the chromatography tank with freshly prepared NH_4HCO_3 solution. Add a final volume of ~150 mL for chromatography. Make sure that the meniscus of the solution is lower than the origin of sample application on the TLC plate.
9. Place the TLC plate in the chromatography tank such that the resin does not contact the side of the tank. Cover the tank and develop until the solvent front is ~2.5 cm from the top of the plate. This may take 2–4 h depending on the source and lot of TLC plate employed (*see* **Note 8**).
10. Remove and dry the TLC plate as in **step 7**. This will require air-drying for 30–60 min or by gentle treatment with a hair dryer a few minutes.
11. Wrap the plate with plastic wrap, and expose it to autoradiography film for 30–60 min. The developed film will show a chromatography pattern similar to that depicted in **Fig. 1**. The spots with the highest chromatographic mobility correspond to the free phosphate released by the hydrolysis of ATP.
12. Overlay the TLC plate on the autoradiogram. The plate and autoradiogram can be aligned easily on a light box using the points of sample application and the standards as markers. Mark the locations of the standards and free phosphate in each sample.
13. Cut out sections of the TLC plate that contain the reaction standards and free phosphate, and place them in scintillation vials (*see* **Note 9**). Add scintillation fluid to cover the TLC plate sections and determine the radioactivity in each by scintillation counting.
14. As an alternative to **items 12–14** under **Subheading 2.**, determine the radioactivity by phosphor-imager analysis.
15. Subtract the free phosphate at time = 0 from each sample, and normalize to mol of ATP hydrolyzed/time using the standards to determine the specific activity of the reaction mix. Results of a typical assay are shown in **Fig. 2**.

4. Notes

1. It is important to use [γ-^{32}P]ATP preparations that contain low levels (i.e., >2%) of contaminating free phosphate.
2. It is advisable to use pencil and not pen to mark the TLC plates, since ink is chromatographically mobile.
3. TLC plates should be handled with gloves at all times to avoid interference with chromatography.
4. Reactions can be run at temperatures other that those specified. For yeast and *Drosophila* topoisomerase II, temperatures up to 30°C may be used *(**8,10**)*. For topoisomerase II from mammalian species, temperatures up to 37°C may be used *(**17,18**)*.

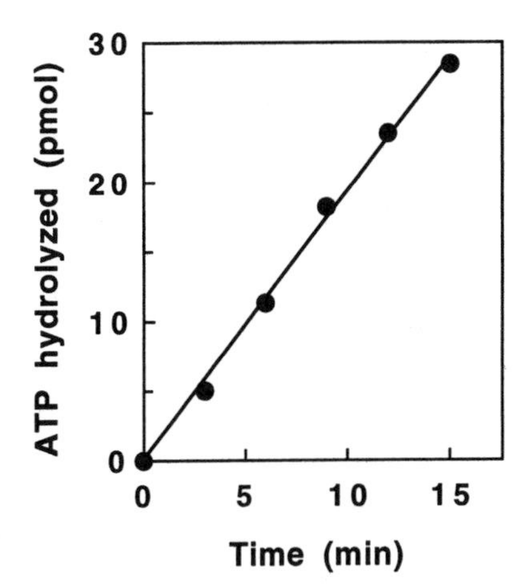

Fig. 2. Typical results of an ATP hydrolysis assay that utilized *Drosophila* topoisomerase II and contained negatively supercoiled pBR322 plasmid DNA.

5. It is important to note that type II topoisomerases are sensitive to the topological state of DNA *(7–9)*. Therefore, if supercoiled plasmid is used with eukaryotic topoisomerase II or relaxed plasmid with DNA gyrase, the plasmid must be present in high enough levels such that the concentration of the initial DNA topoisomer does not change substantially over the course of the reaction.
6. The concentrations of ATP, DNA, and topoisomerase II as well as the final reaction volumes can be changed as needed to optimize individual systems. Although the ATPase activity of topoisomerase II is stimulated 10- to 20-fold by the presence of DNA, enzyme-catalyzed ATP hydrolysis can be measured in the absence of DNA ***(14,19)***. To do so, increase the topoisomerase II concentration in each reaction mixture ~5- to 10-fold.
7. It is advisable to place reaction standards asymmetrically on the plate to help with alignment of the TLC plate and the autoradiogram.
8. Since phosphate interacts with polyethyleneimine, molecules that contain fewer phosphate groups have greater chromatographic mobility. Consequently, free phosphate displays the highest mobility in this aqueous system and ATP the lowest (**Fig. 1**).
9. Plastic-backed TLC plates are required if a scintillation counter is used to analyze samples. It is highly discouraged to attempt scraping and collecting radioactive resin from glass-backed plates.

Acknowledgment

This protocol was developed in part under the auspices of Grant GM33944 from the National Institutes of Health.

References

1. Osheroff, N., Zechiedrich, E. L., and Gale, K. C. (1991) Catalytic function of DNA topoisomerase II. *BioEssays* **13,** 269–273.

2. Watt, P. M. and Hickson, I. D. (1994) Structure and function of type II DNA topoisomerases. *Biochem. J.* **303,** 681–695.
3. Wang, J. C. (1996) DNA topoisomerases. *Annu. Rev. Biochem.* **65,** 635–692.
4. Burden, D. A. and Osheroff, N. (1998) Mechanism of action of eukaryotic topoisomerase II and drugs targeted to the enzyme. *Biochem. Biophys. Acta.* **1400,** 139–154.
5. Osheroff, N. (1986) Eukaryotic topoisomerase II. Characterization of enzyme turnover. *J. Biol. Chem.* **261,** 9944–9950.
6. Roca, J. and Wang, J. C. (1992) The capture of a DNA double helix by an ATP-dependent protein clamp: a key step in DNA transport by type II DNA topoisomerases. *Cell* **71,** 833–840.
7. Sugino, A., Higgins, N. P., Brown, P. O., Peebles, C. L., and Cozzarelli, N. R. (1978) Energy coupling in DNA gyrase and the mechanism of action of novobiocin. *Proc. Natl. Acad. Sci. USA* **75,** 4838–4842.
8. Osheroff, N., Shelton, E. R., and Brutlag, D. L. (1983) DNA topoisomerase II from *Drosophila melanogaster*. Relaxation of supercoiled DNA. *J. Biol. Chem.* **258,** 9536–9543.
9. Mizuuchi, K., O'Dea, M. H., and Gellert, M. (1978) DNA gyrase: Subunit structure and ATPase activity of the purified enzyme. *Proc. Natl. Acad. Sci. USA* **75,** 5960–5963.
10. Goto, T. and Wang, J. C. (1982) Yeast DNA topoisomerase II. An ATP-dependent type II topoisomerase that catalyzes the catenation, decatenation, unknotting, and relaxation of double-stranded DNA rings. *J. Biol. Chem.* **257,** 5866–5872.
11. Maxwell, A. (1993) The interaction between coumarin drugs and DNA gyrase. *Mol. Microbiol.* **9,** 681–686.
12. Robinson, M. J., Corbett, A. H., and Osheroff, N. (1993) Effects of topoisomerase II-targeted drugs on enzyme-mediated DNA cleavage and ATP hydrolysis: evidence for distinct drug interaction domains on topoisomerase II. *Biochemistry* **32,** 3638–3643.
13. Sugino, A. and Cozzarelli, N. R. (1980) The intrinsic ATPase of DNA gyrase. *J. Biol. Chem.* **255,** 6299–6306.
14. Maxwell, A. and Gellert, M. (1984) The DNA dependence of the ATPase activity of DNA gyrase. *J. Biol. Chem.* **259,** 14,472–14,480.
15. Ali, J. A., Jackson, A. P., Howells, A. J., and Maxwell, A. (1993) The 43-kilodalton N-terminal fragment of the DNA gyrase B protein hydrolyzes ATP and binds coumarin drugs. *Biochemistry* **32,** 2717–2724.
16. Lindsley, J. E. and Wang, J. C. (1993) On the coupling between ATP usage and DNA transport by yeast DNA topoisomerase II. *J. Biol. Chem.* **268,** 8096–8104.
17. Halligan, B. D., Edwards, K. A., and Liu, L. F. (1985) Purification and characterization of a type II DNA topoisomerase from bovine calf thymus. *J. Biol. Chem.* **260,** 2475–2482.
18. Schomburg, U. and Grosse, F. (1986) Purification and characterization of DNA topoisomerase II from calf thymus associated with polypeptides of 175 and 150 kDa. *Eur. J. Biochem.* **160,** 451–457.
19. Osheroff, N. (1987) Role of the divalent cation in topoisomerase II mediated reactions. *Biochemistry* **26,** 6402–6406.

6

Use of a Real-Time, Coupled Assay to Measure the ATPase Activity of DNA Topoisomerase II

Janet E. Lindsley

1. Introduction

This chapter describes the use of a common spectrophotometric assay for following the rate of ATP hydrolysis as applied to type II DNA topoisomerases. It is called a "coupled assay," because each time an ATP molecule is hydrolyzed, a molecule of NADH is rapidly oxidized; ATP hydrolysis and NADH oxidation are therefore coupled. Although NADH absorbs strongly at 340 nm, the product of its oxidation, NAD^+, does not. Therefore, the rate of ATP hydrolysis can be determined by following the decrease in optical absorbance of the reaction at 340 nm.

1.1. General Description of the Assay

The first reported use of this assay was by Kornberg and Pricer in 1951 *(1)*. The assay makes use of the well-studied properties of the enzymes pyruvate kinase and lactate dehydrogenase, with the substrates MgADP, phospho-(enol)pyruvate (PEP), and NADH. The reactions involved in the assay are illustrated in **Scheme 1** below.

All of the components, the ATPase (topoisomerase II in this case), MgATP (which will be referred to simply as ATP from here on), pyruvate kinase, lactate dehydrogenase, PEP, NADH, buffer, and salts, are mixed together and transferred to a cuvet. As illustrated in this scheme, when ATP is hydrolyzed to ADP and PO_4^{3-}, pyruvate kinase rapidly converts the ADP back to ATP and PEP to pyruvate. The lactate dehydrogenase reduces the pyruvate to lactate while oxidizing NADH to NAD^+. Therefore, for each ATP hydrolyzed, one

From: *Methods in Molecular Biology, Vol. 95: DNA Topoisomerase Protocols, Part II: Enzymology and Drugs*
Edited by N. Osheroff and M.A. Bjornsti © Humana Press Inc., Totowa, NJ

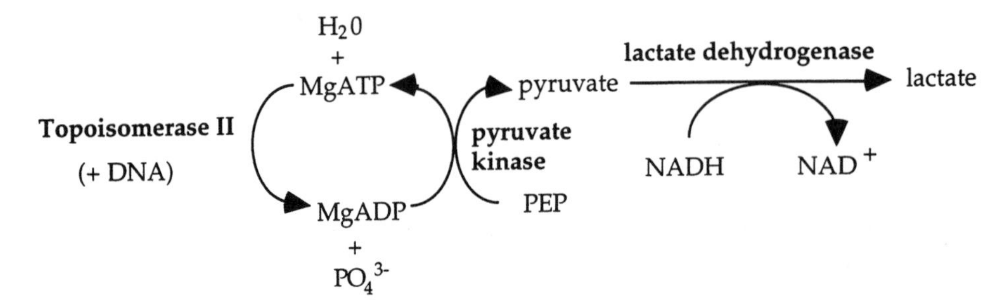

Scheme 1.

NADH is oxidized and the absorbance decreases. In addition, the initial concentration of ATP is maintained throughout the life time of the reaction, because it is rapidly regenerated.

This assay works effectively for several reasons:

1. The equilibria for each of the two coupled reactions favor the desired products as shown in **Scheme 1** *(2)*.
2. Pyruvate kinase and lactate dehydrogenase are very efficient enzymes, and, therefore, their activities do not come close to limiting the overall rate of the reaction. In other words, the coupling enzymes more rapidly couple ADP production to NADH oxidation than topoisomerase II hydrolyzes ATP.
3. NADH, but not NAD^+, absorbs strongly at 340 nm.
4. None of the substrates or products of the coupling system inhibit topoisomerase II activity.
5. The coupling enzymes and substrates are all commercially available and are relatively inexpensive (<$0.20 total/assay using 1996 prices).

1.2. Example of Data Obtained Using the Coupled Assay

The raw data from three ATPase assays of purified yeast DNA topoisomerase II are shown in **Fig. 1**. It was previously determined that the ATP concentration needed to give one-half the maximal reaction rate for *Saccharomyces cerevisiae* topoisomerase II is 0.13 m*M* in the presence of a saturating concentration of DNA *(3)*.

1.3. Advantages of Using a Coupled Assay

1.3.1. Scientific Advantages

1. During a reaction time-course, the spectrophotometer can take tens or hundreds of absorbance readings. By contrast, in standard discontinuous assays, rarely are more than 5–10 aliquots analyzed for ADP concentration during a time course. This means that a coupled assay is generally both more accurate and more precise than a manually performed assay.

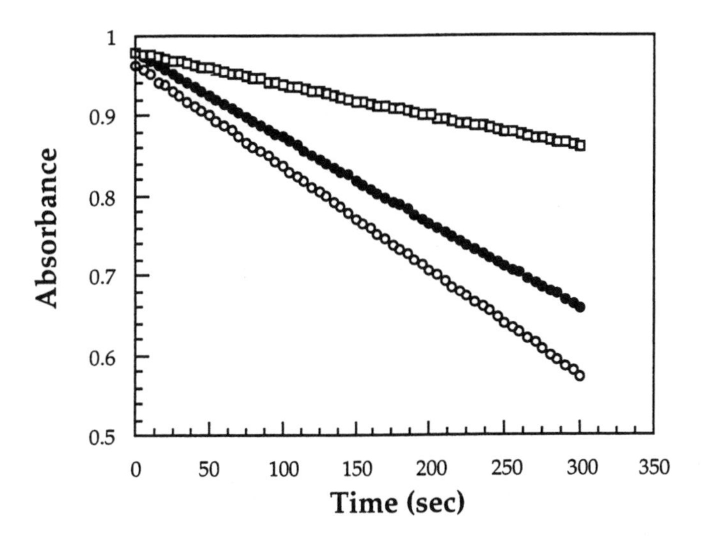

Fig. 1. Example of coupled ATPase assay data. These reactions contained: 50 m*M* potassium-HEPES, pH 7.5, 150 m*M* potassium acetate, 8 m*M* magnesium acetate, 5 m*M* β-mercaptoethanol, 250 μg/mL bovine serum albumin, 2 m*M* PEP, 0.16 m*M* NADH, 5 U of pyruvate kinase, 8 U of lactate dehydrogenase, 50 n*M* *S. cerevisiae* topoisomerase II (dimer), 25 μ*M* (base pair) of plasmid DNA and the indicated concentration of ATP. The reaction temperature was 30°C. An Aviv model 118DS UV-Vis spectrophotometer with a thermojacketed cuvet holder and attached lauda RC6 (Brinkmann) water bath was used for data collection. Absorbance data were collected every 5 s at 340 nm and transferred into KaleidoGraph 2.1 (Abelbeck Software) for plotting. ○ 2.0 m*M* ATP, ● 0.5 m*M* ATP, □ 0.1 m*M* ATP.

2. There is no change in [ATP] with time. Since ATP is constantly being regenerated from ADP and PEP by pyruvate kinase, the starting concentration of ATP remains constant during the reaction. This means that as long as the topoisomerase remains active and PEP/NADH remain plentiful, the amount of ATP hydrolyzed will increase linearly with time. This is not the case in discontinuous ATPase assays where the ATP concentration decreases during the reaction. As the [ATP] drops during the discontinuous reaction, so will the rate of ATP hydrolysis. Data analysis is simpler if the reaction proceeds at a constant rate.

1.3.2. Technical/Practical Advantages

1. The activity can be visualized in real time, and final results can be obtained rapidly. Within seconds of an ATP being hydrolyzed, an NADH is oxidized and the absorbance drops. Therefore, immediately after a reaction is initiated, the researcher can watch its progress. The reaction is generally followed for between 5 and 30 min, depending on the ATPase activity present, and the rate can be calculated immediately afterward. If a spectrophotometer with multiple cuvet holders is used, several reactions can be followed simultaneously.

2. The assay is technically easy to perform and does not use radioactive substrates. Once the reaction is initiated, the spectrophotometer and associated computer collect, store, and process the data. Therefore, the researcher does not need to be present while the assay is ongoing and does not need to enter data into a computer.

1.4. Disadvantages of Using a Coupled Assay

1. A relatively large amount of enzyme is required. For a 1-mL reaction containing 50 n*M* (dimer concentration) topoisomerase II, 16.4 µg of purified yeast topoisomerase II are required/reaction. In order to detect an absorbance decrease of 0.1 OD U in a 1-cm pathlength cuvet, 16 µM ATP must be hydrolyzed (0.1 OD U = [6.22 mM^{-1}/cm^{-1}] [1 cm] [0.016 m*M*]). Although under optimal reaction conditions, 50 n*M* of yeast topoisomerase II will hydrolyze 16 µ*M* ATP within 1 min *(3)*, one often wants to measure the rate under suboptimal conditions. Smaller amounts of topoisomerase can be used if:
   a. 1-cm pathlength cuvets holding volumes smaller than 1 mL can be used;
   b. The reaction is followed for longer times; or
   c. The oxidation of NADH is followed by fluorescence instead of absorbance.

   In this last case, because a significant change in NADH fluorescence can be detected when only 100 n*M* ATP is oxidized, one should be able to decrease the topoisomerase concentration by at least 10-fold. However, the author has never used fluorescence detection in this assay and refers the interested reader to **ref.** ***(4)***.
2. The coupled assay can only be used under conditions in which pyruvate kinase and lactate dehydrogenase are highly active. Therefore, before testing the effects of inhibitors, pH, ionic strength, and so forth, on the ATPase activity of topoisomerase II, one must be sure that these conditions do not inhibit the coupling enzymes. In addition, NADH is unstable at pH values below 7.0 and spontaneously oxidizes. Therefore, the researcher should be careful always either to buffer the reactions above pH 7.0 or determine the spontaneous rate of NADH oxidation at the pH used, and subtract this rate from the determined rate of ATP hydrolysis.

2. Materials

2.1. The Spectrophotometer

The ideal UV-visible spectrophotometer for these assays will be computer-controlled, have a thermojacketed, multiple-cuvet holder, and maintain a stable baseline. However, although most of these features improve the accuracy or convenience of the assay, only the last is essential. If the spectrophotometer is connected to a computer, the absorbance data can be directly transferred to spread sheet and graphics programs. A thermojacketed cuvet holder is important, because the rate of topoisomerase II-catalyzed ATP hydrolysis, like all other enzymatic reactions, varies with temperature. Finally, a multiple-cuvet holder allows several, generally six, reactions to be followed simultaneously.

UV-visible spectrophotometers made by many companies fit all of the above criteria.

2.2. Biochemicals and Enzymes for the Coupling System

2.2.1. Phospho(enol)pyruvate

A 500-mg bottle of the trisodium salt (Sigma #P-7002) contains enough PEP for 1000 × 1 mL reactions.

2.2.2. NADH

For convenience, vials with preweighed quantities of NADH are available (Sigma, St. Louis, MO, cat. #340-102). A vial with 2 mg contains enough NADH for 20 × 1 mL reactions. We normally add 9.4 mg of the PEP to a 2-mg vial of NADH. Both chemicals are then resuspended together in 200 µL of 50 m*M* Tris-HCl, pH 8.0. This makes a 100× stock solution containing 16 mM NADH and 200 m*M* PEP (*see* **Note 1**).

2.2.3. Pyruvate Kinase and Lactate Dehydrogenase

These two enzymes can be purchased as a mixture stored in 50% glycerol (Sigma #P-0294). A 5-mL bottle has enough enzyme for approx 750 × 1 mL reactions.

2.3. ATP

Starting with the disodium salt, a stock solution of buffered ATP is made by dissolving 61 mg into 1 mL of double-distilled water. The solution is neutralized by adding Tris base. Although the concentration of this solution should be approx 100 m*M*, the exact concentration is determined by absorbance at 259 nm, where the extinction coefficient for ATP is 0.154 mM^{-1}/cm. Small aliquots of the ATP stock solution are stored frozen for no longer than 1 yr.

2.4. DNA Topoisomerase II

We use *S. cerevisiae* DNA topoisomerase II overexpressed in and purified from the same organism, as previously described *(5)*. Since a relatively large amount of pure enzyme is needed for this assay, use of an overexpression system is recommended (*see* **Note 2**).

2.5. DNA

Topoisomerase II is a DNA-stimulated ATPase *(6)*. The yeast enzyme is stimulated by approx 10-fold under optimal reaction conditions by a ratio of ≥200:1 DNA base pairs:topoisomerase II dimer *(3)*. For these experiments, plasmid DNA was isolated from *Escherichia coli* by the CsCl banding method

(7). Neither the sequence nor the topology of the plasmid has a significant effect on the ATPase rate *(3)*.

2.6. Buffer

A suitable reaction buffer must contain Mg^{2+} (5–10 m*M*), K^+ (100–150 m*M*), and a pH buffer. Bovine serum albumin (Sigma, fraction V) can also be included to increase protein stability. We generally use a 5X reaction buffer that when diluted to 1X consists of: 50 m*M* potassium-HEPES, pH 7.5, 150 m*M* potassium acetate, 8 m*M* magnesium acetate, 5 m*M* β-mercaptoethanol (added fresh daily), and 250 µg/mL bovine serum albumin (*see* **Note 3**).

2.7. Cuvets

Disposable plastic 1-mL cuvets with a 1-cm pathlength (VWR, S. Plainfield, NJ, cat. #S-7360-10, <$0.07 each) work well.

3. Methods

3.1. The Assay

1. Mix the following together in a microfuge tube to give the indicated final concentrations in a total volume of 1 mL:
   1X Buffer (**Subheading 2.6.**);
   1X PEP + NADH (= 2 m*M* PEP and 0.16 m*M* NADH);
   5 U of pyruvate kinase;
   8 U of lactate dehydrogenase; and
   Topoisomerase II, and DNA if desired.
2. Gently mix the ingredients (no vortexing), and put at the desired reaction temperature for 5 min.
3. Transfer the mixture to a cuvet, and add ATP to initiate the reaction (*see* **Note 4**). To mix the ATP thoroughly with the other components, cover the cuvet with parafilm and gently invert several times.
4. Place the cuvet in the thermojacketed cuvet holder, and activate the spectrophotometer to collect data at 340 nm. Program the spectrophotometer to take readings every 1–10 s depending on the preferred number of final data points (*see* **Note 5**).
5. Calculate the rate of ADP production, and therefore ATP hydrolysis, using a derivation of the Beer-Lambert law:

$$\text{Rate of ATP hydrolysis} = \frac{\Delta[\text{ADP}]}{\Delta\text{time}} = \frac{-\Delta(\text{Abs}_{340})/\Delta\text{time}}{(\text{cuvet pathlength in cm})\ (6.22\ \text{m}M^{-1}/\text{cm})} \qquad (1)$$

   The rate of change of absorbance is calculated by a least-squares fit of the linear portion of the data. When the pathlength of the cuvet is 1 cm, this rate is simply divided by –6.22 to give the ATP hydrolysis rate in m*M*/s or min.

3.2. Controls

Because the coupled assay indirectly measures ATP hydrolysis, control reactions must regularly be performed. Whenever the reaction conditions are varied (change of pH, temperature, researcher, addition of an inhibitor, and so on), the researcher must show that the rate of absorbance decrease is an accurate reflection of the rate of ATP hydrolyzed. The first control is to double the concentration of topoisomerase II in the reaction. This should double the reaction rate. If it does not, then either the assay is being done improperly, there is an inhibitor/activator in the topoisomerase preparation, or the topoisomerase concentration initially used is so low that the enzyme is unstable. In neither case should the data collected be trusted until the situation is resolved. The second control is to double the concentrations of pyruvate kinase and lactate dehydrogenase, and separately, PEP. These should have no effect on the observed reaction rate. If both controls consistently give the desired result, as they should, then the rate NADH oxidation observed will directly correspond to the rate of ATP hydrolyzed.

4. Conclusions

The coupled assay described here has been used to study the activity of many ATPases. However, it has only fairly recently been applied to topoisomerase II studies *(3,8)*. Its technical ease as well as its increased accuracy and precision should make it a popular assay among researchers interested in determining the ATPase activities of their topoisomerase II preparations.

5. Notes

1. NADH is unstable at pH < 7.0. Therefore, if a PEP/NADH stock solution is made, be certain that the trisodium form of PEP is used and check the final pH carefully. This stock solution should be made up daily for best results.
2. Remember that cells contain many ATPases. To ensure that the ATPase activity being measured is actually catalyzed by topoisomerase II in our protein preparations, we have also measured the ATPase activity from an ATPase mutant topoisomerase II, Gly144→Ile, prepared in an identical manner to the wild-type enzyme *(3)*. The Gly144→Ile preparation has <1% of the ATPase activity of the wild-type preparation. Therefore, by analogy, >99% of the ATPase activity detected in our wild-type preparations is catalyzed by topoisomerase II itself.
3. Pyruvate kinase requires potassium and magnesium for maximal activity, and is inhibited somewhat by calcium and sodium *(2)*. In addition, heavy metals (Pb^{2+}, Hg^{2+}, Cd^{2+}, Cu^{2+}, Fe^{2+}) have been found to inhibit lactate dehydrogenase *(2)* and should therefore be avoided.

4. Depending on the experimental goals, reactions can also be initiated by adding topoisomerase II to a mixture already containing ATP.
5. The initial absorbance reading should be approx 1. If the spectrophotometer used does not give readings linear with NADH concentration up to an absorbance of 1 OD U, then use a lower starting concentration of NADH that is within the linear range. The absorbance should decrease linearly with time until nearly all of the NADH is oxidized (absorbance of <0.2). One exception to this linear decrease has been found. When the slowly hydrolyzed or nonhydrolyzed analogs of ATP, adenosine 5'-thiotriphosphate (ATPγS) and 5'-Adenylyl-β,γ-imidodiphosphate (AMPPNP or ADPNP, respectively) are added as competitive inhibitors, the rate of absorbance decrease slows with time ***(9)***, Lindsley and Wang, unpublished results. We believe that this represents a complex, slow-binding inhibition of the enzyme.

Acknowledgments

I am grateful to Timothy Harkins and Rachel Tennyson for critical reviewing and helpful comments on this manuscript. This work was supported by the Lucille P. Markey Charitable Grant for Biophysics (M. C. Rechsteiner, P. I.) and a grant from the US Public Health Service (GM 51194).

References

1. Kornberg, A. and Pricer, W. E., Jr. (1951) Enzymatic phosphorylation of adenosine and 2,6-diaminopurine riboside. *J. Biol. Chem.* **193,** 481–495.
2. Adams, H. (1965) Adenosine-5'-diphosphate and adenosine-5'-monophosphate in *Methods of Enzymatic Analysis* (Bergmeyer, H. U., ed.), Academic, New York, pp. 573–577.
3. Lindsley, J. E. and Wang, J. C. (1993) On the coupling between ATP usage and DNA transport by yeast DNA topoisomerase II. *J. Biol. Chem.* **268,** 8096–8104.
4. Lowry, O. H. and Passonneau, J. V. (1972) *A Flexible System of Enzyme Analysis*, Academic, New York, pp. 4–20.
5. Worland, S. T. and Wang, J. C. (1989) Inducible overexpression, purification, and active site mapping of DNA topoisomerase II from the yeast *Saccharomyces cerevisiae*. *J. Biol. Chem.* **264,** 4412–4416.
6. Osheroff, N., Shelton, E. R., and Brutlag, D. L. (1983) DNA Topoisomerase II from *Drosophila melanogaster*: Relaxation of Supercoiled DNA. *J. Biol. Chem.* **258,** 9536–9543.
7. Sambrook, J., Fritsch, E. F., and Maniatis, T. (1989) Purification of plasmid DNA, in: *Molecular Cloning: A Laboratory Manual*, 2nd ed. Cold Spring Harbor Laboratory Press, Cold Spring Harbor, NY, p. 1.42.
8. Tamura, J. K. and Gellert, M. (1990) Characterization of the ATP binding site on *Escherichia coli* DNA gyrase. *J. Biol. Chem.* **265,** 21,342–21,349.
9. Tamura, J. K., Bates, A. D., and Gellert, M. (1992) Slow Interaction of 5'-Adenylyl-β,γ-imidodiphosphate with *Esherichia coli* DNA gyrase. *J. Biol. Chem.* **267,** 9214–9222.

7

Analysis of Topoisomerase–DNA Interactions by Electrophoretic Mobility Shift Assay

Stewart Shuman

1. Introduction

DNA topoisomerases break and rejoin DNA strands through a covalent protein–DNA intermediate. The reaction chemistry involves nucleophilic attack by a tyrosine moiety of the enzyme on the phosphodiester backbone of DNA to form a phosphotyrosyl linkage to one (in the case of type I enzymes) or both (type II enzymes) of the DNA strands. Although the distribution of topoisomerase cleavage sites in DNA is nonrandom, the principles governing cleavage site choice are not fully understood. The question of how topoisomerases recognize their DNA cleavage sites is of definite interest, insofar as site specificity may have implications for topoisomerase action in vivo, and because the topoisomerase–DNA complex is the pharmacological target of many antimicrobial and anticancer drugs. An extensive battery of footprinting techniques has been applied to analyzing the topoisomerase–DNA interface in selected model systems, including deoxyribonuclease protection, base-specific modification protection, base-specific modification interference, phosphate modification interference, and site-specific photocrosslinking. This chapter focuses on the use of the electrophoretic mobility shift assay (EMSA) to study topoisomerase binding to DNA.

Since its introduction in 1981 ***(1,2)***, EMSA has become one of the most widely used methods to study the binding of proteins to specific sites on DNA. The technique is remarkably simple to perform. Briefly, a radiolabeled DNA ligand (duplex or single-stranded) is incubated in solution with the protein of interest. The reaction mixture is then electrophoresed through a native polyacrylamide gel. The mobility of the free DNA during native gel electrophoresis

From: *Methods in Molecular Biology, Vol. 95: DNA Topoisomerase Protocols, Part II: Enzymology and Drugs*
Edited by N. Osheroff and M.A. Bjornsti © Humana Press Inc., Totowa, NJ

is dictated by its chain length (exclusive of bending effects on gel mobility), such that longer DNAs migrate more slowly. The presence of bound protein or proteins on the DNA will retard its mobility relative to unbound DNA and shift the labeled DNA upward in the gel. The positions of the bound and free DNAs are discerned by autoradiographic exposure of the gel. Quantifying the extent of overall binding (i.e., the decrement in free DNA as a function of protein concentration) and determining the distribution of the DNA ligand among multiple discrete protein–DNA complexes can be performed by using a Phosphorimager. EMSA offers several advantages over alternative methods such as nitrocellulose filter binding and solution DNase footprint titration. These include: (1) the ability to distinguish and physically separate protein–DNA complexes containing different sets of polypeptides, and (2) the capacity to assay simultaneously binding to more than one DNA probe, provided that the DNAs can be electrophoretically resolved from each other (e.g., on the basis of chain length). EMSA data are amenable to rigorous determinations of binding affinity and detection of cooperativity effects ***(3)***.

Despite these desirable features, EMSA has not achieved wide use in the topoisomerase field. Most investigators rely instead on DNA cleavage assays to detect the covalent binding of topoisomerase at specific DNA target sites (*see* Chapters 10–11). Unlike the cleavage assays, EMSA measures total binding—covalent plus noncovalent—of topoisomerase to DNA. The combined use of EMSA and equilibrium cleavage assays can provide a thorough description of the binding affinity and the internal cleavage-religation equilibrium (K_{eq}) at a single topoisomerase target site. For example, EMSA is employed routinely in the author's laboratory, in conjunction with strand cleavage assays, to study site-specific binding of the vaccinia virus topoisomerase to duplex DNA and to DNA cruciform structures ***(4–7)***. EMSA has been used by other investigators to analyze duplex DNA binding by DNA gyrase ***(8,9)***. DiGate's laboratory has employed EMSA to demonstrate binding of *Escherichia coli* topoisomerase III to single-stranded DNA ***(10,11)***. The methods outlined below are those we use for the vaccinia virus topoisomerase, a member of the eukaryotic type I topoisomerase family ***(12,13)***. These procedures should be broadly applicable to other topoisomerases, subject to experimental caveats enumerated below.

2. Materials

2.1. Choice of the DNA Ligand

The critical prerequisite for successful use of EMSA is the identification of a specific DNA binding site for the topoisomerase being studied. Candidate topoisomerase binding sites can be defined by mapping sites of topoisomerase-induced DNA strand cleavage within duplex or single-stranded DNAs, e.g., in plasmid DNA or within specific DNA elements thought to be sites of topoisomerase action. The extent of cleavage at any given binding site is a function of the affinity of the topoisomerase for that site and the internal cleavage-religation equilibrium at that site. It is therefore advisable to conduct topoisomerase titration experiments in order to gauge the order of cleavage site occupancy on the DNA. Once high-affinity sites are mapped, then the dimensions of the DNA site should be estimated by sequential truncations of the cleavage substrate on either side of the site of covalent adduct formation. The ideal probe for EMSA is a DNA fragment that has a single topoisomerase cleavage site (i.e., a unique site of scission on one strand for type I enzymes or a unique site of staggered duplex DNA cleavage for the type II enzymes). The fragment size itself is not critical, since duplex DNAs as large as 500 bp are amenable to polyacrylamide gel EMSA. One advantage of using smaller probes is that there is less chance of detecting the binding of contaminating proteins in the enzyme preparation, either at nonspecific sites or at specific sites fortuitously present in the ligand. It is useful also to have on hand a mutated version of the intended DNA ligand in which the topoisomerase cleavage/binding site is altered or deleted in such a way as to render the DNA incapable of being cleaved by the topoisomerase. This DNA serves as a valuable reagent to evaluate the specificity of the noncovalent binding interaction detected by EMSA. In the same vein, it is very instructive to prepare a mutated version of the topoisomerase in which the active site tyrosine is substituted conservatively by phenylalanine. Such mutants are inactive in covalent catalysis, but may retain the specificity of the wild-type enzyme in noncovalent binding to DNA. By studying the binding properties of the active site mutant, the experimenter can focus exclusively on the determinants of noncovalent interaction with the DNA ligand.

We have followed this strategy to the letter in defining the cleavage specificity of the vaccinia virus topoisomerase. Vaccinia topoisomerase forms a covalent adduct with duplex plasmid DNA at sites containing a conserved pentapyrimidine sequence 5'-(C/T)CCTT$^{\downarrow}$ in the scissile strand ***(14)***. The T$^{\downarrow}$ residue is linked via a 3' phosphodiester bond to Tyr-274 of the 314 amino acid vaccinia protein ***(15)***. A Tyr-274→Phe substitution at the active site abolishes covalent catalysis. However, the Phe-274 mutant protein retains the ability to bind noncovalently to the CCCTT motif, as detected by DNase footprinting ***(16)*** and by EMSA ***(6)***. EMSA analysis of the binding of Topo(Phe-274) to CCCTT-containing duplexes of varying lengths shows that formation of a gel-stable noncovalent complex is optimal when the duplex ligand contains at least 12 bp of DNA upstream and 12 bp downstream of the scissile bond ***(6)***. Therefore, the minimal effective ligand for EMSA studies of the vaccinia topoisomerase is a 24-bp CCCTT-containing duplex. Mutating the CCCTT element to ACACA abolishes the ability of the wild-type enzyme to form a covalent adduct on the 24-mer duplex. The ACACA-containing 24-mer binds with only 7- to 10-fold lower affinity to the topoisomerase than does the CCCTT-containing DNA *(6)*. This result demonstrates that the stringent specificity of the topoisomerase in DNA cleavage is exerted primarily at the chemical step rather than at the level of noncovalent binding.

2.2. Purity of the Topoisomerase

The quality of the binding data obtained with the EMSA technique will inevitably be influenced by the purity of the topoisomerase preparation used in the assay. It is common practice to assay site-specific DNA binding by transcription factors using crude cellular extracts or in vitro translation extracts as the source of the specific DNA binding protein; in such cases, a specific gel shift can be detected when a nonspecific competitor DNA, e.g., a homopolymer like poly(dI-dC), is included in the binding reaction mixtures to suppress binding of the labeled probe by proteins other than the transcription factor of interest. The author would not recommend this practice for EMSA studies of the topoisomerases. The best approach is to purify the enzyme as closely as possible to homogeneity and to study its binding to a defined DNA probe without including any competing DNA in the reaction mixture. Techniques for expressing recombinant topoisomerases from cloned genes or cDNAs and for their subsequent purification by standard or affinity-chromatography steps are described in later chapters in this volume. The vaccinia topoisomerase can be overexpressed in *E. coli* and recovered in homogeneous form with yields of 15–20 mg of topoisomerase/L of bacterial culture *(4,17)*.

2.3. Gel and Buffer System

The buffers used in EMSA are generally of low ionic strength. This is predicted to increase DNA binding affinity and therefore stabilize the protein–DNA complexes during the electrophoresis procedure. We have used two different buffer systems—either 22.5 m*M* Tris borate, 0.625 m*M* EDTA (0.25X TBE) or 6.7 m*M* Tris-HCl, 3.3 m*M* NaOAc, 0.01 m*M* EDTA—to study the binding of vaccinia topoisomerase to duplex DNA. Although either buffer is satisfactory to resolve protein-bound and free DNAs, we find that smearing of the DNA bands is less of a problem with 0.25X TBE *(4)*. A 6% polyacrylamide gel provides excellent separation of vaccinia topoisomerase–DNA complexes from unbound duplex DNA. Investigators using EMSA to study other topoisomerases may want to test both gel buffer systems and to vary the concentration of acrylamide in the gel (as well as the acrylamide to bisacrylamide ratio) in order to optimize the system for their needs.

2.4. Reagents and Equipment

1. T4 polynucleotide kinase.
2. [γ^{32}P]ATP.
3. 10X TBE: 108 g Tris base, 55 g boric acid, 9.3 g EDTA, deionized water to 1 L.
4. Acrylamide stock solution: 30% acrylamide, 0.8% bisacrylamide (150 g acrylamide, 4 g *N,N'*-methylene bis-acrylamide, deionized water to 500 mL).
5. Vertical slab gel electrophoresis apparatus (e.g., BRL Model V16) and power supply.
6. Vacuum gel dryer with heating unit.
7. X-ray film, cassettes, and developer.

3. Methods

3.1. Labeling the DNA Probe

1. In our studies of the vaccinia topoisomerase, we prepare the duplex DNA ligand by annealing two complementary DNA oligonucleotides, one of which is 5' ^{32}P-labeled. The sequence of the scissile strand includes the pentapyrimidine recognition motif CCCTT, which is normally placed centrally within the oligonucleotide for EMSA and footprinting studies. The oligonucleotides, which have 5'-OH termini, are synthesized by automated DMT-phosphoramidite chemistry. The sizes of the CCCTT-containing duplexes used in our EMSA studies vary from 24–60 bp.
2. The 5'-end of the CCCTT-containing strand is labeled by phosphoryl transfer from [γ^{32}P]ATP catalyzed by T4 polynucleotide kinase (*see* **Note 1**). The scissile strand is labeled so that the duplex ligand used to study total binding of DNA to protein by EMSA can also be used to assay covalent binding to the topoisomerase. Covalent binding is assayed by label transfer to the 33-kDa topoisomerase to form a protein–DNA adduct that is stable to SDS-PAGE ***(16)***.
3. The kinase reaction mixture (25 µL) contains 50 m*M* Tris-HCl (pH 8.0), 10 m*M* dithiothreitol, 10 m*M* $MgCl_2$, 0.1 m*M* ATP, 100 µCi [γ^{32}P]ATP, T4 polynucleotide kinase (20 U, Bethesda Research Laboratories), and 20 µ*M* (500 pmol) of DNA oligonucleotide (DNA is quantitated by A_{260}). After incubation for 60 min at 37°C, the reaction is quenched by addition of EDTA to 20 m*M* and formamide to 20% (v/v).
4. The 5'-labeled DNA is freed of protein and radioactive nucleotide by electrophoresis through a 15% polyacrylamide gel (15% acrylamide, 0.4% bisacrylamide; 15 × 17 cm gel; 0.8-mm thickness) in 1X TBE buffer. Full-sized labeled oligonucleotide is localized by autoradiographic exposure of the wet gel. The labeled DNA is recovered from an excised gel slice by soaking the slice in 0.4 mL H_2O in a 1.5 mL microcentrifuge tube for 8–12 h. The aqueous solution is

removed, and an aliquot (1 μL) is counted in a liquid scintillation counter. The molar concentration of the oligonucleotide is then calculated based on the specific activity (cpm/pmol) of the ATP in the kinase reaction mixture.

5. Hybridization of labeled scissile strand to a complementary oligonucleotide (present in two- to fourfold molar excess) is performed in 0.2 *M* NaCl by heating the mixture to 65°C, followed by slow cooling to room temperature. The annealed substrate is stored at 4°C prior to use in DNA binding reactions.

3.2. Native Polyacrylamide Gel

1. A 6% native polyacrylamide gel in 0.25X TBE is prepared in a 50-mL plastic centrifuge tube by mixing 8 mL of 30% acrylamide, 0.8% bis-acrylamide stock solution with 1 mL of 10X TBE, and adjusting the volume to 40 mL with deionized water. (This is just enough to pour one native gel.) Then add 0.4 mL of 10% ammonium persulfate, and mix by inverting the tube. This is followed by addition of 40 μL of *N,N,N',N'*,-tetramethylethylenediamine (TEMED) with mixing by inversion of the tube. The acrylamide solution is poured gently into a 15 × 17 cm (W × L), 1.5-mm thick slab gel. After any fine air bubbles are allowed to escape, the plates are placed horizontally and the comb is inserted. After polymerization, the comb is removed and the gel is clamped in place onto the gel box.
2. The gel running buffer, 0.25X TBE is prepared by mixing 25 mL of 10X TBE in 1 L of deionized water. The upper and lower reservoirs are filled, and the sample wells are cleaned out by injection of running buffer with needle and syringe.
3. The gel is then electrophoresed for ~1 h at 100 V. The binding reaction can be performed while the gel is being prerun.

3.3. Binding Reaction and EMSA

1. Binding reaction mixtures (20 μL) containing 50 m*M* Tris HCl, pH 8.0, 0.5 pmol of 5' ^{32}P-labeled CCCTT-containing duplex DNA, and purified vaccinia topoisomerase are incubated for 5 min at 37°C (*see* **Note 2**). The mixture is adjusted to 10% glycerol and then applied immediately to the prerun native polyacrylamide gel. Note that the wells should be cleaned out again just before sample loading. The addition of 5–10% glycerol facilitates application of the samples to the gel. (Alternatively, to reduce manipulations and pipeting steps, glycerol can be included in the binding reaction mixture prior to the addition of enzyme.).
2. The samples are then electrophoresed through the native polyacrylamide gel at 100 V for 3 h. A mixture of xylene cyanol and bromophenol blue dyes is applied to the extreme right or left lane to monitor the progress of the electrophoretic run
3. After electrophoresis is completed, the plates are separated, and the gel is transferred to a double layer of Whatman 3MM filter paper. After applying an overlay of plastic wrap, the gel is dried under vacuum on a heated (80°C) platform gel dryer.
4. The results are visualized by autoradiographic exposure of the dried gel. In the example shown in **Fig. 1**, vaccinia topoisomerase binds to a 30-bp CCCTT-containing DNA probe to form a single topoisomerase–DNA complex

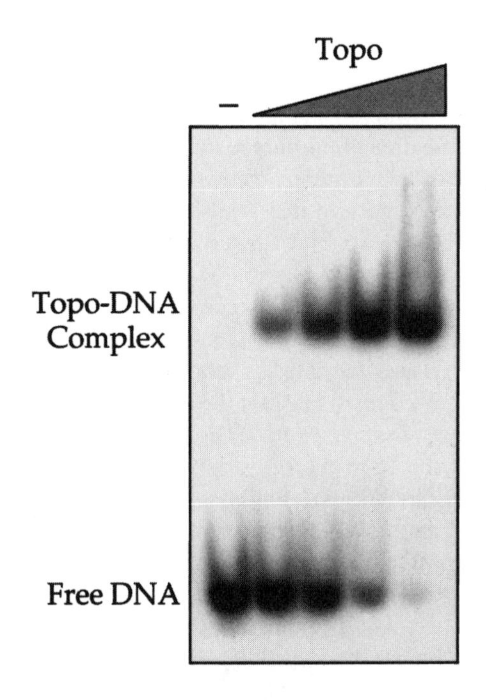

CGTGTCGCCCTTATTCCGATAGTGACTACA
GCACAGCGGGAATAAGGCTATCACTGATGT

Fig. 1. Reaction mixtures (20 µL) containing 50 m*M* Tris HCl, pH 8.0, 0.5 pmol of a 30-bp duplex DNA (5' ^{32}P-labeled on the CCCTT-containing strand) and increasing amounts of vaccinia topoisomerase (0.25, 0.5, 1.0, or 2.5 pmol) were incubated at 37°C for 5 min, and then subjected to native gel electrophoresis as described in the text. An autoradiogram of the gel is shown. The structure of the DNA ligand is depicted below the gel with the scissile bond indicated by an arrowhead.

of retarded electrophoretic mobility (*see* **Notes 3** and **4**). The fraction of the DNA bound increases with the concentration of input topoisomerase; nearly all DNA is bound at 125 n*M* enzyme, i.e., at a 5:1 ratio of protein to DNA (*see* **Note 5**).

5. The most convenient way to quantitate the EMSA data is to scan the dried gel using a Phosphorimager. If this technology is not available, excised gel slices containing bound and free DNA from each reaction can be quantitated by liquid scintillation counting. Alternatively, the X-ray film can be scanned with a densitometer.

4. Notes

1. An alternative method to prepare DNA labeled uniquely at the 5'-end of one strand of the duplex is to PCR-amplify a DNA segment containing the

topoisomerase binding site from a plasmid DNA template using oligonucleotide primers, one of which has been 5'-labeled with [γ^{32}P]ATP. Although it is not necessary to gel-purify the oligonucleotide primer from the kinase reaction in this case, native gel purification of the duplex PCR product is suggested. Labeled probe can also be prepared by traditional methods in which DNA ends generated by restriction endonuclease digestion are 3'-end-labeled by fill-in with DNA polymerase and [α^{32}P]dNTPs or else 5'-end-labeled with T4 kinase after having been dephosphorylation by alkaline phosphatase.

2. Proper choice of binding reaction conditions can enhance the experimenter's ability to detect a discrete topoisomerase–DNA complex. In fact, conditions optimal for topoisomerase activity, as determined by relaxation of supercoiled DNA, may not be well suited to EMSA. This is because the rate-limiting step in DNA relaxation, at least for the type I enzymes under conditions of DNA excess, is the dissociation of the enzyme from relaxed plasmid DNA product. The marked stimulation of DNA relaxation by salt and magnesium under multiple turnover conditions can be attributed to an acceleration of the topoisomerase off-rate ***(18–20)***. Thus, inclusion of salt and/or magnesium in the binding reaction may actually reduce the yield of a gel-shifted complex. This is what we observe experimentally for the vaccinia topoisomerase.
3. The standard EMSA technique focuses only on the fate of the labeled nucleic acid probe; the electrophoretic behavior of the unbound protein is usually ignored. Often, the unbound protein does not migrate into the native polyacrylamide gel under EMSA electrophoresis conditions. (The free protein can be located simply by fixing the native gel and staining it with Coomassie brilliant blue dye. Alternatively, the gel contents can be transferred to a membrane and immunoblotted.) The binding of the negatively charged DNA ligand increases the electrophoretic mobility of the protein, causing a protein shift from the well into the gel. This is so for the vaccinia topoisomerase; indeed, we find that shortening the length of duplex DNA probe (i.e., which makes the DNA less negatively charged) actually slows the electrophoretic mobility of the topoisomerase–DNA complex ***(6)***.
4. In performing gel-shift assays, one often finds that some or even all of the DNA probe becomes shifted to the sample well when protein is included in the reaction mixture. Shift to the sample well indicates that the DNA is bound, but provides no sense of the number and distribution of protein–DNA complexes formed. Therefore, this EMSA finding is really no more informative than results obtained from filter binding assays. Should the "well-shift" problem occur during EMSA studies with DNA topoisomerases, the question becomes: does the well shift genuinely reflect the binding of a single topoisomerase molecule to a single binding site on DNA? If it does, then the topoisomerase–DNA complex may be either too large to electrophorese into the gel or insufficiently negatively charged. Increasing the ratio of acrylamide to bis-acrylamide in the gel may circumvent the problem of protein–DNA complex size. Other options are to vary the length

of the DNA ligand to alter the net charge (*see* **Note 3**), to run the gel for longer times, or to switch to a native agarose gel in lieu of polyacrylamide.

5. A well shift may also result from the formation of protein–DNA aggregates. This tends to occur when relatively large amounts of protein are added to the binding reaction compared to input DNA. Even with a homogenous preparation of vaccinia topoisomerase, the inclusion of a 10-fold or greater molar excess of enzyme over probe causes the discrete protein–DNA complex to smear upward in the native gel. Ultimately, as protein concentration is increased further, all the probe winds up in the well. This occurs even with a 24-bp DNA ligand, which we know from DNase footprinting accommodates just one molecule of the topoisomerase bound directly on the DNA *(16)*. This suggests that the smear and well shift at high protein concentrations are caused by interactions between DNA-bound enzyme and free protein. If aggregation is suspected during EMSA trials using other topoisomerases, it is advisable to back-titrate the amount of input protein to a point where a only a fraction of the input probe is bound. This should diminish self-aggregation and lessen the potential problem of DNA binding by contaminants in the topoisomerase preparation. Another approach to counteract the tendency to form nonspecific aggregates is to increase the stringency of the binding reaction condition, e.g., by increasing ionic strength. Addition of nonspecific competitor is another option.

References

1. Garner, M. M. and Revzin, A. (1981) A gel electrophoresis method for quantifying the binding of proteins to specific DNA regions: application to components of the Escherichia coli lactose operon regulatory system. *Nucleic Acids Res.* **9,** 3047–3060.
2. Fried, M. and Crothers, D. M. (1981) Equilibrium and kinetics of lac repressor-operator interactions by polyacrylamide gel electrophoresis. *Nucleic Acids Res.* **9,** 6505–6525.
3. Senear, D. F. and Brenowitz, M. (1991) Determination of binding constants for cooperative site-specific protein-DNA interactions using the gel-mobility shift assay. *J. Biol. Chem.* **266,** 13,661–13,671.
4. Morham, S. G. and Shuman, S. (1992) Covalent and noncovalent DNA binding by mutants of vaccinia DNA topoisomerase I. *J. Biol. Chem.* **267,** 15,984–15,992.
5. Wittschieben, J. and Shuman, S. (1994) Mutational analysis of vaccinia virus DNA topoisomerase defines amino acid residues essential for covalent catalysis. *J. Biol. Chem.* **269,** 29,978–29,983.
6. Sekiguchi, J. and Shuman, S. (1994) Requirements for noncovalent binding of vaccinia topoisomerase I to duplex DNA. *Nucleic Acids Res.* **22,** 5360–5365.
7. Sekiguchi, J., Seeman, N. C., and Shuman, S. (1996) Resolution of Holliday junctions by eukaryotic DNA topoisomerase I. *Proc. Natl Acad. Sci. USA* **93,** 785–789.

8. Maxwell, A. and Gellert, M. (1984) The DNA dependence of the ATPase activity of DNA gyrase. *J. Biol. Chem.* **259,** 14,472–14,480.
9. Yang, Y. and Ames, G. F. (1988) DNA gyrase binds to a family of prokaryotic repetitive extragenic palindromic sequences. *Proc. Natl. Acad. Sci. USA* **85,** 8850–8854.
10. Zhang, H. L. and DiGate, R. J. (1994) The carboxyl-terminal residues of Escherichia coli DNA topoisomerase III are involved in substrate binding. *J. Biol. Chem.* **269,** 9052–9059.
11. Zhang, H. L., Malpure, S., and DiGate, R. J. (1995) Escherichia coli DNA topoisomerase III is a site-specific DNA binding protein that binds asymmetrically to its cleavage site. *J. Biol. Chem.* **270,** 23,700–23,705.
12. Shuman, S. and Moss, B. (1987) Identification of a vaccinia virus gene encoding a type I DNA topoisomerase. *Proc. Natl. Acad. Sci. USA* **84,** 7478–7482.
13. Caron, P. R. and Wang, J. C. (1994) Alignment of primary sequences of DNA topoisomerases. *Adv. Pharmocol.* **29B,** 271–297.
14. Shuman, S. and Prescott, J. (1990) Specific DNA cleavage and binding by vaccinia virus DNA topoisomerase I. *J. Biol. Chem.* **265,** 17,826–17,836.
15. Shuman, S., Kane, E. M., and Morham, S. G. (1989) Mapping the active-site tyrosine of vaccinia virus DNA topoisomerase I. *Proc. Natl. Acad. Sci. USA* **86,** 9793–9797.
16. Shuman, S. (1991) Site-specific interaction of vaccinia virus topoisomerase I with duplex DNA. *J. Biol. Chem.* **266,** 11,372–11,279.
17. Shuman, S., Golder, M., and Moss, B. (1988) Characterization of vaccinia virus DNA topoisomerase I expressed in *Escherichia coli. J. Biol. Chem.* **263,** 16,401–16,407.
18. McConaughy, B. L., Young, L. S., and Champoux, J. J. (1981) The effect of salt on the binding of the eucaryotic DNA nicking-closing enzyme to DNA and chromatin. *Biochim. Biophys. Acta* **655,** 1–8.
19. Stivers, J. T., Shuman, S., and Mildvan, A. S. (1994) Vaccinia DNA topoisomerase I: Single-turnover and steady-state kinetic analysis of the DNA strand cleavage and ligation reactions. *Biochemistry* **33,** 327–339.
20. Sekiguchi, J. and Shuman, S. (1994) Stimulation of vaccinia DNA topoisomerase by nucleoside triphosphates. *J. Biol. Chem.* **269,** 29,760–29,764.

8

Filter Binding Assays for Topoisomerase–DNA Complexes

Joaquim Roca

1. Introduction

1.1. Fundamentals of the Filter Binding Assays

Protein–nucleic acid interactions have been studied by density shift gradients, affinity columns, and gel retardation. Some of these methods are tedious or have severe limitations in quantitative analysis. Filter binding assays are fast, sensitive, and suitable for physical chemical measurements *(1,2)*. Filter binding techniques are based on the observation that proteins, either free or complexed with nuclei acids, bind quantitatively to certain materials in conditions that free RNA and DNA are not retained. Experimental data can thus be obtained from analysis of the nucleic acids retained in the filter, or found in filtrates and washes. Because protein-adsorbing materials can differ in their efficiency, capacity, and flow characteristics, filters from the same lot should be used in serial experiments, and the optimal filtration conditions need to be empirically determined in each case.

Filters of polyvinylcloride *(3,4)*, nitrocellulose *(5)*, and borosilicate *(6–8)* have been used in the study of topoisomerase–DNA complexes. In this chapter, a simple filter binding assay, adapted from Thomas et al. *(2)*, is described for the study of topoisomerase–DNA complexes. This procedure uses borosilicate filters (glass fibers) in a spin filtration setting, and allows qualitative and quantitative analysis of topoisomerase-free and topoisomerase-bound DNA species by gel electrophoresis.

From: *Methods in Molecular Biology, Vol. 95: DNA Topoisomerase Protocols, Part II: Enzymology and Drugs*
Edited by N. Osheroff and M.A. Bjornsti © Humana Press Inc., Totowa, NJ

1.2. Types of Topoisomerase–DNA Complexes

For most proteins that interact with DNA, a common question is whether a protein recognizes a defined chromosomal sequence. In case of topoisomerase–DNA interactions, two additional questions become often more relevant: the structure of the bound DNA and the nature of the attachment itself. Topoisomerases have different DNA binding affinities regarding DNA structure. Prokaryotic type I topoisomerases have high affinity for single-stranded DNA, whereas type II topoisomerases can interact with two or more double-stranded DNA segments simultaneously. Therefore, the conformation of DNA and the local concentration of DNA segments are parameters to consider in topoisomerase–DNA binding experiments. Additionally, three kinds of attachments can be found between topoisomerases and DNA: noncovalent, covalent, and a topological link (a protein clamp around DNA). Below 50 m*M* salt, most type I and type II topoisomerases have a low DNA dissociation rate. Higher salt concentrations progressively dissociate all noncovalent topoisomerase–DNA interactions, and only covalent interactions and topological links remain stable over 1 *M* salt. Covalent interactions are associated to DNA strand breakage by the topoisomerase and become irreversible on denaturation of the enzyme. Topological links, instead, only exist in DNA molecules without free ends (i.e., circular molecules) and are destroyed on protein denaturation. Filter binding assays should be designed to distinguish among these three kinds of protein–DNA interactions.

2. Materials

2.1. Filter Units

Whatman GF/C microfiber glass filters and standard 1.5-mL disposable Eppendorf centrifuge tubes with concave caps (Sarstedt # 65.697) are needed to construct the filter units. Other type of tubes with concave caps can be used, as long as the unit can be placed in a fixed-angle centrifuge rotor and the cap concavity can hold 100 μL.

2.2. Topoisomerases

Type I or type II topoisomerases can be purified from eukaryotic or bacterial cells. Working stocks of the enzymes can be stored at –20°C with 20% glycerol during several days without lost of specific activity. The presence of salt (200–600 m*M*) in the store buffer is recommended to avoid protein aggregation.

2.3. DNA Substrates

According to the purpose of each experiment, several DNA or different conformers of a particular DNA molecule can be used in a topoisomerase binding assay. DNA isolated by standard protocols should have sufficient purity for

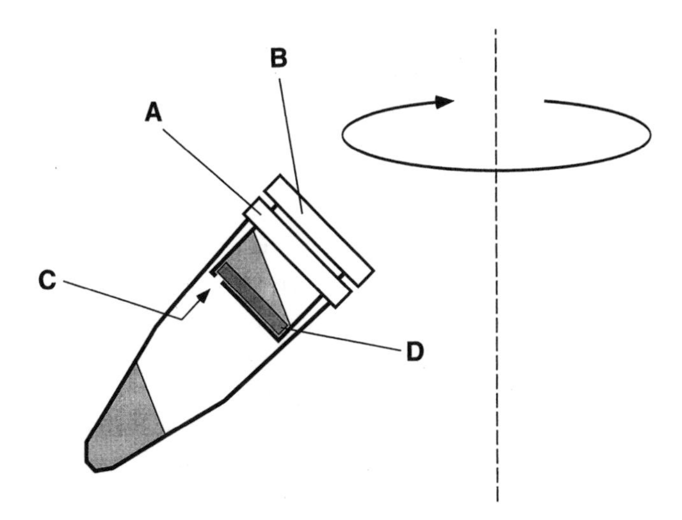

Fig. 1. Setting for spin filtration through glass fiber filters. **(A)** Regular 1.5-ml microcentrifuge tube. **(B)** Push-in cap with a concavity deep enough to hold at least 100 µL. **(C)** Small hole pierced with a needle at the bottom edge of the cap concavity. **(D)** Whatman GF/C glass fiber filter 6 mm in diameter.

use directly in the reactions. DNA molecules can be labeled to facilitate quantitative analysis.

2.4. Solutions

1. 1X topoisomerase–DNA reaction buffer without salt (to be adjusted).
2. 1X topoisomerase–DNA reaction buffer with salmon sperm DNA (100 µg/mL).
3. 2 *M* NaCl.
4. 0.1% SDS.

3. Methods

3.1. Construction of the Filter Units

Filters 6 mm in diameter are cut with a cork borer from standard size Whatman GF/C microfiber glass filters. Each filter is placed in the bottom of a concave cap of a microfuge tube, in which a small drainage hole has been pierced through near the lowest perimeter of the concavity. As is illustrated in **Fig. 1**, when the filtration unit is placed in a fixed-angle microfuge rotor, the drainage hole in the cap concavity should be away from the center of rotation.

3.2. Binding of Topoisomerase to DNA

DNA and topoisomerase can be mixed in any of the buffers currently used for activity assays, but to ensure a homogeneous distribution of the protein, it is recommended to start the mixture in a high salt buffer (over 150 m*M*), and then dilute the sample and adjust the desired final salt concentration. The amount of total protein assayed should not exceed the filter capacity (*see* **Note 1**), and the stoichiometry of protein and DNA needs to be adjusted to ensure a correct interpretation of the results (*see* **Note 2**).

3.3. Filter Binding Assay

Once the reaction is done, the mixture (up to 100 μL) is placed on the filter unit. Filtrates and washes with increasing salt concentrations (50–100 μL each) can now be recovered from the centrifuge tubes following brief spins of 10–20 s at 500*g* (*see* **Note 3**). A final wash with 0.1% SDS will dissociate any of the protein–DNA complexes still retained in the filter.

In order to minimize background, before loading a reaction sample, each filter can be prewashed by filtering 100 μL of 100 μg/mL salmon sperm DNA in the corresponding reaction buffer (*see* **Note 4**).

3.4. Analysis of the Results

DNA from the filtrates and washes can be precipitated and analyzed by gel electrophoresis. Samples can be also directly electrophoresed if volumes and salt concentration are normalized by dilution. DNA is visualized by ethidium staining of the gel, or transferred to a membrane and probed. Relative amounts of DNA forms can be now determined by a variety of methods. Scintillation counting or phospho-image analysis are the best choice to obtain accurate data.

Figure 2 describes an experiment using this filter binding technique. The purpose of the experiment was to quantify the single-step DNA transport efficiency by yeast topoisomerase II on binding of AMPPNP *(8)*. The substrate was a purified single-DNA topoisomer for which ΔLK was close to 0 in the reaction conditions. Because AMPPNP addition converts topoisomerase II in a closed clamp around DNA, filtration was done directly in high salt, and the DNA molecules topologically linked to the topoisomerase were then recovered from the filter by an SDS wash. The results of the experiment were interpreted as follows: In agreement with an enzyme/DNA molar ratio of 0.5, <30% of DNA was retained in the filter as a high salt-resistant DNA–protein complex, and therefore, most DNA molecules retained had only one bound enzyme. As expected for topologicaly linked complexes, only circular DNA species were retained in the filter and were recover in the SDS wash. Because the reacted single DNA topoisomer had an ΔLk close to 0, DNA transport by the topoisomerase resulted both in an increase or decrease of the initial Lk (*see* **Note 5**). DNA transport efficiency by the topoisomerase can be calculated from the population of molecules that underwent Lk change, relative to the total closed-circular molecules bound by the enzyme.

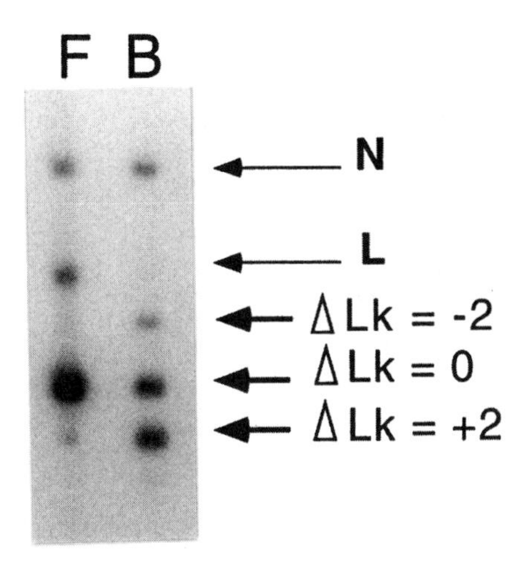

Fig. 2. Single-step DNA transport efficiency by yeast topoisomerase II assayed by filter binding. Approximately 100 fmol of a single 4-kb DNA topoisomer were reacted with 50 fmol of yeast DNA topoisomerase II in a 40-µL reaction mixture containing 50 m*M* Tris-HCl, pH 8.0, 1 m*M* EDTA, 150 m*M* KCl, 8 m*M* $MgCl_2$, 7 m*M* 2-SHEtOH, and 100 µg/mL BSA. AMPPNP was added to 2-m*M* final concentration, and after incubation at 30°C for 5 min, 40 µL of 2 *M* NaCl were added, and the mixture was filtered using the setting depicted in **Fig. 1** and described in the **Subheading 3.** DNA recover from the filtrate (80 µL) and eluted from the filter by 0.1% SDS wash (80 µL) was directly analyzed by agarose-gel electrophoresis. The gel was blotted and probed with DNA obtained by random priming. The topoisomer bands in the topoisomerase-free and topoisomerase-bound fractions were quantitated by phosphor-imaging analysis. Lane F: filtrate. Line B: 0.1% SDS wash. Some nicked and linearized molecules, denoted by (N) and (L), respectively, were generated during the purification of the single DNA topoisomer, which is denoted by (ΔLk = 0). The position of DNA species generated in the reaction with altered ΔLk values (-2 and +2) are also indicated.

4. Notes

1. Reaction samples with high protein content may saturate the filter, and most of the protein–nucleic acid complexes may not be retain. A Whatman GF/C glass fiber filter 6 mm in diameter is saturated with 25 µg of protein.
2. The bulk of the filter-bound DNA represents molecules with at least one bound protein. The average number of bound protein per DNA molecule (i) can be calculated from the Poisson relation (i) = - ln x , where x is the fraction of total DNA in the filtrate. For example, if 63% of DNA is retained in the filter and 37% is protein-free DNA, the average number of protein molecules bound to each DNA molecule in the initial mixture is – ln 0.37 or 1.

3. Spins should drain most of the liquid from the cap in few seconds. In several ways, the filtration step can alter the result. For example by removing complexes during a very slow filtration, the filters may pull the equilibrium in direction of complex formation. Binding to the filter could also change the characteristics of topoisomerase–DNA complexes formed in solution.
4. The no-specific retention of protein-free nucleic acids, especially double-stranded DNA, is very low in filter assays. No more than 3% of no-specific binding is reported with nitrocellulose *(1)* or glass fiber filters *(2)*. Background can be further minimized by washing the filters with a competitor nucleic acid prior filtration of the sample.
5. Note that no molecules with altered Lk are found in the filtrate. That is an indication of a good retention efficiency of the filter. Retention efficiency can be easily tested when filtering DNA–topoisomerase complexes, because the topoisomerase usually changes some characteristic of the bound DNA. If retention is efficient, altered DNA molecules should not be found in the protein-free DNA filtrates.

Acknowledgment

I am grateful to James C. Wang for his support during the preparation of this chapter.

References

1. Riggs, A. D., Suzuki, H., and Bourgeois, S. (1970) Lac repressor-operator interaction: equilibrium studies. *J. Mol. Biol.* **48,** 67–83.
2. Thomas, C. A., Saigo, K., McLeod, E., and Ito, J. (1979) The separation of DNA segments attached to proteins. *Anal. Biochem.* **93,** 158–166.
3. Pommier, Y., Mindford, J. K., Schwartz, R. E., Zwelling, L. A., and Kohn, K. W. (1985) Effects of the DNA intercalators 4'-(9-Acridinylamino) methanesulfon-m-aniside and 2 methyl-9-hydroxyellipticinium on topoisomerase II mediated DNA strand cleavage and strand passage. *Biochemistry* **24,** 6410–6416.
4. Pommier, Y., Kerrigan, D., and Kohn, K. W. (1989) Topological complexes between DNA and topoisomerase II and effects of polyamines. *Biochemistry* **28,** 995–1002.
5. Manden, K. R., Steward, L., and Champoux, J. J. (1995) Preferential binding of human topoisomerase I to superhelical DNA. *EMBO J.* **14,** 5399–5409.
6. Roca, J. and Wang, J. C. (1992) The capture of a DNA double helix by an ATP-dependent protein clamp: a key step in DNA transport by type II DNA topoisomerases. *Cell* **71,** 833–840.
7. Roca, J., Ishida, R., Berger, J. M., Andoh, T., and Wang, J. C. (1994) Antitumor bisdioxipiperazines inhibit yeast DNA topoisomerase II by trapping the enzyme in the form of a closed protein clamp. *Proc. Natl. Acad. Sci. USA* **91,** 1781–1785.
8. Roca, J. and Wang, J. C. (1996) The probabilities of supercoil removal and decatenation by yeast DNA topoisomerase II. *Genes to Cells* **1,** 17–27.

9

DNA Topoisomerase I-Mediated Nicking of Circular Duplex DNA

James J. Champoux

1. Introduction

Type I topoisomerases catalyze the reversible nicking of duplex DNA (for review, *see* **ref. *1***). In the nicked state, the enzyme is attached to the DNA by way of a phosphodiester bond between a tyrosine residue in the protein and the end of the broken strand. The polarity of attachment depends on the source of the enzyme; the bacterial enzyme is linked to the 5'-end, whereas the eukaryotic enzyme is attached to the 3'-end of the DNA.

Mechanistic and specificity studies often require that one experimentally interrupt the nicking-closing cycle of the topoisomerase-catalyzed reaction to trap the covalent intermediate. For *Escherichia coli* topoisomerase I and single-stranded substrates, the addition of denaturants, such as alkali or SDS, results in strand breakage and formation of the covalent complex *(2,3)*. These same denaturants will trap the covalent complex on double-stranded circular DNA provided one starts with a negatively superhelical substrate *(4)*. Similarly, covalent complexes can be trapped for the eukaryotic topoisomerase I using either double-stranded or single-stranded DNAs by stopping the reactions with SDS, alkali, or low pH *(5,6)*. Since the eukaryotic enzyme is less specific with regard to the topological state of its substrate, the starting duplex DNA need not be supercoiled. As will be shown below, prolonged incubation of the human topoisomerase I with duplex DNA also leads to a low level of spontaneous abortive nicking. The use of special substrates that cause suicide cleavage of the DNA and the formation of the covalent complexes are discussed in Chapters 11–12 and will not be considered here.

From: *Methods in Molecular Biology, Vol. 95: DNA Topoisomerase Protocols, Part II: Enzymology and Drugs*
Edited by N. Osheroff and M.A. Bjornsti © Humana Press Inc., Totowa, NJ

A convenient procedure for detecting the enzyme–DNA covalent complex is to start with a supercoiled or relaxed closed-circular duplex DNA (form I DNA) and analyze the products of the cleavage reaction by electrophoresis through an agarose gel *(7)*. Although the protein–DNA complexes can be visualized directly owing to their slower mobilities relative to the substrate DNA, the bands are quite diffuse and difficult to quantitate, owing in part to the fact that more than one protein molecule can become attached to each DNA. A more sensitive and quantitative approach is to treat the enzyme–DNA complexes with proteinase K to remove all but a few amino acids of the bound protein, converting them to a uniform population of molecules that comigrate with nicked circles (form II). Although these nicked molecules are well resolved from any remaining superhelical form I DNA in an agarose gel, they do migrate close to the population of relaxed topoisomers (form I_r). The overlap of the relaxed product with the form II DNA is not as serious a problem with *E. coli* topoisomerase I, since the final set of topoisomers still retain some negative supercoils reflecting the inability of the enzyme to relax the DNA completely (*see* Chapter 1) *(8)*. A method is described here for detecting topoisomerase-nicked circular DNAs both before and after protease treatment, which takes advantage of electrophoresis in the presence of ethidium bromide to separate the nicked circles from the relaxed topoisomers.

2. Materials

2.1. Closed-Circular DNA

The closed-circular substrate can either be a small viral DNA, such as SV40, or a bacterial plasmid DNA. For ease of handling and gel analysis, it is convenient if the DNA has a size in the range of 2.5–7 kb (*see* **Note 1**). Standard isolation protocols should yield DNA of sufficient purity for use directly in the breakage assay. Since one is looking for the conversion of supercoiled circles to nicked circles, it is important that the starting material contain <~10% nicked circles. To achieve this low level of contaminating form II DNA, it is sometimes necessary to purify the DNA further by CsCl equilibrium centrifugation in the presence of ethidium bromide. Care should be taken to ensure that all of the ethidium bromide is extracted from the DNA prior to use in the assay.

2.2. Topoisomerase I

The type I topoisomerase can be purified from either a bacterial or a eukaryotic cell type. Since any contaminating endonuclease activity will generate form II DNA that will comigrate with the nicked covalent complexes after protease treatment, it is important to use a highly pure enzyme (*see* **Note 2**).

2.3. Agarose-Gel Electrophoresis

1. 10X TBE buffer: 0.89 *M* Tris-borate, 20 m*M* EDTA, pH 8.0.
2. 1X TBE buffer with ethidium bromide: 89 m*M* Tris-borate, pH 8.0, 2 m*M* EDTA, 0.25 µg/mL ethidium bromide, made up just prior to use.
3. 3X gel-loading buffer: 30 m*M* EDTA, 2.5% SDS, 0.1% bromphenol blue, 12% Ficoll 400.

3. Methods

3.1. Topoisomerase Reactions

The efficiency of trapping the covalent complex is relatively low, and thus, one commonly uses high molar ratios of enzyme to DNA to observe measurable levels of cleavage (*see* **Note 3**). In general, the processivity of a topoisomerase decreases as the salt concentration is increased. Accordingly, to observe the maximum level of breakage when stopping a reaction with a protein denaturant, such as SDS, it is necessary to use the lowest salt concentration that is compatible with activity. The extent of breakage by the eukaryotic enzyme can be enhanced by the use of a variety of poisons that slow the rate of closure ***(9,10)***. The optimal detergent concentration for efficient breakage needs to be empirically determined for the particular enzyme of interest (*see* **Note 4**).

3.2. Agarose-Gel Electrophoresis

1. Cast either a vertical or horizontal 1% agarose gel in 1X TBE with a total length of at least 10 cm and a thickness of 3–5 mm (*see* **Note 5**).
2. Prepare the samples for electrophoresis by adding 1/2 volume of 3X gel-loading buffer.
3. Load the samples and carry out the electrophoresis at ~1 V/cm for 16 h. The time and voltage should be adjusted for the size of the DNA substrate.
4. Carefully remove the gel from the electrophoresis apparatus, and stain the gel for 30–60 min in 1X TBE buffer with ethidium bromide. Thicker gels may require a longer staining time.
5. Return the gel to the electrophoresis apparatus, and continue the electrophoresis for ~1 h at 3 V/cm. The time should be adjusted to give the desired separation between the form II DNA and the form I_r DNA.
6. Transfer the gel to a UV light box, and either record the image by photography or by use of a digitized scanner or video camera.

3.3. Analysis of Results

Figure 1 shows the results of an agarose gel analysis of human topoisomerase I-nicked SV40 DNA. The reactions were stopped by the addition of SDS to trap the covalent intermediate (lanes 9–12) or high salt to dissociate

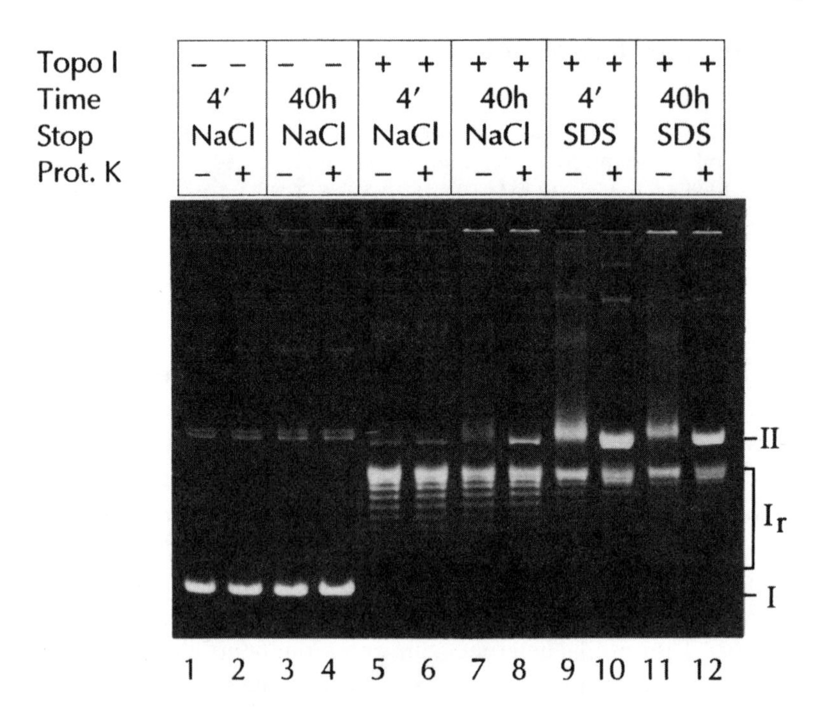

Fig. 1. Agarose-gel electrophoresis of human topoisomerase I-nicked DNA. Form I SV40 DNA (98 µg/mL final concentration) was incubated at room temperature (~22°C) with human topoisomerase I (225 µg/mL final concentration) in 20 m*M* Tris-HCl, pH 7.5, 50 m*M* NaCl, 1 m*M* EDTA for either 4 min or 40 h as indicated (final volume 140 µL). For the samples shown in lanes 5–8, 8-µL aliquots of the reaction were stopped by the addition of NaCl to 1.0 *M* and incubated further at room temperature for 5 min to dissociate any noncovalently bound enzyme prior to addition of the gel-loading buffer. For the samples shown in lanes 9–12, the covalent complexes in a 8-µL aliquot were trapped by the addition of an equal volume of 1% SDS prior to the addition of the gel-loading buffer. The enzyme was omitted from the controls shown in lanes 1–4 and the samples treated as indicated. One-half of each sample was treated with proteinase K (100 µg/mL final concentration) (even numbered lanes) before loading onto the gel. In all cases, an aliquot corresponding to 0.2 µg of DNA was loaded onto a 1% agarose gel and the electrophoresis carried out as described in the text. The mobilities of the supercoiled substrate (I), the nicked circles (II), and the relaxed topoisomers (I_r) are shown along the right side of the figure.

any noncovalently bound protein prior to the analysis (lanes 5–8). Regardless of the incubation time, the SDS stop procedure produced a spectrum of DNA–protein complexes that migrated slower than form II DNA, and that collapsed to the position of the nicked circles after treatment with proteinase K (lanes 9–12). Without the second electrophoresis in the presence of ethidium bromide, the form II and form I_r DNAs would have overlapped, precluding any conclusions about the amount of form II DNA in the samples. For the 4-min incubation time, the sample stopped with high salt and treated with proteinase K (lane 6) showed that none of the observed nicking resulted from a contaminating nuclease, since the amount of form II DNA remained identical to that in the control samples (lanes 1–4). Prolonged incubation with the enzyme (40 h) did, however, produce a significant increase in nicked DNA when the reaction was similarly treated with high salt and proteinase K (lane 8). This nicking was likely because of spontaneous breakage without reclosure by the enzyme, since the sample that was not treated with protease (lane 7) showed the smear of slower migrating species diagnostic of topoisomerase I–DNA covalent complexes. Such spontaneous breakage by eukaryotic type I topoisomerase on a duplex substrate has not been previously reported (*see* **Note 6**).

The relative amounts of form II and form I_r DNAs can be determined by a variety of methods. By digitizing the image using a CCD camera or scanner, quantitation can easily be accomplished using readily available software. Alternatively, a photographic negative can be scanned using a densitometer. If radioactively labeled DNA is available, the DNA bands can be excised from the gel and quantitated with a scintillation counter.

The sequence specificity for both the bacterial and eukaryotic type I topoisomerases is sufficiently weak that a DNA molecule with >2.5 kb will likely contain hundreds of potential break sites ***(3,11,12)***. One can therefore calculate the average number of breaks per molecule (b) by substituting the fraction of unbroken molecules (f) for the zero term of the Possion distribution [$b = -\ln(f)$]. In practice, this relationship is useful up to an average of ~4 breaks/mol, since above that number, the amount of remaining form I_r falls below the level that one can reliably measure.

4. Notes

1. Closed-circular DNAs in this size range often contain detectable amounts of supercoiled dimer DNA that migrates near the position of the nicked circles (*see* **Fig. 1**, lanes 1–4). Since relaxation of the dimers during the course of the topoisomerase reaction reduces their mobility, they do not interfere with the visualization of the topoisomerase-nicked DNA.
2. As described in **Subheading 3.3.**, a contaminating endonuclease will produce nicked form II DNA that does not contain bound protein and that persists when the reaction is stopped with high salt (~1 *M*).
3. For eukaryotic type I topoisomerases, molar ratios of enzyme to DNA of 10–100 are typically used to observe SDS-induced breakage in the absence of an enhancing agent, such as camptothecin.
4. The agarose-gel analysis described here for cleavage of duplex DNA can easily be adapted to detect cleavage of single-stranded circle DNA ***(6)***. However, in this

case, multiple cleavage events produce a smear of fragments with lengths less than unit linears.

5. Although vertical gels are somewhat more difficult to run, in our experience, the topoisomer bands are sharper than on a horizontal gel.
6. To detect the low level of spontaneous breakage, it was necessary to incubate the enzyme with the DNA at very high protein to DNA ratios. Under the conditions employed here, we have noticed that the complexes tend to precipitate *(13)*. The relationship between this precipitation and the observed spontaneous breakage remains to be determined.

Acknowledgments

I am grateful to Lance Stewart and Sharon Schultz for helpful comments during the preparation of the manuscipt. This work was supported by NIH grant GM-49156.

References

1. Champoux, J. J. (1990) Mechanistic aspects of type-I topoisomerases, in *DNA Topology and Its Biological Effects* (Cozzarelli, N. R. and Wang, J. C., eds.), Cold Spring Harbor Laboratory Press, Cold Spring Harbor, NY, pp. 217–242.
2. Depew, R. E., Liu, L. F., and Wang, J. C. (1978) Interaction between DNA and *Escherichia coli* protein ω. Formation of a complex between single-stranded DNA and ω protein. *J. Biol. Chem.* **253,** 511–518.
3. Kirkegaard, K., Plugfelder, G., and Wang, J. C. (1984) The cleavage of DNA by type-I DNA topoisomerases. *Cold Spring Harbor Symp. Quant. Biol.* **49,** 411–419.
4. Liu, L. F. and Wang, J. C. (1979) Interaction between DNA and *Escherichia coli* DNA topoisomerase I. Formation of complexes between the protein and superhelical and nonsuperhelical duplex DNAs. *J. Biol. Chem.* **254,** 11,082–11,088.
5. Champoux, J. J. (1977) Strand breakage by the DNA untwisting enzyme results in covalent attachment of the enzyme to DNA. *Proc. Natl. Acad. Sci. USA* **74,** 3800–3804.
6. Been, M. D. and Champoux, J. J. (1980) Breakage of single-stranded DNA by rat liver nicking-closing enzyme with the formation of a DNA-enzyme complex. *Nucleic Acids Res.* **8,** 6129–6142.
7. Champoux, J. J. and Aronoff, R. (1989) The effects of camptothecin on the reaction and the specificity of the wheat germ type I topoisomerase. *J. Biol. Chem.* **264,** 1010–1015.
8. Wang, J. C. (1971) Interaction between DNA and an *Escherichia coli* protein ω. *J. Mol. Biol.* **55,** 523–533.
9. Hsiang, Y.-H., Hertzberg, R., Hecht, S., and Liu, L. F. (1985) Camptothecin induces protein-linked DNA breaks via mammalian DNA topoisomerase I. *J. Biol. Chem.* **260,** 14,873–14,878.
10. Chen, A. Y., Yu, C., Gatto, B., and Liu, L. (1993) DNA minor groove-binding ligands: a different class of mammalian DNA topoisomerase I inhibitors. *Proc. Natl. Acad. Sci. USA* **90,** 8131–8135.

11. Tse, Y.-C., Kirkegaard, K., and Wang, J. C. (1980) Covalent bonds between protein and DNA. Formation of phosphotyrosine linkage between certain DNA topoisomerases and DNA. *J. Biol. Chem.* **255,** 5560–5565.
12. Been, M. D., Burgess, R. R., and Champoux, J. J. (1984) Nucleotide sequence preference at rat liver and wheat germ type 1 DNA topoisomerase breakage sites in duplex SV40 DNA. *Nucleic Acids Res.* **12,** 3097–3114.
13. Stewart, L., Ireton, G. C., and Champoux, J. J. (1996) The domain organization of human topoisomerase I. *J. Biol. Chem.* **271,** 7602–7608.

10

Analysis of Cleavage Complexes Using Reactive Inhibitor Derivatives

Kenneth N. Kreuzer, Catherine H. Freudenreich, and Yves Pommier

1. Introduction

With recent progress in determining the crystal structures of type I and type II topoisomerases (topo I and topo II, respectively) *(1,2)*, an understanding of the binding of DNA-cleavage-inducing inhibitors at the atomic level is within reach. A very useful approach for investigating the nature of the inhibitor binding site is the crosslinking of reactive inhibitor derivatives to the DNA or protein within the cleavage complex. In this chapter, we describe methods that have been used to crosslink a 4'-(9-acridinylamino)methanesulfon-*m*-anisidide (*m*-AMSA) derivative to the DNA within a topo II cleavage complex and a camptothecin (CPT) derivative to the DNA within a topo I cleavage complex. In both cases, the derivative reacted only with the base pairs immediately adjacent to the scissile phosphodiester bond and only in the presence of topoisomerase, providing strong and direct evidence that the DNA substrate forms an important component of the inhibitor binding site.

1.1. Photoactivation of an m*-AMSA Derivative*

Denny's group *(3)* first synthesized the photoactive 4'-(3-azido-9-acridinylamino)methanesulfon-*m*-anisidide (3-azido-AMSA) (*see* **Fig. 1** for structure), and Byrn and colleagues *(4)* analyzed the photochemical, physical, and biological properties of this compound. Importantly, DNA binding by *m*-AMSA and 3-azido-AMSA were very similar, but 3-azido-AMSA could be photoactivated to react with DNA, mononucleosides, or dinucleotides *(4)*.

Freudenreich and Kreuzer *(5)* used the photoactive derivative to localize the inhibitor binding site in a topo II cleavage complex. First, they found that in

From: *Methods in Molecular Biology, Vol. 95: DNA Topoisomerase Protocols, Part II: Enzymology and Drugs*
Edited by N. Osheroff and M.A. Bjornsti © Humana Press Inc., Totowa, NJ

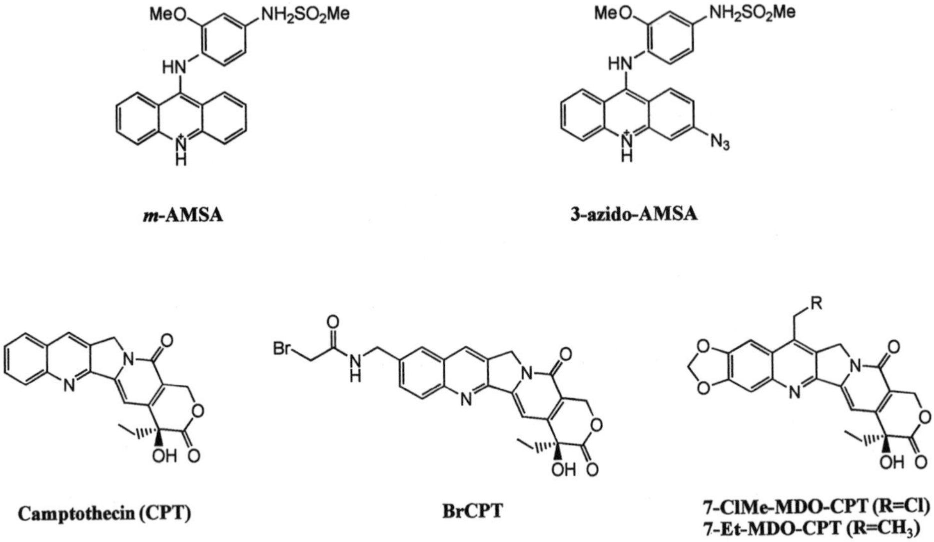

m-AMSA 3-azido-AMSA

Camptothecin (CPT) BrCPT 7-ClMe-MDO-CPT (R=Cl)
7-Et-MDO-CPT (R=CH_3)

Fig. 1. Structure of *m*-AMSA, CPT, and reactive derivatives.

the absence of light, 3-azido-AMSA induces cleavage complexes with similar potency and at the same DNA cleavage site as does *m*-AMSA, arguing that the photoactive agent is an accurate probe for analyzing inhibitor binding. The location of the inhibitor within the cleavage complex was then investigated by photoactivating cleavage complexes formed with 3-azido-AMSA, an oligonucleotide substrate, and the phage T4 topo II (*see* **Subheading 2.**). Covalent adducts along the DNA substrate were localized by both primer-extension analysis and piperidine-induced cleavage of the DNA backbone. The inhibitor reacted with four nucleotide residues in the DNA substrate, and each of these was within one of the two base pairs that flank the scissile bond on each strand (**Fig. 2A**, reactive bases are boxed).

The precise geometry of the reactive bases, kitty-corner and 5' with respect to the scissile bond position, is apparently imposed by the orientation of the inhibitor and the cleaved DNA, because substitution of the +1 base pair with T/A (top/bottom) or the −1 base pair with A/T (top/bottom) did not result in reactivity of the dT residue (even though dT is reactive in the other locations; *see* **Note 1**). Considering the fact that many topo II inhibitors (including *m*-AMSA) are DNA intercalators, the inhibitor binding site could consist of an intercalation site between the two base pairs that flank each cleaved phosphodiester bond. Given the locations of the bases that react with 3-azido-AMSA, it is also conceivable that the two broken ends of the DNA are separated and the central 4-base stagger is single-stranded at the time of the induced photoreaction.

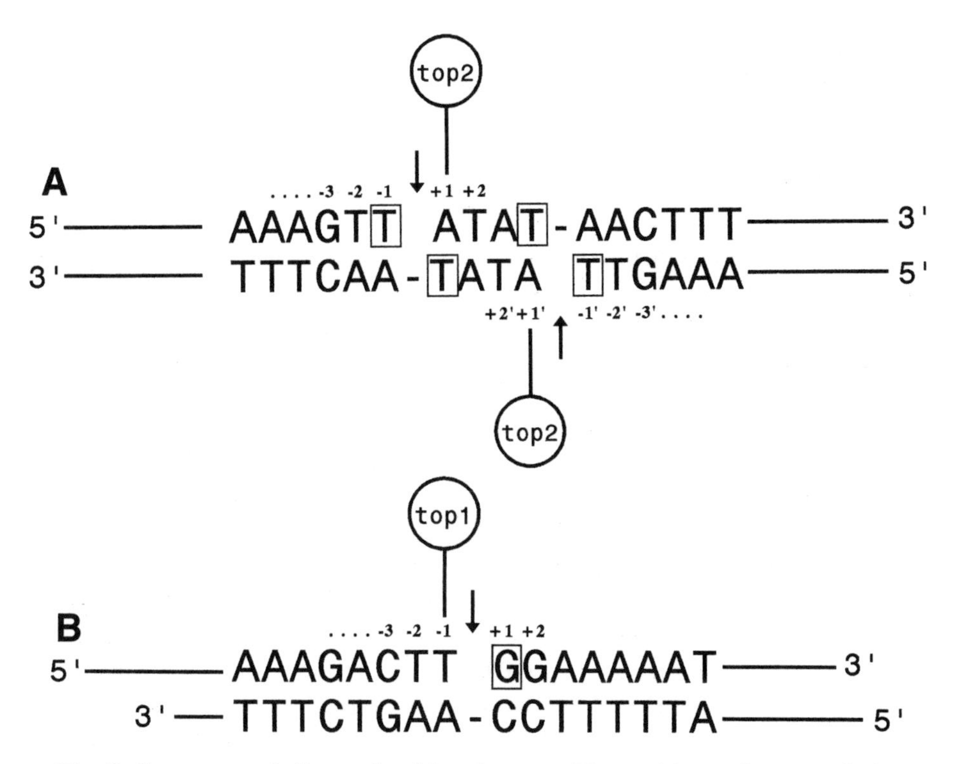

Fig. 2. Sequences of oligonucleotide substrates. The positions of enzyme-induced DNA cleavage are indicated by arrows (the dash across from the arrow indicates that the opposite strand is intact). The positions of covalently attached enzyme are indicated by the lollipops labeled "top1" or "top2", and the reactive bases are boxed. The sequences in the flanking regions of both oligonucleotides are not shown, but rather indicated by solid lines. In **A**, the lines represent: 10 bp to the left and 12 bp to the right of the indicated sequence. In **B**, the lines represent: 6 bases on top (5'-strand) and 2 bases on bottom (3'-strand) to the left; 11 bases on top (3'-strand) and 14 bases on bottom (5'-strand) to the right.

1.2. Alkylation by CPT Derivatives

Three different alkylating derivatives of CPT, 7-ClMe-MDO-CPT (7-chloromethyl-10,11-methylenedioxycamptothecin), 7-Et-MDO-CPT (7-ethyl-10,11-methylenedioxycamptothecin), and BrCPT (10-bromoacetamidomethylcamptothecin), have been used to analyze cleavage complexes involving topo I *(6,7)* (*see* **Fig. 1** for structures). CPT induces cleavage complexes preferentially at DNA sites with a dG residue at the 5'-terminus of the break (the +1 position; *see* **Fig. 2B**), raising the possibility of a specific interaction between CPT and the dG residue adjacent to the scissile bond ***(8–10)***.

Pommier et al. *(6)* investigated this possibility by forming cleavage complexes with 7-ClMe-MDO-CPT (or 7-Et-MDO-CPT), an oligonucleotide substrate, and mammalian topo I (*see* **Subheading 3.**). When the labeled substrate was analyzed by gel electrophoresis, a specific alkylation product was readily detected. By treating the alkylated DNA with formic acid and piperidine (to cleave the DNA backbone at the modified base), the alkylation product was demonstrated to result from reactivity at a dG or dA residue in the +1 position, immediately adjacent to the scissile bond (**Fig. 2**; *see* **Note 1**). These results directly localize the CPT derivative to the immediate neighborhood of the 5'-residue in the cleavage complex, and strongly support the model that cleavage site selection is modulated by interaction of CPT with that base residue. Reactivity was detected only when the topoisomerase was present, and was not detected with either an active site tyrosine mutant, or a CPT-resistant mutant (which does not induce cleavage complexes with the inhibitor) *(6)*.

Hertzberg et al. *(7)* used radiolabeled BrCPT, with the reactive group at the 10 rather than the 7 position (*see* **Fig. 1**), to analyze the CPT binding site within the topo I cleavage complex. In this case, the CPT derivative reacted covalently with the enzyme and not the DNA substrate. The precise amino acid residue(s) of the topo I that reacted with BrCPT was not determined. The results of the two CPT studies together *(6,7)* provide the strongest evidence for the hypothesis that topoisomerase inhibitors bind at the enzyme–DNA interface (*also see* ***9***).

1.3. Perspectives and Future Directions

Reactive inhibitor derivatives are clearly useful in analyzing the inhibitor binding site within the cleavage complex. As already discussed, results with reactive inhibitors have localized inhibitor binding sites to the immediate vicinity of the cleaved phosphodiester bond within the cleavage complex. Crosslinking can also provide a finer level of resolution. For example, the orientation of the inhibitor within the cleavage complex could be more accurately assigned by determining the precise nature of the covalent DNA adduct (e.g., by mass spectrometry). Even without such information, molecular modeling can be attempted. For instance, modeling of the 7-ClMe-MDO-CPT alkylation product shows a preferred configuration with CPT stacked between the +1 and −1 base pairs flanking the cleavage site. In this model, the concave arch of CPT faces the DNA major groove, and the E (lactone) ring is oriented toward the DNA break with the 20(S)hydroxyl interacting with topo I asparagine-722 (next to the catalytic tyrosine-723, and mutated in the CPT-resistant topo I, Asn722Ser *[11]*).

The interactions of reactive inhibitors with DNA within the cleavage complex can also be compared to the interactions that occur between drug and naked DNA. Regarding topo I, CPT has been shown to interact preferentially with dG residues within naked single- or double-stranded DNA, consistent with its interaction with dG residues within the DNA-topo I cleavage complex ***(12)***. Regarding topo II, the relationship between inhibitor interaction with naked DNA and interaction with DNA within the cleavage complex has yet to be analyzed.

Additional reactive derivatives can presumably be utilized very effectively. Possible topoisomerase inhibitors that might alkylate cleavage complexes include members of the acridine, terpenoide, anthraquinone, and anthracycline families ***(13–16)***. Perhaps the most useful approach would be to place the reactive group at a variety of positions within one particular inhibitor, systematically probing contacts of the drug with DNA and enzyme. For instance, in the case of CPT, protein contacts would be expected when the reactive group is appended to the concave region of the drug. In the case of topo II inhibitors, protein contacts might be expected with substitutions on the anilino group of *m*-AMSA, on the pendant ring of epipodophyllotoxins or azatoxins, or on the A ring of anthracyclines. In a related vein, it would be very informative to locate the precise amino acid residue that is modified by reactive inhibitors in the cleavage complex.

Finally, we would like to note that covalent linkage of the inhibitor to the protein or DNA substrate apparently results in a nonreversible complex ***(6,7,14,15)*** (Freudenreich and Kreuzer, unpublished results). The generation of nonreversible complexes may extend the potency of topoisomerase inhibitors. For example, an important limitation of camptothecins is the rapid reversal of cleavage complexes on drug removal, and cytotoxicity is strongly dependent on long drug exposures. The 7-ClMe-MDO-CPT derivative alleviates this problem. In a therapeutic sense, a single dose of a reactive inhibitor would be expected to produce persistent cleavage complexes and marked cytoxicity. Preliminary results indicate that 7-ClMe-MDO-CPT is markedly more effective than other camptothecins in cell-culture experiments (Valenti and Pommier, unpublished results). Furthermore, 7-ClMe-MDO-CPT is much more cytotoxic in DNA-replication-inhibited cells (e.g., treated with aphidicolin) than is CPT (Valenti and Pommier, unpublished results).

2. Materials

1. Oligonucleotides: Synthetic oligonucleotides are purchased from commercial sources and should be purified to eliminate shorter products that contribute to background. A strong cleavage site for the relevant topoisomerase is designed into the oligonucleotide so that a large majority of the signal will reflect a single enzyme recognition site. For the phage T4 topo II, the results of a mutational analysis of a moderately strong cleavage site ***(17)*** were used to design a stronger cleavage site for crosslinking experiments ***(5)***. For topo I experiments, several

oligonucleotides containing a single strong cleavage site have been described, including one that corresponds to the SV40 replication origin ***(9)*** and another derived from tetrahymena ribosomal DNA ***(18,19)***. In the latter case, the sequence was modified by changing the +1 base to a G to optimize topo I-specific DNA cleavage enhanced by CPT ***(20)***.

2. Labeling reagents: T4 polynucleotide kinase and terminal deoxynucleotidyl transferase are purchased from standard commercial sources, and γ-^{32}P]ATP, [α-^{32}P]ddATP, and [α-^{32}P]cordycepin 5'-triphosphate are available from Amersham Pharmacia (Piscataway, NJ) and/or DuPont NEN (Boston, MA). Spin columns (e.g., Sephadex G-25, DNA-grade; Amersham Pharmacia) are used for eliminating unincorporated label.
3. Topoisomerase: Purification of the phage T4 topo II is described in Chapter 18 of this volume. Calf thymus topo I can be purchased from Gibco/BRL (Grand Island, NY) or DNA Technology Incorporation (Gaithersburg, MD), human placenta topo I from Topogen, Inc. (Columbus, OH), and wheat germ topo I from Epicentre Technologies or Promega (Madison, WI). The base sequence selectivity of calf thymus and wheat germ enzymes is well conserved, with only minor differences ***(10)***.
4. Reactive *m*-AMSA derivative: A stock solution of 3-azido-AMSA (kindly provided by Stephen R. Byrn, Purdue University, West Lafayette, IN) is dissolved in 100% ethanol at a maximum concentration of 600 µg/mL (1.38 m*M*). This stock solution is kept in the dark at –20°C, and aliquots are taken out in near darkness to avoid degradation of the compound by ambient light. Under these conditions, a stock solution typically retains near-maximal reactivity for about 2 mo. Further dilutions are made in water immediately before use. The exact concentration of 3-azido-AMSA is determined by spectrophotometry, with molar extinction coefficients of: 10,900 at 352 nm and 9510 at 432 nm (in EtOH); 9550 at 353 nm and 8760 at 433 nm (in water) ***(4)***. A 15-W GE F15T8 black light source (peak wavelength 360–370 nm with no emission below 300 nm) or light source with similar spectral characteristics is used for photocrosslinking (3-azido-AMSA absorbs maximally at about 350 nm ***[4]***).
5. Reactive CPT derivatives: 7-ClMe-MDO-CPT and 7-Et-MDO-CPT were synthesized by Monroe E. Wall and Mansukh Wani (Research Triangle Park, NC) according to published procedures ***(21,22)***. Both drugs are dissolved in dimethyl sulfoxide at 10 m*M* and stored in aliquots at –20°C. Further dilutions are made in water immediately before use.
6. Reagents for piperidine reactions: Cleavage of modified bases is induced by piperidine (freshly diluted to 1 *M* from 10.1 *M* stock), with or without prior treatment with formic acid.
7. Reagents for primer extension analysis: The following reagents are needed for primer-extension analysis of covalently modified substrates: buffered phenol, 5'-end-labeled primer (complementary to the appropriate region of the substrate), Klenow polymerase (Boehringer Mannheim, Indianapolis, IN), 10X Klenow

buffer (500 m*M* Tris-Cl, pH 7.6, 100 m*M* $MgCl_2$, 2 m*M* dithiothreitol), and 1 m*M* deoxyribonucleotide mix (250 μ*M* of each nucleotide).

8. Gel electrophoresis reagents: Formamide loading solution contains 98% (v/v) formamide, 10 m*M* Na_2EDTA, bromophenol blue (1 mg/mL), and xylene cyanol (1 mg/mL). Gels contain the desired concentration of acrylamide, 7 *M* urea, and TBE (89 m*M* Tris base, 89 m*M* boric acid, 2.5 m*M* Na_2EDTA); TBE is used as running buffer.

3. Methods

3.1. Photocrosslinking of m-AMSA Derivative to Topo II Cleavage Complex

1. Duplex oligonucleotide substrates are produced by mixing top- and bottom-strand oligonucleotides at equimolar concentration in 1X TE (10 m*M* Tris, pH 7.6, 1 m*M* EDTA) and slowly cooling from 65 to 4°C. Reactions that will be analyzed directly or subjected to piperidine cleavage utilize double-stranded substrates with one strand end-labeled. The 5'-end is labeled by polynucleotide kinase in the presence of [γ-^{32}P]ATP, or the 3'-end is labeled by terminal deoxynucleotide transferase with [α-^{32}P]ddATP. If cleavage of the two DNA strands is to be compared quantitatively, they should be labeled separately and brought to the same specific activity before annealing (the efficiencies of labeling of different ends can vary). An unlabeled duplex substrate is used for topoisomerase cleavage reactions that will be analyzed by primer extension.
2. Topoisomerase reactions are performed in T4 topo II reaction buffer (40 m*M* Tris-HCl pH 7.6, 60 m*M* KCl, 10 m*M* $MgCl_2$, 0.5 m*M* Na_2EDTA, 0.5 m*M* ATP, bovine serum albumin at 30 μg/mL). A typical reaction contains 1–2 pmol each of DNA substrate and T4 topo II (10–20 μL reaction volume).
3. Immediately prior to use, an aliquot of the 3-azido-AMSA stock is diluted with water to 10 times the final drug concentration, keeping light exposure to a minimum. One-tenth vol of the diluted drug is added to the topoisomerase reactions (still in the dark; 10 μ*M* [final concentration] 3-azido-AMSA gives maximal reactivity).
4. Photoactivation reactions are placed directly under the UV light with microcentrifuge tube lids open. A distance of 4 cm from the 15-W GE F15T8 light source to the reaction liquid provides an energy of 3200 μW/cm^2 at 360–370 nm (emission energy measured with a UV meter). The reactions are done either at room temperature or in a water bath at constant temperature. Under these conditions, an exposure time of 10–30 min (depending on the age of the drug stock) gives maximal reactivity of 3-azido-AMSA with the DNA substrate. Control reactions without photoactivation should remain covered with tinfoil, away from the UV light source.
5. All reactions are stopped by the addition of SDS to 1% and proteinase K to 1 mg/mL, followed by incubation at 30°C for 30 min. If desired, topo II-induced cleavage

can also be reversed by incubation at 65°C for 10 min or addition of EDTA to 50 m*M* before addition of SDS/proteinase K (*see* **Note 2**).

6. Reaction products destined for primer-extension analysis are extracted once with buffered phenol (to eliminate proteinase K) and purified by passing through 0.6-mL Sephadex G-25 (Amersham Pharmacia) spin-columns (to eliminate unreacted drug and SDS). One-fourth of each reaction product (usually 0.5 pmol DNA) is annealed to equimolar 5'-end-labeled primer of either the bottom- or top-strand sequence in 1X Klenow buffer containing 100 μ*M* deoxyribonucleotide mix (10 μL total volume). Klenow enzyme (1 U) is added and the extension reaction is incubated for 5 min at 24°C (*see* **Notes 3** and **4**). The reaction is terminated by adding 6 μL formamide loading solution, and the reaction products are analyzed by polyacrylamide gel electrophoresis according to standard procedures. Sequencing standards should be generated using the chain-termination method *(23)*.
7. If labeled reaction products are to be analyzed by piperidine-induced cleavage, add 100 μL freshly diluted 1 *M* piperidine to the purified products (typically 1 pmol in 10 μL), and incubate at 90°C for 30 min. The piperidine is then removed by three successive lyophilizations, resuspending in 30 μL water after the first and 20 μL water after the second. After the third lyophilization, each sample is resuspended in 10 μL formamide loading solution, and the reaction products are analyzed by polyacrylamide gel electrophoresis according to standard procedures (incomplete removal of piperidine results in smeary bands). Sequencing standards should be generated using the Maxam-Gilbert method *(23,24)*.

3.2. Alkylation of Topo I Cleavage Complex by Camptothecin Derivative

1. Single-stranded oligonucleotides are labeled at either their 5'-end (using T4 polynucleotide kinase and [γ-^{32}P]ATP) or their 3'-end [using terminal deoxynucleotidyl transferase and [α-^{32}P]cordycepin 5'-triphosphate; *see* ***9,20***). Duplex substrates are produced by annealing the labeled strand with an excess of the complementary strand in 20 μL annealing buffer (10 m*M* Tris-HCl pH 7.8, 100 m*M* NaCl, 1 m*M* EDTA). The annealing mixture is heated to 95°C for 5 min, slowly chilled to 20°C, and then centrifuged through a Quick Spin G-25 Sephadex column (to remove excess single-stranded oligonucleotides) and stored at 4°C.
2. Reactions are performed in 10 μL topo I reaction buffer (10 m*M* Tris-HCl, pH 7.5, 50 m*M* KCl, 5 m*M* $MgCl_2$, 0.1 m*M* EDTA, 15 μg/mL bovine serum albumin) containing substrate oligonucleotide (approx 50–200 fmol/reaction), topo I (10 U), and either 7-Cl-Me-MDO-CPT or 7-Et-MDO-CPT. Reactions are stopped by adding either SDS or NaCl (0.5% and 0.25 M, respectively, as final concentrations). Proteinase K (0.5 mg/mL, final concentration) is then added to the reaction mixtures and proteolysis carried out for 60 min at 50°C. Proteolysis is halted by the addition of 36 μL formamide loading solution.
3. Formic acid and piperidine reactions are performed according to standard procedures *(23,24)*.

4. The reaction products are separated by electrophoresis through 25% sequencing gels (25% acrylamide, 0.1 % bis-acrylagel [National Diagnostics, Manville, NJ], 7 *M* urea, TBE) at 40 V/cm for 2–3 h (50°C). After electrophoresis, the gels are transferred to Whatmann 3MM paper sheets and dried under vacuum. Imaging and quantification are performed using a PhosphorImager (Molecular Dynamics) or autoradiography.

4. Notes

1. One caution in interpreting results with reactive inhibitors is the possibility that all DNA bases (or amino acid residues) are not equally receptive to covalent addition. For example, 3-azido-AMSA crosslinked to dT, dG, and dC, but not to a dA residue, at the appropriate location (complement of +1 in **Fig. 2A**) of the cleavage complexes with the T4 topo II *(5)*. In addition, a dA residue at the +1 position in a topo I cleavage complex (top strand in **Fig. 2B**) was alkylated by 7-ClMe-MDO-CPT more efficiently than a dG residue *(6)*.
2. Reversal of the cleavage reactions can be used to eliminate intense cleavage bands, to distinguish cleavage bands from nonreversible photoproducts, to visualize drug binding sites that lie past topoisomerase cleavage sites, and to evaluate changes in reversibility dependent on covalent drug binding.
3. The optimal temperature and time for Klenow extension reactions vary with the template and primer used. The temperature should be high enough to avoid secondary structure of the template and yet low enough to allow efficient primer annealing. Extending for the minimal effective time is helpful in minimizing extra base addition (*see* **Note 4**).
4. Polymerases lacking a 3' → 5' exonuclease (such as Sequenase) will usually add an extra untemplated base to the end of the primer extension, complicating analysis of the site of covalent drug attachment *(25)*. Klenow, which has 3' → 5' exonuclease, is therefore the enzyme of choice for the extension reaction. Unfortunately, Klenow will also add an extra base in some situations, particularly when the template DNA is short *(5,26)* (C. Freudenreich and K. Kreuzer, unpublished results).

Acknowledgments

K. N. K. and C. H. F. thank Stephen Byrn for his generous gift of 3-azido-AMSA. Research in the lab of K. N. K. is supported by Grant CA60836 from the National Cancer Institute. Y. P. thanks Monroe Wall and Mansukh Wani for synthesizing the camptothecin derivatives, and Kurt Kohn for support and suggestions.

References

1. Wigley, D. B. (1995) Structure and mechanism of DNA topoisomerases. *Annu. Rev. Biophys. Biomol. Struct.* **24,** 185–208.
2. Berger, J. M., Gamblin, S. J., Harrison, S. C., and Wang, J. C. (1996) Structure and mechanism of DNA topoisomerase II. *Nature (Lond.)* **379,** 225–232.

3. Denny, W. A., Cain, B. F., Atwell, G. J., Hansch, C., Panthananickal, A., and Leo, A. (1982) Potential antitumor agents. 36. Quantitative relationships between experimental antitumor activity, toxicity, and structure for the general class of 9-anilinoacridine antitumor agents. *J. Med. Chem.* **25,** 276–315.
4. Shieh, T.-L., Hoyos, P., Kolodziej, E., Stowell, J. G., Baird, W. M., and Byrn, S. R. (1990) Properties of the nucleic acid photoaffinity labeling agent 3-azidoamsacrine. *J. Med. Chem.* **33,**1225–1230.
5. Freudenreich, C. H. and Kreuzer, K. N. (1994) Localization of an aminoacridine antitumor agent in a type II topoisomerase-DNA complex. *Proc. Natl. Acad. Sci. USA* **91,** 11,007–11,011.
6. Pommier, Y., Kohlhagen, G., Kohn, K. W., Leteurtre, F., Wani, M. C., and Wall, M. E. (1995) Interaction of an alkylating camptothecin derivative with a DNA base at topoisomerase I-DNA cleavage sites. *Proc. Natl. Acad. Sci. USA* **92,** 8861–8865.
7. Hertzberg, R. P., Busby, R. W., Caranfa, M. J., Holden, K. G., Johnson, R. K., Hecht, S. M., and Kingsbury, W. D. (1990) Irreversible trapping of the DNA topoisomerase I covalent complex. Affinity labeling of the camptothecin binding site. *J. Biol. Chem.* **265,** 19,287–19,295.
8. Porter, S. E. and Champoux, J. J. (1989) The basis for camptothecin enhancement of DNA breakage by eukaryotic topoisomerase I. *Nucleic Acids Res.* **17,** 8521–8532.
9. Jaxel, C., Capranico, G., Kerrigan, D., Kohn, K. W., and Pommier, Y. (1991) Effect of local DNA sequence on topoisomerase I cleavage in the presence or absence of camptothecin. *J. Biol. Chem.* **266,** 20,418–20,423.
10. Tanizawa, A., Kohn, K. W., and Pommier, Y. (1993) Induction of cleavage in topoisomerase I cDNA by topoisomerase I enzymes from calf thymus and wheat germ in the presence and absence of camptothecin. *Nucleic Acids Res.* **21,** 5157–5156.
11. Fujimori, A., Harker, W. G., Kohlhagen, G., Hoki, Y., and Pommier, Y. (1995) Mutation at the catalytic site of topoisomerase I in CEM/C2, a human leukemia cell resistant to camptothecin. *Cancer Res.* **55,** 1339–1346.
12. Leteurtre, F., Fesen, M., Kohlhagen, G., Kohn, K. W., and Pommier, Y. (1993) Specific interaction of camptothecin, a topoisomerase I inhibitor, with guanine residues of DNA detected by photoactivation at 365 nm. *Biochemistry* **32,** 8955–8962.
13. Gourdie, T. A., Valu, K. K., Gravatt, G. L., Boritzki, T. J., Baguley, B. C., Wakelin, L. P. G., Wilson, W. R., Woodgate, P. D., and Denny, W. A. (1990) DNA-directed alkylating agents. 1. Structure-activity relationships for acridine-linked aniline mustards: Consequences of varying the reactivity of the mustard. *J. Med. Chem.* **33,** 1177–1186.
14. Kawada, S., Yamashita, Y., Fujii, N., and Nakano, H. (1991) Induction of a heat-stable topoisomerase II-DNA cleavable complex by nonintercalative terpenoides, terpentecin and clerocidin. *Cancer Res.* **51,** 2922–2925.
15. Kong, X.-B., Rubin, L., Chen, L.-I., Ciszewska, G., Watanabe, K. A., Tong, W. P., Sirotnak, F. M., and Chou, T.-C. (1992) Topoisomerase II-mediated DNA

cleavage activity and irreversibility of cleavable complex formation induced by DNA intercalator with alkylating capability. *Mol. Pharmacol.* **41,** 237–244.

16. Arcamone, F. (1994) Design and synthesis of anthracycline and distamycin derivatives as new, sequence-specific, DNA-binding pharmacological agents. *Gene* **149,** 57–61.
17. Freudenreich, C. H. and Kreuzer, K. N. (1993) Mutational analysis of a type II topoisomerase cleavage site: Distinct requirements for enzyme and inhibitors. *EMBO J.* **12,** 2085–2097.
18. Svejstrup, J. Q., Christiansen, K., Andersen, A. H., Lund, K., and Westergaard, O. (1990) Minimal DNA duplex requirements for topoisomerase I-mediated cleavage *in vitro. J. Biol. Chem.* **265,** 12,529–12,535.
19. Kjeldsen, E., Svejstrup, J. Q., Gromova, I. I., Alsner, J., and Westergaard, O. (1992) Camptothecin inhibits both the cleavage and religation reactions of eukaryotic DNA topoisomerase I. *J. Mol. Biol.* **228,** 1025–1030.
20. Tanizawa, A., Kohn, K. W., Kohlhagen, G., Leteurtre, F., and Pommier, Y. (1995) Differential stabilization of eukaryotic DNA topoisomerase I cleavable complexes by camptothecin derivatives. *Biochemistry* **43,** 7200–7206.
21. Sawada, S., Nokata, K., Fureta, T., Yokokura, T., and Miyasaka, T. (1991) Chemical modification of an antitumor alkaloid, camptothecin. Synthesis and antitumor activity of 7-C-substituted camptothecins. *Chem. Pharm. Bull.* **39,** 2574–2580.
22. Luzzio, M. J., Besterman, J. M., Emerson, D. L., Evans, M. G., Lackey, K., Leitner, P. L., McIntyre, G., Morton, B., Myers, P. L., Peel, M., Sisco, J. M., Sternbach, D. D., Tong, W.-Q., Truesdale, A., Uehling, D. E., Vuong, A., and Yates, J. (1995) Synthesis and antitumor activity of novel water-soluble derivatives of camptothecin as specific inhibitors of topoisomerase I. *J. Med. Chem.* **38,** 395–401.
23. Ausubel, F. M., Brent, R., Kingston, R. E., Moore, D. D., Seidman, J. G., Smith, J. A., and Struhl, K. *Current Protocols in Molecular Biology*. John Wiley, New York.
24. Maxam, A. and Gilbert, W. (1980) Sequencing end-labeled DNA with base-specific chemical cleavages. *Methods Enzymol.* **65,** 499–560.
25. Clark, J. M., Joyce, C. M., and Beardsley, G. P. (1987) Novel blunt-end addition reactions catalyzed by DNA polymerase I of *Escherichia coli. J. Mol. Biol.* **198,** 123–127.
26. Freudenreich, C. H. and Kreuzer, K. N. (1994) Differentially labeled mutant oligonucleotides for analysis of protein-DNA interactions. *BioTechniques* **16,** 104–108.

11

Uncoupling of Topoisomerase-Mediated DNA Cleavage and Religation

Anni H. Andersen, Kent Christiansen, and Ole Westergaard

1. Introduction

DNA topoisomerase I and II catalyze topological changes of DNA by transient breakage of the nucleic acid backbone. Topoisomerase I transiently cleaves one DNA strand and allows the passage of single-stranded DNA through the generated nick, whereas topoisomerase II introduces a transient double-stranded break and permits passage of another double helix through the gap *(1–4)*. Owing to the physiological importance of the topoisomerase-mediated cleavage reactions, they have been intensively studied by use of an in vitro assay involving treatment with a strong protein denaturant *(5,6)*. On this treatment, the cleavage–religation cycle of the enzyme is interrupted, and abortive topoisomerase–DNA complexes are formed, where the enzyme is covalently linked to the cleaved DNA. Topoisomerase I is covalently linked to the newly generated 3'-end, whereas a subunit of topoisomerase II is covalently linked to each of the generated 5'-ends. Several studies have utilized the detergent method to define the sites of interaction between DNA and topoisomerase I and II in vitro as well as in vivo *(7,8)*. One of the major problems in these studies has been ascribed to the involvement of SDS in trapping cleavage complexes, as the detergent denatures the enzyme and thus prevents a subsequent study of the topoisomerase-mediated religation reaction. To overcome this problem, different attempts have been performed to separate the cleavage and religation half-reactions. Thus, circular or linear single-stranded DNA has been used as substrate for both topoisomerase I and II. The generated cleavage complexes can catalyze recircularization or ligation to double-stranded DNA containing either a 3'-overhang, a 5'-overhang, or a blunt end *(9–12)*. However, the

From: *Methods in Molecular Biology, Vol. 95: DNA Topoisomerase Protocols, Part II: Enzymology and Drugs*
Edited by N. Osheroff and M.A. Bjornsti © Humana Press Inc., Totowa, NJ

cleavage complexes generated with single-stranded DNA are heterogenous in DNA sequence and structure, making these systems inadequate for detailed studies of the ligation reaction.

We describe in this chapter an assay that allows a total uncoupling of the cleavage and religation half-reactions by use of partially double-stranded substrates, the so-called suicidal cleavage substrates, which contain a strong recognition site for the topoisomerase enzyme of interest. The advantage of suicide substrates is that topoisomerase–DNA cleavage complexes can be trapped in the absence of denaturing agents, as topoisomerase-mediated cleavage releases the noncovalently bound DNA end, normally involved in religation. Since denaturing agents are absent from the reaction mixture, the topoisomerase enzyme covalently linked to the cleaved DNA is kinetically competent and able to mediate ligation, whenever an appropriate ligation substrate is added. The ligation reaction can be characterized as intra- or intermolecular depending on the DNA substrate of choice. The assays, as described below, involve suicidal cleavage followed by intramolecular DNA ligation, and methods are provided for both topoisomerase I and II. The substrate specificities for intermolecular ligation are discussed under **Subheading 4.** The assays are optimized for mammalian topoisomerase I and II enzymes, but identical observations have been obtained with enzymes purified from a wide variety of eukaryotic sources.

2. Materials

2.1. Topoisomerase I

1. Purified eukaryotic topoisomerase I.
2. Oligonucleotides. A 38-mer (OL19, 5' GCCTGCAGGTCGACTCTAGAGGATCTAAAAGACTTAGA 3') and a 77-mer (OL6, 5' ATTCGAGCTCGGTACCCGGGGATCTTTTTAAAAATTTTTCTAAGTCTTTTAGATCCTCTAGAGTCGACCTGCAGGC 3') are used to generate the cleavage substrate, and a 40-mer (OL18, 5' AGAAAAATTTTTAAAAAGATCCCCGGGTACCGAGCTCGA 3') is used as the ligation substrate. The oligonucleotides are synthesized on a DNA synthesizer model 381A from Applied Biosystems.
3. T4 polynucleotide kinase (Boehringer Mannheim).
4. [γ-^{32}P] ATP (4 µCi/mL; SA 7000 Ci/mmol; ICN).
5. 100 m*M* ATP (Pharmacia).
6. 10X kinase buffer: 500 m*M* Tris-HCl, pH 7.6, 100 m*M* $MgCl_2$, and 20 m*M* dithiothreitol.
7. 0.2 *M* EDTA, pH 8.0.
8. 10X Topoisomerase I reaction buffer: 100 m*M* Tris-HCl, pH 7.5, 500 m*M* NaCl, 50 m*M* $MgCl_2$, 50 m*M* $CaCl_2$.
9. TE buffer: 10 m*M* Tris HCl, pH 8, 0.1 m*M* EDTA.

10. Sephadex G50 column: Sephadex G50 coarse (Pharmacia) is equilibrated in 10 m*M* Tris-HCl, pH 7.5, 0.1 m*M* EDTA (TE-buffer) overnight and applied to a 1-mL syringe blocked with a siliconiced glass bead. When the syringe is filled with Sephadex, it is centrifuged at 1000*g* for 3 min to remove excess buffer. The DNA is applied to the column and is collected after another round of centrifugation.
11. 25:24:1 (v/v/v) Phenol/chloroform/isoamyl alcohol (Merck) is prepared with liquefied redistilled phenol supplied with 8-hydroxyquinoline to 1 g/L and made buffered with 3 vol of 50 m*M* Tris HCl, pH 8.
12. 4 *M* NaCl.
13. 80 and 96% Ethanol.
14. Trypsin (Sigma, type XIII) is prepared as a 10 mg/mL stock in 1 m*M* HCl and stored at –20°C.
15. 2X Denaturing polyacrylamide gel-loading buffer: deionized formamide (Merck) supplied with 0.05% (w/v) bromphenol blue and 0.05% (w/v) xylene xyanol.
16. 12% Denaturing polyacrylamide gel: The gel is prepared from 12% (w/v) acrylamide (2X, Serva), 0.6% (w/v) *N,N'*-methylene bis-acrylamide (2X, Serva), and 7.5 *M* urea (Merck) in 89 m*M* Tris base, 89 m*M* boric acid, and 2 m*M* EDTA, pH 8.3 (TBE) supplied with 0.03% (w/v) ammonium peroxodisulfate (Merck) and 0.1% (v/v) *N,N,N',N'* Tetramethylenediamide (Merck).
17. 5X Native polyacrylamide gel-loading buffer: 30% glycerol, 0.05% (w/v) bromphenol blue, and 0.05% (w/v) xylene xyanol.
18. 12% Native polyacrylamide gel: The gel is prepared of 12% (w/v) acrylamide (2X, Serva), 0.6% (w/v) *N,N'*-methylene bis-acrylamide (2X, Serva) in 89 m*M* Tris base, 89 m*M* boric acid, and 2 m*M* EDTA, pH 8.3 (TBE), supplied with 0.03% (w/v) ammonium peroxodisulfate (Merck) and 0.1% (v/v) *N,N,N',N'*-Tetramethylenediamide (Merck).
19. Fuji medical X-ray film.

2.2. Topoisomerase II

1. Purified eukaryotic topoisomerase II.
2. Oligonucleotides: The substrate for the topoisomerase II-mediated cleavage reaction is generated from the two oligonucleotides 5' ATGAGCGCATTGTTAG 3' (16-mer, top strand) and 5' AATCTAACAATGCGCTCATCGTCATCCT 3'(28-mer, bottom strand). The oligonucleotide used for the ligation reaction is the 45-mer 5' GGCTCCCACGACGGCCAGTGCCAAGCTTGCATGAGG ATGACGATG 3'. The oligonucleotides are synthesized on a DNA synthesizer model 381A from Applied Biosystems. The bottom strand oligonucleotide is synthesized with a 3' end modification consisting of the amino link -O-PO_2-O-CH_2-CHOH-CH_2-NH_2 (**Note 14**).
3. 10X Hybridization buffer: 400 m*M* Tris-HCl, pH 7.5, 200 m*M* $MgCl_2$, 500 m*M* NaCl.
4. 10X Sequenase buffer: 240 m*M* Tris-HCl, pH 7.5, 120 m*M* $MgCl_2$, 300 m*M* NaCl, 60 m*M* dithiothreitol.

5. T7 DNA polymerase (Sequenase, version 2.0; Amersham Corp.).
6. [α-^{32}P]dATP, 3000 Ci/m*M*ol (Amersham Corp.).
7. Sephadex G50 column: *see* **Subheading 2.1., item 10.**
8. 80 and 96% Ethanol.
9. 10X Topoisomerase II cleavage buffer: 100 m*M* Tris-HCl, pH 7, 1 m*M* EDTA, 25 m*M* $MgCl_2$, 25 m*M* $CaCl_2$.
10. 4 *M* NaCl.
11. 10% SDS.
12. Proteinase K (Merck): A 10 mg/mL stock is prepared in H_2O and stored at –20°C
13. TE buffer: 10 m*M* Tris-HCl, pH 8, 0.1 m*M* EDTA.
14. 2X Denaturing polyacrylamide gel-loading buffer, 12% denaturing polyacrylamide gel, and X-ray film: *see* **Subheading 2.1., items 15**, **16**, and **19**, respectively.

3. Methods

3.1. Topoisomerase I

1. Prepare stock solutions of oligonucleotides: All oligonucleotides are diluted in TE buffer to 5 pmol/µL (**Notes 1–4**).
2. The 38-mer (OL19) is radioactive-labeled at the 5'-end prior to hybridization by incubation of 10 pmol of the oligonucleotide with 20 U of T4 polynucleotide kinase in kinase buffer (50 m*M* Tris-HCl, pH 7.6, 10 m*M* $MgCl_2$, and 2 m*M* dithiothreitol) supplied with 4 µCi/mL [γ-^{32}P] ATP in a 50 µl reaction volume at 37°C for 30 min. The 77-mer (OL6) is cold-phosphorylated by incubation of 100 pmol of the oligonucleotide with 20 U of T4 polynucleotide kinase in 50 µl of kinase buffer supplied with 100 µ*M* ATP at 37°C for 30 min (**Note 5**). The kinase reactions are stopped by addition of 10 m*M* EDTA.
3. Phosphorylated oligonucleotides are freed of protein and nucleotide by centrifugal gel filtration using a 1-mL Sephadex G-50 column followed by extraction with 1 vol phenol/chloroform/isoamyl alcohol (25:24:1). After adjustment of NaCl to 300 m*M*, 3 vol 96% ethanol are added, and the sample is placed at –20°C overnight to precipitate the DNA. Following centrifugation for 10 min at 15.900*g* (Eppendorf centrifuge), the DNA pellet is washed with 1 mL 80% ethanol. The precipitated DNA is finally dissolved in 10 µL 10 m*M* Tris-HCl, pH 8.0, 0.1 m*M* EDTA (TE buffer).
4. Hybridization of the 38-mer (OL19) and the 77-mer (OL6) occurs after mixing of 10 pmol OL19 (10 µL) with 10 pmol OL6 (10 µL) by heating the mixture to 90°C followed by a slow cooling to room temperature.
5. To purify the hybridized DNA (**Note 6**), 5 µL of 5X native polyacrylamide gel-loading buffer are added to the hybridization mixture, and the sample is subjected to electrophoresis on a 12% native polyacrylamide gel at room temperature. The band representing the double-stranded DNA is localized by autoradiography and excised from the gel. The DNA is subsequently eluted by soaking the gel slice in 10 m*M* Tris-HCl, pH 7.5, 5 m*M* $MgCl_2$, and 5 m*M* $CaCl_2$ (topoisomerase I reaction buffer) overnight at 4°C. Gel pieces are removed by centrifugation. By this procedure, 50–90% of the labeled DNA is recovered from the gel.

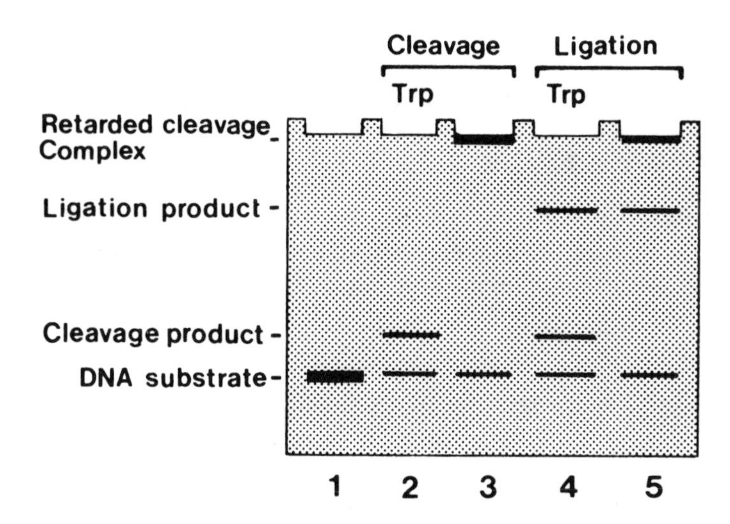

Fig. 1. Migration of topoisomerase I-generated cleavage and ligation products on a denaturing polyacrylamide gel. Lane 1, DNA substrate. Lanes 2 and 3, topoisomerase I cleavage assay terminated with and without trypsin digestion, respectively. Lanes 4 and 5, topoisomerase I ligation assay terminated with and without trypsin digestion, respectively.

6. The topoisomerase I-mediated DNA cleavage reaction is started by incubating 10–100 fmol of the purified double-stranded DNA with topoisomerase I (the molar ratio of enzyme:DNA has to be approx 60:1) in a total volume of 50 μL in topoisomerase I reaction buffer. Incubation occurs for 30 min at 30°C (**Note 7**). To prevent further DNA cleavage, NaCl is added to a final concentration of 300 m*M*. To ascertain the amount of cleaved complexes after 30 min, 3 vol of 96% ethanol can be added to a control sample.
7. The topoisomerase I DNA ligation reaction is initiated by adding a 10- to 50-fold molar excess (relative to the cleavage substrate) of the 40-mer (OL18) to the NaCl-treated sample, and incubation takes place for 30 min at 30°C. The ligation reaction is subsequently terminated by addition of 3 vol of 96% ethanol. The ethanol-precipitated DNA is redissolved in 20 μL of 10 m*M* Tris-HCl, pH 8.0, 0.1 m*M* EDTA (TE-buffer) containing 1 mg/mL trypsin, and incubated at 37°C for 1 h.
8. After trypsin digestion (**Note 8**), the sample is mixed with one vol of denaturing polyacrylamide gel loading buffer, heated to 90°C for 2 min and applied to 12% denaturing gel electrophoresis followed by autoradiography. A schematic example of a cleavage/ligation experiment is shown in **Fig. 1**. Topoisomerase I-mediated DNA cleavage requires bipartite interaction with a duplex region encompassing the cleavage site (region A) and another duplex region on the side holding the 5'-OH end generated by cleavage (region B) (**Fig. 2** and **Note 3**). The fundamental advantage of using defined DNA substrates to analyze separately the cleavage and the intramolecular ligation reactions of topoisomerase I is dia-

A B

5'- - - A A A A G A C T T A G A A A A T T T T T A A A A A - - - 3'
-9 -8 -7 -6 -5 -4 -3 -2 -1 1 2 3 4 5 6 7 8 9 10 11 12 13 14 15 16
3'- - - T T T T C T G A A T C T T T T T A A A A A T T T T - - - 5'

Fig. 2. The bipartite minimal duplex region required for topoisomerase I-mediated cleavage. The bipartite minimal duplex region is shaded. The positive and negative numbers in the DNA sequence are relative to the cleavage site, which is marked with an arrow.

grammed in **Fig. 3**. Generation of DNA insertions and DNA deletions during intramolecular ligation is discussed in **Note 9** and diagrammed in **Fig. 4**. Intermolecular ligation (**Fig. 5**) as well as use of camptothecin in topoisomerase I-mediated cleavage/ligation are discussed in **Notes 10** and **11**, respectively.

3.2. Topoisomerase II

1. Prepare solutions of the DNA oligonucleotides. All oligonucleotides are diluted in TE buffer to 5 pmol/μL (*see* **Notes 12–14**).
2. 10 pmol of the 16-mer oligonucleotide (top strand) is hybridized with 10 pmol of the complementary 28-mer (bottom strand) in 40 m*M* Tris-HCl, pH 7.5, 20 m*M* $MgCl_2$, and 50 m*M* NaCl (hybridization buffer) in a 55 μL reaction volume. The mixture is heated to 70°C for 2 min and cooled slowly to room temperature.
3. After hybridization, 5 pmol of the hybrid DNA are labeled at the 3'-end of the top strand by incubation of the DNA with 1 U of T7 DNA polymerase in 24 m*M* Tris-HCl, pH 7.5, 12 m*M* $MgCl_2$, 30 m*M* NaCl, and 6 m*M* dithiothreitol (Sequenase buffer), supplied with 0.2 μ*M* [α^{32}P]dATP in a 30-μL reaction volume for 5 min at room temperature. Since cold deoxynucleotides are absent from the reaction mixture, labeling results in an extension of the top strand with only one nucleotide.
4. The labeled DNA is freed of protein and nucleotide by centrifugal gel filtration using a 1-mL Sephadex G-50 column followed by extraction with 1 vol phenol/chloroform/isoamyl alcohol (25:24:1). The DNA is subsequently precipitated from the water phase by addition of 3 vol of 96% ethanol. The sample is placed at –20°C overnight, and after centrifugation for 10 min at 15.900*g* (Eppendorf centrifuge), the DNA pellet is washed with 1 mL 80% ethanol. The precipitated DNA is finally dissolved in 100 μL 10 m*M* Tris-HCl, pH 8.0, and 0.1 m*M* EDTA (TE buffer).
5. The topoisomerase II-mediated cleavage reaction is initiated by incubation of 0.1 pmol of labeled DNA with a 25–50 molar excess of topoisomerase II in 10 m*M* Tris-HCl, pH 7.0, 0.1 m*M* EDTA, 2.5 m*M* $MgCl_2$, and 2.5 m*M* $CaCl_2$ (topoisomerase II cleavage buffer) at 30°C for 60 min in a total volume of 50 μL (*see* **Note 15**). The cleavage reaction is stopped by adding NaCl to 0.4 *M*. To show the amount of cleavage at the start of ligation, a control sample can be prepared, where the cleavage reaction is stopped by adding SDS to 1%.

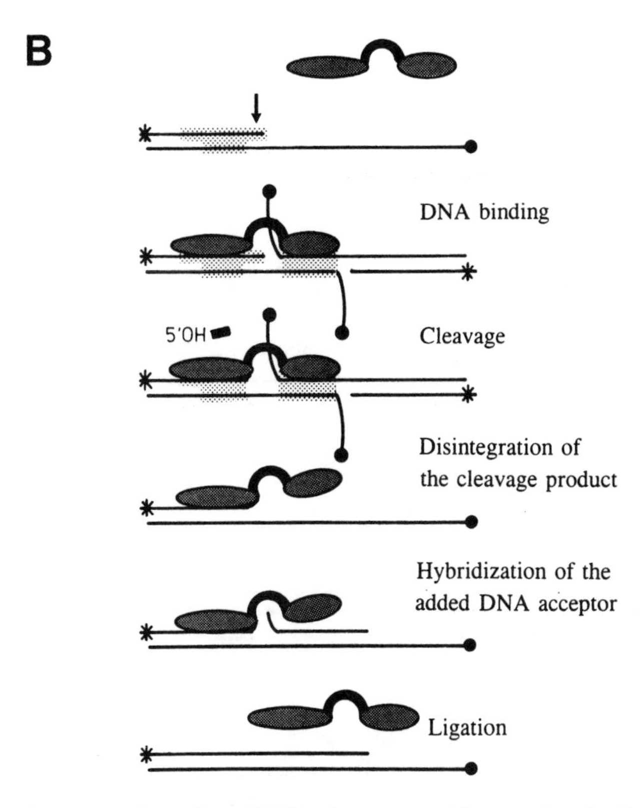

Fig. 3. Topoisomerase I-mediated DNA cleavage and intramolecular ligation. Panel **A**, the DNA sequence of the cleavage substrate. Panel **B**, schematic diagram for topoisomerase I-mediated cleavage and intramolecular ligation of a DNA strand capable of hybridizing to the noncleaved strand. Topoisomerase I is depicted as a monomeric enzyme containing two DNA binding domains separated by a flexible protein hinge, exclusively to illustrate one of the more obvious possibilities. Covalent linkage between topoisomerase I and DNA is indicated by an arc. Asterisk and filled circle denote radioactive 5'-end-labeling and "cold" 5'-end-phosphorylation, respectively. Other labeling is the same as in **Fig. 2**.

6. The ligation reaction is started by the addition of 15 pmol of the 45-mer ligation substrate to the salt-treated sample (**Note 16**). Incubation is continued for 60 min at 30°C, and the reaction is stopped by the addition of SDS to 1%.

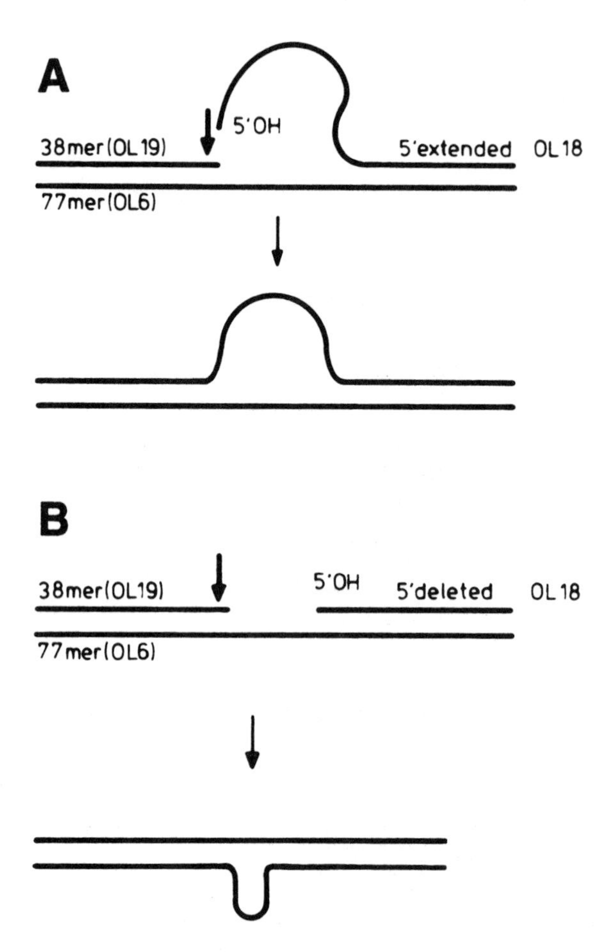

Fig. 4. Topoisomerase I-mediated cleavage/ligation of displaced overhangs (panel **A**) and gaps (panel **B**) resulting in DNA insertions and deletions, respectively. The topoisomerase I cleavage site is indicated by an arrow.

7. To degrade topoisomerase II in the formed cleavage complexes, 500 µg/mL proteinase K is added, and the mixture is incubated at 37°C for 2 h (**Note 17**). The DNA is precipitated with 96% ethanol and redissolved in 10 µL of 10 m*M* Tris-HCl, pH 7.5, and 0.1 m*M* EDTA (TE buffer). The sample is then mixed with 1 vol of deionized formamide, 0.05% bromphenol blue, and 0.05% xylene xyanol, heated to 90°C for 1 min and subjected to 12% denaturing gel electrophoresis followed by autoradiography. A schematic illustration of a cleavage/ligation experiment is shown in **Fig. 6**, and the individual steps in the reactions are diagrammed in **Fig. 7**.

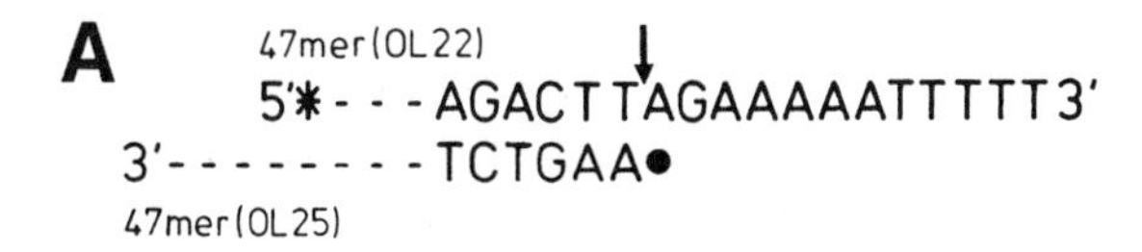

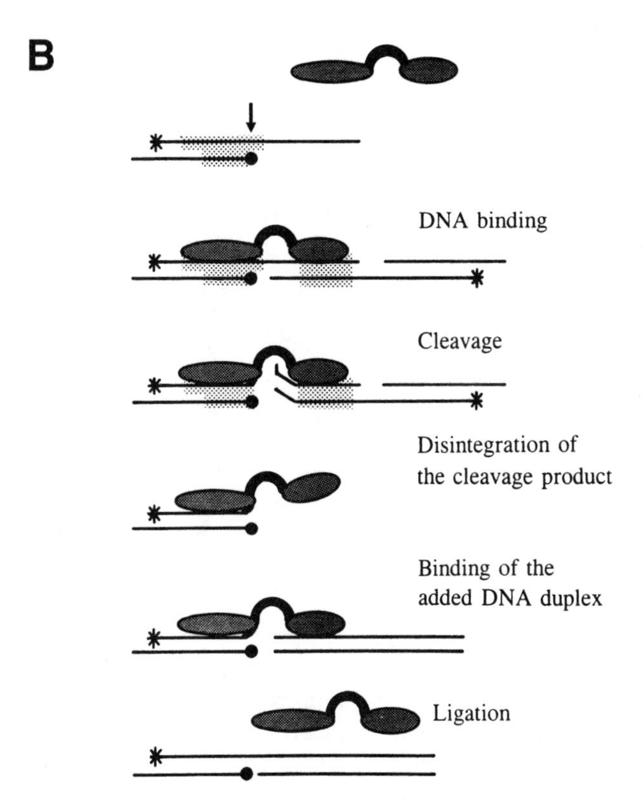

Fig. 5. Topoisomerase I-mediated DNA cleavage and intermolecular ligation. Panel **A**, the DNA sequence of the cleavage substrate. Panel **B**, schematic diagram for topoisomerase I-mediated cleavage and intermolecular ligation of a blunt-ended DNA duplex. Labeling is the same as in **Fig. 3**.

4. Notes

4.1. Topoisomerase I

1. The oligonucleotides used in the topoisomerase I reactions are chosen as they form a very strong recognition sequence for eukaryotic topoisomerase I. The sequence is the hexadecameric topoisomerase I recognition sequence found in the DNase I hypersensitive regions 5' and 3' to the rRNA genes in *Tetrahymena* and *Dictyostelium (7,13–15)*.

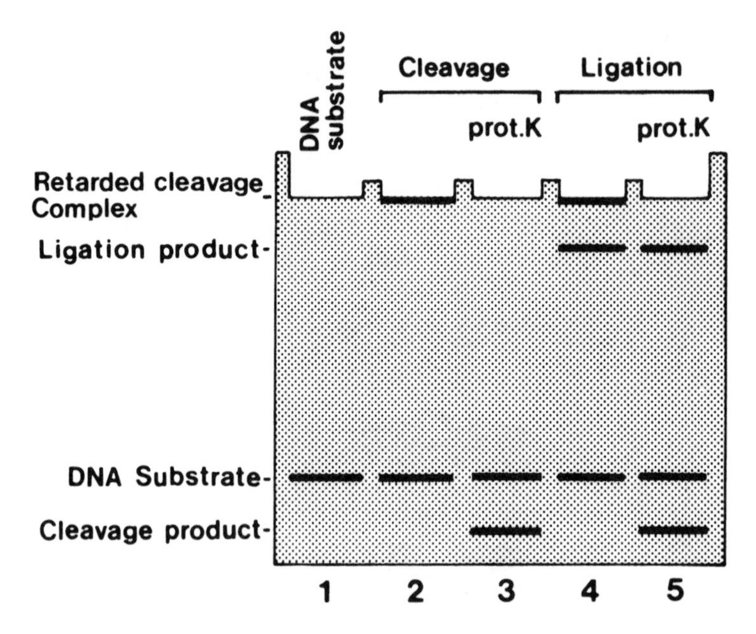

Fig. 6. Schematic illustration of the migration of topoisomerase II-generated cleavage and ligation products. Lane 1, DNA substrate. Lanes 2 and 3, topoisomerase II cleavage assay terminated without and with proteinase K digestion, respectively. Lanes 4 and 5, topoisomerase II ligation assay terminated without and with proteinase K digestion, respectively.

2. A practical advantage can be obtained by a reduction in the length of the 77-mer (OL6) removing 30 nucleotides from its 5'-end giving the noncleaved strand a size of 47 nucleotides, although this reduction results in a small decrease in the cleavage efficiency ***(16)***. Correspondingly, the 40-mer (OL18) can be exchanged with a dinucleotide 5' AG 3', which is able to hybridize to positions +1 and +2 on the noncleaved strand relative to the cleavage site. This dinucleotide defines the minimal length of the DNA strand that can be ligated to the topoisomerase I–DNA cleavage complex ***(17)***.
3. Topoisomerase I-mediated DNA cleavage requires separate interaction with a duplex region encompassing the cleavage site (region A) and a duplex region located on the site holding the 5'-OH end generated by cleavage (region B) (**Fig. 2**) ***(16)***. The bipartite interaction takes place with the cleavage substrate generated from the 38-mer (OL19) and the 77-mer (OL6), since these oligonucleotides, in addition to hybridization to each other, are able to form dimerized DNA substrates (**Fig. 3**). In the dimerized DNA substrate, both region A and B are present in a double-stranded form. Although the cleavage reaction requires an interaction of topoisomerase I with both region A and B, the ligation reaction does not require any interaction with region B ***(17)***.

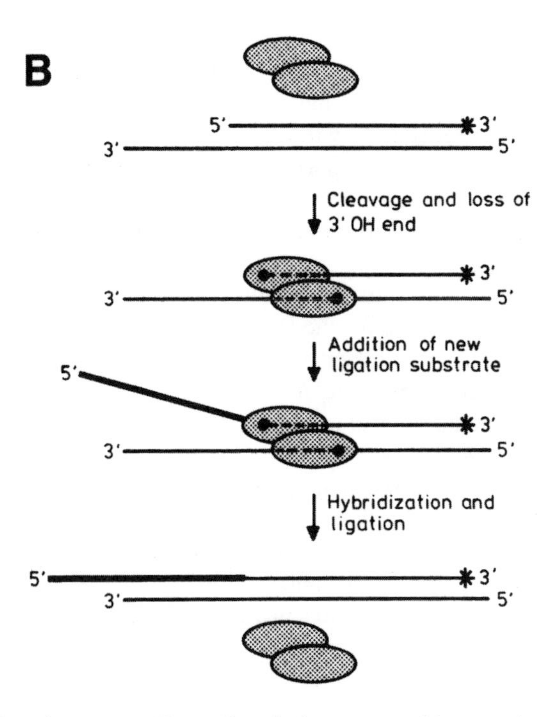

Fig. 7. Topoisomerase II-mediated cleavage and intramolecular ligation. Panel **A**, the DNA sequence of the suicide cleavage substrate. Panel **B**, schematic illustration of the individual steps in the topoisomerase II-mediated DNA cleavage and intramolecular ligation reactions. An asterisk denotes radioactive labeling, topoisomerase II is depicted as shaded ovals, and the ligation substrate is shown as a thick line.

4. Before use of oligonucleotides, the crude synthesis products have to be checked for the presence of incomplete synthesized oligonucleotides. A small aliquot of each oligonucleotide is 5'-radiolabeled by T4-polynucleotide kinase, and the sample is subjected to 12% denaturing polyacrylamide gel electrophoresis. If the full-length oligonucleotide is present in a background of smaller oligonucleotides, it has to be purified further. For this purpose, 100–150 µg DNA are loaded in a 5-cm wide slot on a 12% denaturing sequencing gel (1.5 m*M* thick). After electrophoresis, the gel is stained for 30 min in an aqueous solution of ethidium bro-

mide (2 µg/mL). The DNA band is visualized by illumination with UV light. The full-length oligonucleotide is then excised from the gel and eluted by incubation of the gel slice in 0.5 mL of 0.5 *M* ammonium acetate, 0.01 *M* magnesium acetate, 0.1% SDS, and 0.1 m*M* EDTA for 12 h at 37°C. Gel pieces are removed by centrifugation. The DNA is freed of ethidium bromide by extracting the solution twice with equal volumes of *n*-butyl alcohol, and it is concentrated by ethanol precipitation.

5. If cold-phosphorylation of the 77-mer (OL6) is omitted, a substantial fraction of the cleavage complexes will ligate to the 5'-end of this oligonucleotide, and the actual ligation to the 40-mer (OL18) will be lowered.
6. Omission of 38-mer/77-mer duplex purification before topoisomerase I-mediated cleavage results in a slight decrease in DNA cleavage, and ligation may occur to 5'-OH ends generated from cleavage of partially hybridized oligonucleotides.
7. The chosen incubation times and temperatures for topoisomerase I-mediated cleavage and ligation are optimal for the DNA substrates used here, but if other substrates are used, it has to be taken into consideration that the reactions are both time- and temperature-dependent ***(18)***.
8. Following trypsin digestion of human topoisomerase I–DNA complexes, a trypsin-resistant heptadpeptide is still covalently attached to the 3'-end of the cleaved DNA ***(19)***. The peptide retards the migration of the band resulting from DNA cleavage of the 38-mer/77-mer duplex to a position above the uncleaved substrate. Omission of trypsin digestion will prevent the cleavage product from entering the gel, and only uncleaved substrate and ligated material will enter the gel (**Fig. 1**).
9. Suicidal DNA substrates can be used to study the involvement of topoisomerase I in illegitimate recombination events. Thus, the enzyme will introduce DNA insertions or deletions depending on the ligation substrate added after cleavage complexes are formed from incubation of topoisomerase I with the 38-mer/77-mer duplex (**Fig. 4**) ***(20)***. DNA insertion occurs if the ligation substrate forms a displaced single-stranded overhang on hybridization to the cleavage complex. In this case, the 5'-OH end of the ligation substrate will be ligated to the cleaved 3'-end of the scissile strand, and the resulting ligation product will contain extra, outlooping nucleotides on this strand (**Fig. 4A**). This DNA ligation reaction will only take place provided that the incoming ligation substrate is able to hybridize to part or all of the 5'-overhang of the noncleaved strand in the cleavage complex. The efficiency of the ligation reaction is increased considerably if the two nucleotides at the 5'-end of the ligation substrate are 5' AG 3', since they will hybridize to the +1 and +2 positions on the noncleaved strand and hold the displaced strand in correct position for the ligation reaction (**Fig. 4A**). DNA deletions are formed if the ligation substrate on hybridization to the 5'-overhang of the noncleaved strand leaves a gap immediately downstream of the cleavage position. During and after the ligation reaction, the noncleaved strand is here believed to loop out

as seen in **Fig. 4B**. The illegitimate recombination reactions of topoisomerase I are less efficient than ligation events, where the ligation substrate is fully complementary to the noncleaved strand.

10. The method described above is concerned with the possibility of topoisomerase I mediating suicidal cleavage followed by intramolecular ligation. This type of ligation requires a hybridization of the incoming ligation substrate to the noncleaved strand in the cleavage complex. Depending on the DNA substrate chosen for uncoupling the cleavage and ligation half-reactions, the enzyme is also able to perform intermolecular ligation (**Fig. 5**). The latter involves ligation of two separate double-stranded molecules by topoisomerase I covalently attached to a blunt-ended DNA molecule. The substrates involved in these reactions are the 47-mer (OL22, 5' GCCTGCAGGTGACTCTA-GAGGATCTAAAA GACTTAGAAAAATTTTT 3') as the scissile strand and the 47-mer (OL25, 5' AAGTCTTTTAGATCCTCTAGAGTCGACCTGCAGGCATGCAAGCT-TGG3') as the noncleaved strand. The latter is cold-phosphorylated at the 5'-end to block ligation of this DNA strand to the cleaved DNA. Hybridization between OL22 and OL25 generates a DNA substrate with a 12-nucleotide-long 3'-overhang of the scissile strand, and a 5'-end of the noncleaved strand ending at the cleavage site. On topoisomerase I-mediated cleavage, an enzyme–DNA complex is generated, in which topoisomerase I is covalently attached to a blunt end. After addition of excess amounts of another blunt-ended DNA molecule, the covalently linked topoisomerase I will mediate ligation of the two duplexes. Ligation is here independent of the sequence and length of the added duplex, whereas single-stranded DNA is not ligated to the cleaved DNA strand ***(17)***.
11. The possibility of uncoupling the DNA cleavage and religation reactions of DNA topoisomerase I is very useful concerning studies of topoisomerase I inhibitors, such as camptothecin. Camptothecin inhibits the religation reaction of topoisomerase I, but leaves the cleavage reaction unaffected ***(18)***. Owing to the observations of preferential formation of camptothecin-trapped topoisomerase I–DNA cleavage complexes at sites containing guanine immediately 3' to the cleavage position ***(21,22)***, studies of camptothecin effects can benefit from the use of a slightly modified DNA substrate, which on cleavage allows ligation of a dinucleotide 5' GC 3' within region A in the absence of other DNA contacts. Here the 47-mer (OL22) is used as the scissile strand, and a 37-mer (OL35, 5' GCAAGTCTTTTAGATCCTCTAGAGTCGACCTGCAGGC 3') is used as the noncleaved strand. Hybridization will result in a branched DNA molecule having a protruding scissile strand with the sequence 5'-AGA-5-T_5 3' and a protruding noncleaved strand with the sequence 3'-CG 5' just downstream of the cleavage site. On cleavage, NaCl is added to a concentration of 1 *M* to terminate the cleavage reaction and avoid intermolecular DNA binding of topoisomerase I. After addition of different concentrations of camptothecin, ligation can be initiated by addition of the dinucleotide 5' GC 3' ***(23)***.

4.2. Topoisomerase II

12. The DNA substrate used for topoisomerase II contains a strong recognition sequence for the enzyme *(24,25)*. The cleavage substrate has a 5'-recessed top strand with only three nucleotides 5' to the cleavage position (**Fig. 7**). It is otherwise identical to the 28-bp DNA region, which has been shown to constitute the total binding region of the enzyme *(26)* and to give optimal cleavage with topoisomerase II in vitro *(25)*.
13. The cleavage substrate has a 3-bp duplex region 5' to the cleavage position on the top strand. If substrates are used containing a longer duplex region (4 or 5 bp) 5' to the cleavage position, the cleavage level will decrease considerably *(25)*.
14. If the 28-mer bottom strand is synthesized without an aminolink at the 3'-end, its 3'-OH end will take part in ligation, thereby decreasing the actual level of ligation to the 45-mer ligation substrate.
15. To optimize for cleavage products in experiments, where only the topoisomerase II-mediated cleavage half-reaction has to be studied, covalent topoisomerase II–DNA complexes can be recovered from a phenol/water interphase after phenol extraction. Cleavage products are first obtained as described in **Subheading 3.2., step 5**. After stopping the cleavage reaction by SDS or NaCl, 1 vol of phenol/chloroform/isoamyl alcohol (25:24:1) is added. The sample is shaken for few minutes and centrifuged at 15.900*g* in an Eppendorf centrifuge for 5 min. The water phase is removed, and the interphase as well as the organic phase are washed with 1 mL of 0.6 *M* NaCl. Following one further round of shaking and centrifugation, the water phase is again removed, and cleavage products are precipitated from the interphase by adding 3 vol of 96% ethanol. Cleavage products are redissolved in 10 µL of 10 m*M* Tris-HCl, pH 7.5, 0.1 m*M* EDTA (TE buffer), containing 500 µg/mL proteinase K. After proteinase K digestion for 2 h at 37°C, the sample is mixed with 1 vol of deionized formamide, 0.05% bromophenol blue, and 0.05% xylene xyanol, heated to 90°C for 1 min and subjected to 12% denaturing gel electrophoresis followed by autoradiography.
16. The method described above is concerned with the possibility of topoisomerase II to perform suicidal cleavage followed by intramolecular ligation. This type of ligation requires that the ligation substrate is single-stranded and able to hybridize to the 28-mer bottom strand, since hybridization ensures a higher local concentration of the ligation substrate in the vicinity of the covalently attached enzyme *(27,28)*. In addition to intramolecular ligation, topoisomerase II can mediate intermolecular ligation. In this case, the DNA substrate used for the cleavage reaction is made up of four oligonucleotides as shown in **Fig. 8**. The generation of this substrate is described by Schmidt et al. *(28)*. The top strand has a gap, lacking a nucleotide at position –4 relative to the cleavage position. The bottom strand has a nick at the position normally cleaved by topoisomerase II. Upon cleavage of this substrate, a cleavage complex is generated, where one subunit of topoisomerase II has cleaved the top strand and become covalently linked to the generated 5'-end, whereas cleavage of the bottom strand is made

```
5'-*GGCTCCCACGACGGCCAGTGCCAAGCTTGCATGAGGATGAC(dd)ATG|AGCGCATTGTTAGATTGATGACTCT*AT(dd)-3'
          3'-aGCCGGTCACGGTTCGAACGTACTCCTACTGCTACTCGC|GTAACAATCTAACTACTGAGATAC-5'
                                                   ↑
                                                  nick
```

Fig. 8. DNA substrate used for topoisomerase II-mediated intermolecular ligation. An asterisk denotes radioactive labeling, *a* denotes cordycepin, and C^{dd} and T^{dd} represent dideoxycytosine and dideoxythymidine, respectively. The bottom strand in the substrate has a nick at the position of cleavage as indicated by the arrow.

impossible, since this strand already contains a nick at this position. Cleavage occurs in 10 m*M* Tris-HCl, pH 7.0, 2.5 m*M* $MgCl_2$, 2.5 m*M* $CaCl_2$, and 30 m*M* NaCl. To study intermolecular ligation, 175 m*M* NaCl are added to stop further cleavage followed by a 20-fold molar excess of the ligation substrate of interest, and the mixture is incubated for 1 h at 30°C. Further treatments are as described above for intramolecular ligation. The most efficient ligation substrate during intermolecular ligation is duplex DNA containing a 4-base 5'-overhang, where the sequence of the overhang is complementary to the 4-base stagger in the topoisomerase II–DNA cleavage complex *(28)*. Ligation still takes place although at a reduced level if the ligation substrate is blunt-ended, or if it contains a 5'-overhang containing from 1–7 bases *(28)*.

17. Visualization of cleavage products by autoradiography reveals a major cleavage product with a mobility equivalent to a DNA fragment of approx 15 nucleotides instead of the expected 14 nucleotides. This retardation is caused by a proteinase K-resistant peptide covalently linked to the cleaved DNA. A minor cleavage product migrates to a position equivalent to a DNA fragment of approx 19 nucleotides. This cleavage product contains a longer peptide as a result of partial proteinase K digestion *(27)*. If only the topoisomerase II-mediated ligation reaction has to be studied, proteinase K treatment can be omitted (**Fig. 6**). Owing to the covalently attached enzyme, all cleavage products will then be prevented from entering the gel, and only the substrate and ligation products will be visible on autoradiography. A sole labeling of the ligation product can otherwise be obtained by 5'-radiolabeling of the 45-mer ligation substrate by T4 polynucleotide kinase and [γ^3P]ATP.

Acknowledgments

This work was supported by the Danish Society grants 93-004 and 78-5000, the Danish Center for Human Genome Research, the Danish Natural Science and Medical Research Councils, and the Danish Centre for Molecular Gerontology.

References

1. Vosberg, H.-P. (1985) DNA topoisomerases: enzymes that control DNA conformation. *Curr. Top. Microbiol. Immunol.* **114,** 19–102.

2. Osheroff, N. (1989) Biochemical basis for the interactions of type I and type II topoisomerase with DNA. *Pharmacol. Ther.* **41,** 223–241.
3. Gupta, M. and Pommier, Y. (1995) Eukaryotic DNA topoisomerase I. *Biochim. Biophys. Acta* **1262,** 1–14.
4. Watt, P. M. and Hickson, I. D. (1994) Structure and function of type II DNA topoisomerases. *Biochem. J.* **303,** 681–95.
5. Edwards, K. A., Halligan, B. D., Davis, J. L., Nivera, N. L., and Liu, L. F. (1982) Recognition sites of eukaryotic DNA topoisomerase I: DNA nucleotide sequencing analysis of topoisomerase I cleavage sites on SV40 DNA. *Nucleic Acids Res.* **10,** 2565–2576.
6. Liu, L. F., Rowe, T. C., Yang, L., Tewey, K. M., and Chen, G. L. (1983) Cleavage of DNA by mam*M*alian DNA topoisomerase II. *J. Biol. Chem.* **258,** 15,365–15,370.
7. Bonven, B. J., Gocke, E., and Westergaard, O. (1985) A high affinity topoisomerase I binding sequence is clustered at DNAase I hypersensitive sites in *Tetrahymena* R-chromatin. *Cell* **41,** 541–551.
8. Yang, L., Rowe, T. C., Nelson, E. M., and Liu, L. F. (1985) In vivo mapping of DNA topoisomerase II-specific cleavage sites on SV40 chromatin. *Cell* **41,** 127–132.
9. Been, M. D. and Champoux, J. J. (1981) DNA breakage and closure by rat liver type I topoisomerase: separation of the half-reactions by using a single-stranded substrate. *Proc. Natl. Acad. Sci. USA* **78,** 2883–2887.
10. Halligan, B. D., Davis, J. L., Edwards, K. A., and Liu, L. F. (1982) Intra- and intermolecular strand transfer by HeLa DNA topoisomerase I. *J. Biol. Chem.* **257,** 3995–4000.
11. Gale, K. C. and Osheroff, N. (1990) Uncoupling the DNA cleavage and religation activities of topoisomerase II with a single-stranded nucleic acid substrate: evidence for an active enzyme-cleaved DNA intermediate. *Biochemistry* **29,** 9538–9545.
12. Gale, K. C. and Osheroff, N. (1992) Intrinsic intermolecular DNA-ligation activity of eucaryotic topoisomerase II. Potential roles in recombination. *J. Biol. Chem.* **267,** 12,090–12,097.
13. Christiansen, K., Bonven, B. J., and Westergaard, O. (1987) Mapping of sequence-specific chromatin proteins by a novel method: topoisomerase I on *Tetrahymena* ribosomal chromatin. *J. Mol. Biol.* **193,** 517–525.
14. Andersen, A. H., Gocke, E., Bonven, B., Westergaard, O. (1985) Topoisomerse I has a strong binding preference for a conserved hexadecameric sequence in the promoter region of the rRNA gene from *Tetrahymena pyriformis. Nucleic Acids Res.* **13,** 1543–1557.
15. Ness, P. J., Koller, T., and Thoma, F. (1988) Topoisomerase I cleavage sites identified and mapped in the chromatin of *Dictyostelium* ribosomal RNA genes. *J. Mol. Biol.* **200,** 127–1397.
16. Christiansen, K., Svejstrup, A. B. D., Andersen, A. H., and Westergaard, O. (1993) Eukaryotic topoisomerase I-mediated cleavage requires bipartite DNA interac-

tion. Cleavage of DNA substrates containing strand interruptions implicates a role for topoisomerase I in illegitimate recombination. *J. Biol. Chem.* **268,** 9690–9701.
17. Christiansen, K. and Westergaard, O. (1994) Characterization of intra- and intermolecular DNA ligation mediated by eukaryotic topoisomerase I. Role of bipartite DNA interaction in the ligation process. *J. Biol. Chem.* **269,** 721–729.
18. Svejstrup, J. Q., Christiansen, K., Gromova, I. I., Andersen, A. H., and Westergaard, O. (1991) New technique for uncoupling the cleavage and religation reactions of eukaryotic topoisomerase I. The mode of action of camptothecin at a specific recognition site. *J. Mol. Biol.* **222,** 669–678.
19. Svejstrup, J. Q., Christiansen, K., Andersen, A. H., Lund, K., and Westergaard, O. (1990) Minimal DNA duplex requirements for topoisomerase I-mediated cleavage in vitro. *J. Biol. Chem.* **265,** 12,529–12,535.
20. Christiansen, K. and Westergaard, O. (1993) Involvement of eukaryotic topoisomerase I in illigitimate recombination: Generation of deletions and insertions, in *DNA Repair Mechanisms*, proceedings of the 35th Alfred Benzon Symposium (Bohr, W. A., Wassermann, K., and Krämer, K. H., eds.), Munksgaard, Copenhagen, pp. 361–371.
21. Kjeldsen, E., Mollerup, S., Thomsen, B., Bonven, B. J., Bolund, L., and Westergaard, O. (1988) Sequence-dependent effect of camptothecin on human topoisomerase I DNA cleavage. *J. Mol. Biol.* **202,** 333–342.
22. Jaxel, C., Capranico, G., Kerrigan, D., Kohn, K. W., and Pommier, Y. (1991) Effect of local DNA sequence on topoisomerase I cleavage in the presence or absence of camptothecin. *J. Biol. Chem.* **266,** 20,418–20,423.
23. Christiansen, K. and Westergaard, O. (1996) The effect of camptothecin on topoisomerase I catalysis, in: *The Camptothecins from Discovery to the Patient* (Pantasiz, P., Giovanella, D. C., and Rothenberg, M. L., eds.), Annals of the New York Academy of Sciences, pp. 50–59.
24. Sander, M. and Hsieh, T. (1983) Double strand DNA cleavage by type II DNA topoisomerase from *Drosophila melanogaster*. *J. Biol. Chem.* **258,** 8421–8428.
25. Lund, K., Andersen, A. H., Christiansen, K., Svejstrup, J. Q., and Westergaard, O. (1990) Minimal DNA requirement for topoisomerase II-mediated cleavage in vitro. *J. Biol. Chem.* **265,** 13,856–13,863.
26. Thomsen, B., Bendixen, C., Lund, K., Andersen, A. H., Sørensen, B. S., and Westergaard, O. (1990) Characterization of the interaction between topoisomerase II and DNA by transcriptional footprinting. *J. Mol. Biol.* **215,** 237–244.
27. Andersen, A. H., Sørensen, B. S., Christiansen, K., Svejstrup, J. Q., Lund, K., and Westergaard, O. (1991) Studies of the topoisomerase II-mediated cleavage and religation reactions by use of a suicidal double-stranded DNA substrate. *J. Biol. Chem.* **266,** 9203–9210.
28. Schmidt, V. K., Sorensen, B. S., Sorensen, H. V, Alsner, J. and Westergaard, O. (1994) Intramolecular and intermolecular DNA ligation mediated by topoisomerase II. *J. Mol. Biol.* **241,** 18–25.

12

Synthesis and Use of DNA Containing a 5'-Bridging Phosphorothioate as a Suicide Substrate for Type I DNA Topoisomerases

Alex B. Burgin, Jr.

1. Introduction

Type I DNA topoisomerases change the linking number of supercoiled DNA by carrying out orcheastrated cleavage and ligation reactions involving phosphodiester bonds. Strand cleavage is not the result of phosphodiester hydrolysis, but the result of transesterification of an active-site tyrosine, creating a 3'-DNA covalent intermediate and a 5'-terminus (5'-OH). Strand ligation occurs during a second transesterification event when the 5'-OH displaces the enzyme (*see* **Fig. 1**). A change in DNA linking number results when strand unwinding occurs between these cleavage and ligation reactions.

It has been demonstrated for Vaccinia topoisomerase I *(1)*, and agreed to be generally true for type I topoisomerases that the rate of cleavage (k_c) is much slower than the rate of ligation (k_l). This feature makes mechanistic studies of topoisomerase reactions difficult because the enzyme-imposed equilibrium ($K = k_c/k_l < 1$) favors accumulation of the strand-joining product, which has the same primary structure as the original substrate. Suicide substrates, which shift the equilibrium ($K >> 1$) and thereby trap the enzyme–DNA complex, have therefore been developed. Traditionally, these substrates contain nicks near the site of cleavage so that a small incised fragment is released following strand cleavage, thereby removing the critical 5'-OH necessary for strand ligation *(2)*. However, it has been recently demonstrated that DNA containing a 5'-bridging phosphorothioate linkage (OPS) also acts as a suicide substrate for elementary topoisomerase reactions *(3)*. These modified DNA linkages (**Fig. 1**; $X = S$) are efficient suicide substrates probably because sulfur is a poor

From: *Methods in Molecular Biology, Vol. 95: DNA Topoisomerase Protocols, Part II: Enzymology and Drugs*
Edited by N. Osheroff and M.A. Bjornsti © Humana Press Inc., Totowa, NJ

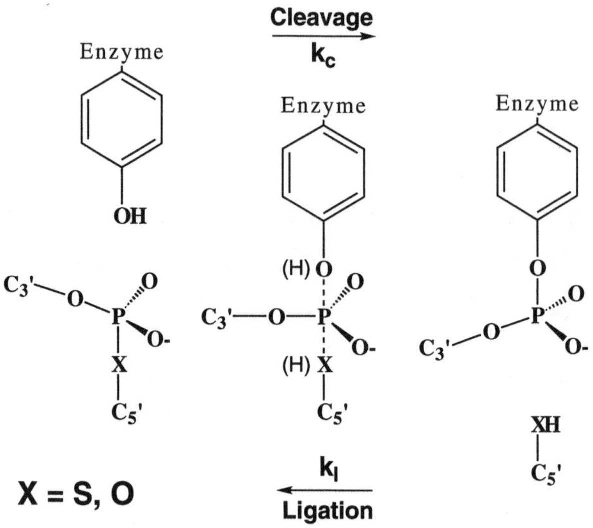

Fig. 1. Rationale for suicide substrate design. A phosphodiester linkage at the site of strand cleavage/ligation and an active site tyrosine are diagrammed on the left; the cleaved DNA and the 3'-phosphotyrosyl enzyme covalent complex are diagrammed on the right. In the unmodified phosphodiester, X represents oxygen; for the 5'-bridging phosphorothioate (OPS DNA), X represents sulfur. Chiral phosphorothioate diesters, in which one of the nonbridging oxygens is replaced by sulfur (not diagrammed), have been extensively used as mechanistic probes in enzymology and should not be confused with the achiral 5'-bridging phosphorothioate diesters described here.

nucleophile at phosphorous *(4)*; the rate of ligation is therefore extremely slow relative to an unmodified substrate ($k_l^O > 100 \times k_l^S$). In addition, the efficiency of these substrates may also benefit from the fact that a thio-anion is a better leaving group than an oxy-anion, and one would therefore expect k_c^S to be $> k_c^O$ *(5)*.

Both kinds of suicide substrates (nicked and OPS) shift the topoisomerase equilibrium by inhibiting the ligation reaction. However, both suicide substrates offer distinct advantages and disadvantages. First, phosphorothioate suicide substrates minimize pleiotropic effects on the cleavage reaction that follow from structural perturbations introduced by the use of nicked suicide substrates. Because no oligonucleotides are lost, the OPS DNA generates product that much more closely mimics that generated under standard conditions. These substrates are therefore likely to be advantageous for the study of drugs or other factors that bind or affect the enzyme–DNA complex. In addition, the OPS modification can presumably be placed within a supercoiled DNA duplex, which is the natural substrate of topoisomerase I. These substrate are therefore better for studying factors that affect enzyme–substrate binding. Finally, the efficiency of nicked suicide substrates relies on the rate of diffusion of the released incised fragment relative to the rate of strand ligation, a ratio that cannot be easily controlled or measured; and it has been demonstrated that the small incised fragment that is releasd by nicked suicide substrates can reattack the enzyme–DNA complex ***(6)***. Nicked suicide substrates, however, do have the distinct advantage that no special chemistry is required, and they can be rapidly synthesized and assembled. Oligonucleotides containg 5'-bridging phosphorothioates are currently not commercially available. The purpose of this chapter is therefore intended to detail the synthetic chemistry of 5'-bridging phosphorothioate-containing DNA oligonucleotides to make their synthesis as facile as possible.

The solid-phase synthesis of 5'-bridging phosphorothioates described below is adapted from the procedure originally described by Mag et al. *(7)*. In this procedure, the sulfur substitution is introduced into the standard oligonucleotide synthesis protocol by the use of a 5'-modified thymidine phosphoramidite [**4**]. The synthesis of this 5'-modified synthon (5'-S-trityl, 3'-*O*-[2-cyanoethyl-*N,N*-diisopropylamino] thymidine) is a three-step synthesis (outlined in **Fig. 2**). In the first step, thymidine [**1**] is converted to 5'-*O*-tosyl-thymidine [**1**]. The 5'-*O*-tosyl modification enables the introduction of a protected 5'-sulfhydryl into the molecule through displacement of the tosylate by the thio-anion of trityl mercaptan. In the third step, the 3'-hydroxyl of 5'-S-trityl thymidine [**3**] is modified so that the molecule can be incorporated into a polynucleotide by automated DNA synthesizers ***(8)*** using standard phosphoramidite chemistry.

2. Materials

1. Thymidine, *p*-toluene sulfonyl chloride, trityl mercaptan, diisopropylethylamine, and 2-cyanoethyl *N,N*-diisopropylchlorophosphoramidite are available from Aldrich Chemicals. Anyhydrous sodium sulfate and sodium bicarbonate are available from Sigma.
2. Anhydrous solvents (pyridine, acetonitrile, methanol, dichloromethane, dimethyl formamide) are available from Aldrich in "sure-seal" bottles; however, these solvents can also be distilled (*see* ***9–11***) to ensure an anhydrous condition (*see* **Note 2**).
3. Nonanhydrous solvents (dichloromethane, acetonitrile, and ethyl acetate) for chromatography and routine manipulations are available from Aldrich.
4. Aluminum-backed silica gel 60 F_{254} thin-layer chromatography (TLC) is available from EM Separation Technologies. Silica Gel (230–400 mesh 60A) for flash chromatography is available from Sigma.
5. Dithiothrietol (DTT) and silver nitrate for deblocking of the 5'-sulfhydryl during oligonucleotide synthesis are available from Aldrich.
6. Oligonucleotide syntheses were performed on an ABI380 or ABI391EP DNA synthesizer. In principle, any automated synthesizer that contains one additional

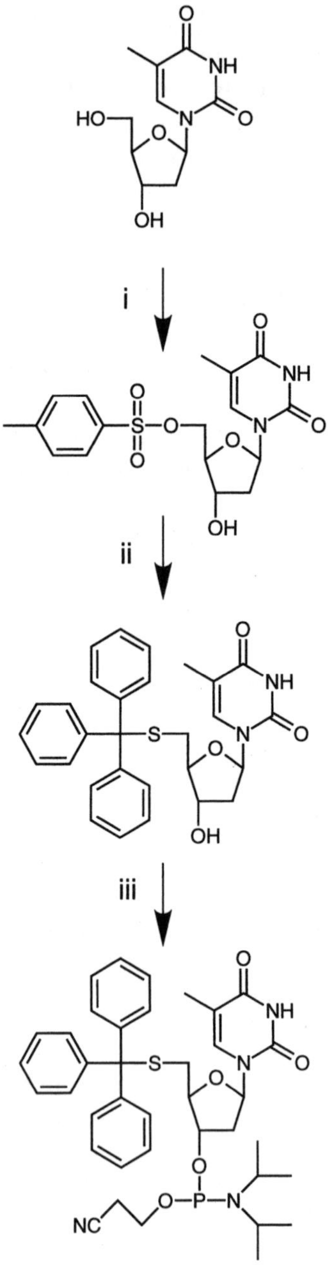

Fig. 2. Synthesis scheme for 5'-S-trityl, 3'-*O*-(2-cyanoethyl-*N,N*-diisopropylamino) thymidine. i. *p*-Toluene sulfonyl chloride. ii. Sodium salt of trityl mercaptan. iii. Diisopropylelthyl amine, 2-cyanoethyl *N,N*-diisopropylchlorophosphoramidite. *See text* for details.

port for the modified phosphoramidite would be suitable. However, we were not successful in synthesizing these oligonucleotides on a Pharmacia Gene Assembler Plus. We do not know why these attempts were unsucessful. However, different reagents (oxidizing, washing solvents, and so forth) are required on this machine and may be the explanation.

7. All solvents and reagents for the ABI synthesizer are available from ABI, except for the β-cyanoethyl DNA phosphoramidites, which is available from Glen Research.

3. Methods

3.1. Synthesis of 5'-O-tosyl thymidine [2]

1. Dissolve 25 g thymidine (100 mmol) in 80 mL of dry pyridine at room temperature with rapid stirring and then cool the solution to 0°C in an ice water bath (*see* **Note 1**).
2. Dissolve 23.75 g *p*-toluene sulfonyl chloride (125 mmol) in 20 mL dry pyridine at room temperature, and then add the solution dropwise (over approx 30 min) through a stop cock-fitted graduated cylinder. Continue stirring at 0°C for 16 h.
3. Monitor the reaction by TLC. Remove ≤0.1 mL using a dry syringe, and add to ethyl acetate:water (0.5 mL, 1:1), mix vigorously, and then add a small sample (~2 μL) of the ethyl acetate (top layer) to the TLC plate. Excess pyridine should be removed by drying the plate *in vacuo* or by a warm air/heat gun. Develop in CH_2Cl_2:CH_3OH (20:1, v/v), and identify product by exposing plate to short-wave (254 n*M*) UV light. The reaction is typically quantitative, and no side products are usually observed.
4. Add 500 mL of water (0°C) to quench the reaction, and then extract with 200 mL of chloroform. Repeat the extraction for a total of three times. Pool the chloroform layers, and then extract the organic phase sequentially with 500 mL 5% $NaHCO_3$ and 500 mL saturated NaCl. Evaporate the organic phase to an oil.
5. Dissolve the oil in 500 mL absolute ethanol by heating to 60°C with rapid stirring (1–4 h). Allow the solution to stand at –20°C overnight, and then collect the crystallized 5'-*O*-tosyl-thymidine [**2**] by filtration. Dry the white solid under oil pump vacuum overnight (typically 90% yield, 36 g) to ensure dryness.

3.2. Synthesis of 5'-S-trityl Thymidine [3]

1. Dissolve 19.8 g (50 mmol) 5'-tosyl-thymidine [**2**] in 150 mL anhydrous DMF at room temperature under argon (*see* **Note 2**).
2. Dissolve 16.56 g (60 mmol) trityl mercaptan in 120 mL anhydrous DMF, and then add 1.42 g sodium hydride (59 mmol) dissolved in 30 mL anydrous DMF under argon. Stir at room temperature for 2 min, cool in an ice water bath, and then add the 5'-tosyl-thymidine solution. Stir at room temperature for 3 h under argon.
3. Monitor the reaction by TLC. Withdraw a small sample (≤ 0.1 mL), and mix it with ethyl acetate and 10% aqueous $NaHCO_3$ (0.5 mL, 1:1), mix vigorously, and then add a sample of the ethyl acetate (top layer) to the TLC plate. Develop in

CH_2Cl_2:CH_3OH (15:1, v/v). The reaction typically yields multiple products. The 5'-S-trityl thymidine [**3**] can be identified by spraying the plate with 60% perchloric acid:ethanol (3:2, v/v) after resolving the reaction components. Products containing the trityl group will immediately turn yellow on spraying. Then place the plate in an 80°C oven for 2–3 min. Products containing a ribose sugar with a free hydroxyl will turn brown on heating. Side products can therefore be identified because they will not turn both yellow on spraying and brown on heating.

4. Evaporate the solution to an oil, and dissolve the residue in a minimal amount of CH_2Cl_2.
5. Purify 5'-S-trityl-thymidine [**3**] by flash chromatography (5% methanol in methylene chloride) 230–400 mesh 60A silica gel. The methodology for flash chromatography is described elsewhere in detail *(7–9)*.
6. The identity of the final purified product should be confirmed by chemical ionization mass spectroscopy and/or ^{1}H-NMR analysis.

3.3. Synthesis of 5'-S-trityl, 3'-O-(2-cyanoethyl-N,N-diisopropylamino)-thymidine [4]

This reaction is extremely moisture-sensitive. The reaction is used in the synthesis of standard DNA phosphoramidites and is also described extensively in **refs.** *7* and *9*.

1. Dissolve 3 g (6 mmol) 5'-S-trityl-thymidine [**3**] in 50 mL dry acetonitrile containing 10% pyridine. Dry the solution, and keep the white powder under an oil pump vacuum overnight. This will help to ensure that the starting material is anhydrous (*see* **Note 2**).
2. Dissolve the 5'-S-trityl-thymidine in 50 mL dry acetonitrile under argon. Add 1.1 mL (6.5 mmol) diisopropylethylamine under argon.
3. Add 1.4 mL (6.5 mmol) 2-cyanoethyl *N,N*-diisopropylchlorophosphoramidite dropwise using a gas-tight syringe while stirring under argon. Stir at room temperature. This reaction is typically fast (<30 min) and quantitative.
4. Monitor the reaction by TLC. Withdraw a small sample (≤mL) using a dry syringe, and mix it with water and ethyl acetate (0.5 ml, 1:1). Mix vigorously, and then add a sample of the ethyl acetate (top layer) to TLC place. Develop in 45:45:10 (v/v) ethylacetate:dichloromethane:triethylamine. The reaction results in two faster migrating spots, since the phosphoramidite is chiral and the chromatography resolves the two stereoisomers (both stereoisomers are used in the synthesis of the oligonucleotide).
5. When the reaction has gone to completion, quench excess phosphitylating reagent by injecting anhydrous methanol (0.2 mL) and continue stirring for 5 min. Dry the solution to an oil, and then as quickly as possible, dissolve the oil into a minimal amount of dichloromethane (approx 5 mL) and purify the 3'-phosphoramidite [**4**] by flash chromatograpy (45:45:10 ethyl acetate: dichloromethane:triethylamine; 230–400 mesh 60A silica gel).

3.4. Solid-Phase Synthesis of Oligonucleotides Containing a 5'-Bridging Phosphorothioate Linkage

1. Typically, 100 mg of the 5'-S-trityl thymidine phosphoramidite [**4**] is weighed out in a oven-dried ABI bottle. Cap the bottle with a rubber septum, and then place a 23-gage needle through the septum. This will help ensure no solid material is lost during the *in vacuo* drying. Place the vented bottle under an oil pump vacuum for 12–24 h to ensure that all water is removed. Resuspend the phosphoramidite in anhydrous acetonitrile (final 0.1 *M* solution), and place onto the ABI DNA synthesizer. This requires that the machine have at least one extra port (five total) for the delivery of the modified phosphoramidite.
2. Program the synthesizer to assemble the oligonucleotide (typically 1-µmol scale) so that the last phosphoramidite added (i.e., 5'-terminal) is the 5'-S-trityl thymidine phosphoramidite. All steps (coupling, capping, oxidizing, and so on) should be left unchanged from ABI recommeded conditions. For example, to synthesize the oligonucleotide (5'-GAGGATCTAAAAGACTTsTGAAAAATTT, where "s" represents the OPS linkage), program the sequence 5'-XGAAAAATTT, where X = 5'-S-trityl phosphoramidite [**4**].
3. The continued synthesis requires removal of the trityl group, so that the resulting 5'-sulfhydryl can be coupled on delivery of the next phosphoramidite. Typically, this is done by treating the 5'-*O*-dimethoxytrityl group with mild acid. The 5'-S-trityl group, however, is not labile to even strong acids, so heavy metal treatment is used to remove the trityl group. Remove the column from the synthesizer and wash with approx 2 mL water. This is most easily done by placing two 10-mL syringes on each side of the column and flushing the wash between the two syringes. Remove the water, and then add 2 mL of freshly prepared 50 m*M* silver nitrate ($AgNO_3$). Flush thoroughly, and let stand in the dark for 10 min. Remove the $AgNO_3$ solution, and wash three times each with 10 mL of water. Then flush thoroughly with 2 mL freshly prepared 50 m*M* DTT. Let stand at room temperature for 20 min. Remove the DTT solution, and wash with 10 mL water. Then quickly flush with 10 mL dry acetonitrile. Sulfhydryls can be very reactive, so it is useful to minimize the time between removal of the DTT (i.e., exposure of 5'-sulfhydryl) and addition of the next phosphoramidite.
4. Immediately place the column back on the machine and continue synthesis. The sequence should be 5'-GAGGATCTAAAAGACT<u>T</u>Y, where Y represents the sequence synthesized in **step 2** above. The first coupling step (*i.e.*, coupling of the underlined T) should be modified in two ways. First, the 5'-trityl group has already been removed, so do not treat the column with acid; this is normally the first step to remove the dimethoxytrityl group present on standard phosphoramidites. Second, because sulfur is a poor nucleophile at phosphorous, the coupling time should be increased to 5 min. All other steps should be left unmodified. The final sequence of the oligonucleotide will be 5'-GAGGATCTAAAAGACTTsTGAAAAATTT. This oligonucleotide can be paired with 5'-AAATTTTTCAAGTCTTTTAGATCCTC (unmodified) to form a topoisomerase I suicide substrate.

5. After synthesis is complete, deprotect the oligonucleotide by adding the resin to 1 mL ammonium hydoxide (28–30%) containing 1 m*M* DTT in a screw-cap tube, and incubate at 55°C for 12–16 h. Cool the tube on ice for 5 min, and then remove the solution and evaporate to a white powder. The final oligonucleotides are relatively stable. However, care should be taken to make sure that the OPS-containing oligonucleotides are not exposed to heavy metals (*see* **Notes 4** and **5**).
6. The phosphorothioate oligonucelotides were typically obtained in 25–50% yield, as compared with unmodified oligonucleotides. This is likely owing to reaction of the 5'-sulfhydryl with the solvent and/or resin.

4. Notes

1. Because only the synthesis of the 5'-modified thymidine synthon is described above, the final suicide substrate must contain thymidine immediately 3' to the topoisomerase cleavage site. Topoisomerase I binding sites, as defined by Sverstrup et al. ***(12)***, contain adenine immediately 3' to the cleavage site. Although we have found no effect of the A to T transversion using nicked suicide substrates, Kuimelis and McLaughlin ***(13)*** have recently described the synthesis of RNA containing 5'-bridging phosphorothioates using a 5'-S-trityl adenine synthon.
2. Care should be taken to keep all vessels, syringes, and so forth, dry. The preparation and handling of moisture-sensitive reagents is available from Aldrich and is described elsewhere ***(8)***.
3. It is also important to note that both 3'-bridging ***(14)*** and 5'-bridging ***(13)*** phosphorothioates are becoming important tools for the study of a variety of phosphoryl transfer reactions in addition to elementary toposomerase reactions ***(3)***.
4. The existence and location of the bridging phosphorothioate linkages can be identified by treating the oligonucleotide with heavy metals. Label the 5'-end of the oligonucleotide using T4 polynucleotide kinase and γ-^{32}P-ATP. Mix the modified oligonucleotide (in water; 50 µL is convenient) with an equal volume of saturated phenyl mercuric acetate (Aldrich) in water (a white precipitate will ocassionally form). Incubate at 65°C for 15 min. Extract three times with an equal volume of ethyl acetate to remove the phenyl mercuric acetate. Add freshly prepared DTT to the aqueous phase (final 10 m*M*), add an equal volume of 10 *M* urea, and then resolve the products by denaturing poolyacrylamide gel electrophoresis.
5. The conditions for 3'-end labeling OPS-containing oligonucleotides using terminal deoxynucleotidyl transferase (TdT) must be modified, since this enzyme uses cobalt as a metal cofactor. The cobalt should be replaced with an equal concentration of magnesium. In addition, cacodylate buffers should be avoided; Tris-HCl at an equal concentration and pH should be substituted. Finally, because the enzyme preparations also contain some cobalt and/or zinc, do not add more than 1 µL (10 U/µL, Pharmacia Biotech.) of enzyme/20 µL final reaction volume. These modifications decrease the efficiency of the TdT. However, by increasing

the time of reaction (typically 1 h, 37°C), 3'-modified product can be obtained with little or no decomposition of the internal OPS linkage.

6. Cleavage of 5'-bridging phosphorothioate DNA by topoisomerase I can be followed using 5'-end-labeled DNA and monitoring accumulation of covalent DNA–protein complexes (10–15% acrylamide [33:1, acrylamide:bis-acrylamide], 0.1% SDS, Tricine gels), or by using 3'-end-labeled DNA and monitoring accumulation of the 3'-fragment by denaturing polyacrylamide gel analysis (*see* **ref. *3*** for details). It important to remember that the 3'-fragment generated by enzyme cleavage contains a 5'-sulfhydryl. If no reducing agent is present, disulfide-linked oligonucleotides will form. Therefore, to monitor cleavage using 3'-end-labeled DNA, 10 m*M* DTT should be present in the gel-loading buffer. The gel buffer should also contain 1 m*M* β-mercaptoethanol.

References

1. Stivers, J. T., Shuman, S., and Mildvan, A. S. (1994) Vaccinia DNA topoisomerase I: single-turnover and steady-state kinetic analysis of the DNA strand cleavage and ligation reactions. *Biochemistry* **33,** 327–339.
2. Svejstrup, J. Q., Christiansen, K., Gromova, I., Andersen, A. H., and Westergaard, O. (1991) New technique for uncoupling the cleavage and religation reactions of eukaryotic topoisomerase I. *J. Mol. Biol.* **222,** 669–678.
3. Burgin, A. B., Huizenga, B. N., and Nash, H. A. (1995) A novel suicide substrate for DNA topoisomerases and site-specific recombinases. *Nucl. Acids Res.* **23,** 2973–2979.
4. Pearson, R. G. (1966) Acids and bases. *Science* **151,** 172–177.
5. Milstein, S. and Fife, T. H. (1967) The hydrolysis of S-aryl phosphorothioates. *J. Am. Chem. Soc.* **89,** 5820–5826.
6. Christiansen, K. and Westergaard, O. (1994) Characterization of intra- and intermolecular DNA ligation mediated by eukaryotic topoisomerase I. *J. Biol. Chem.* **269,** 721–729.
7. Mag, M., Luking, S., and Engels, J. W. (1991) Synthesis and selective cleavage of an oligodeoxynucleotide containing a bridged internucleotide 5'-phosphorothioate linkage. *Nucleic Acids Res.* **19,** 1537–1441.
8. Sproat, B. S. and Gait, M. J. (1985) Solid-phase synthesis of oligodeoxyribonucleotides by the phosphotriester method, in *Oligonucleotide Synthesis: a Practical Approach.* (Gait, M. J., ed.), IRL, Oxford, pp. 83–115.
9. Jones, R. A. (1985) Preparation of protected deoxyribonucleotides, in *Oligonucleotide Synthesis: a Practical Approach* (Gait, M. J., ed.), IRL, Oxford, pp. 23–34.
10. Perrin, D. D. and Armarego, W. L. F., (eds.) (1988) *Purification of Laboratory Chemicals*, 3rd ed. Pergamon.
11. Caruthers, M. H., Geaton, G., Wu, J. V., and Wiesler, W. (1992) Chemical synthesis of deoxyoligonucleotides and deoxyoligonucleotide analogs. *Methods Enzymol.* **211,** 3–19.

12. Svejstrup, J. Q., Christiansen, K., Andersen, A. H., Lund, K., and Westergaard, O. (1990) Minimal DNA duplex requirements for topoisomerase I-mediated cleavage in vitro. *J. Biol. Chem.* **265,** 12,529–12,535.
13. Kuimelis, R. G. and McLaughlin, L. W. (1995) Cleavage properties of an oligonucleotide containing a bridged internucleotide 5'-phosphorothioate RNA linkage. *Nucleic Acids Res.* **23,** 4753–4760.
14. Piccirilli, J. A., Vyle, J. S., Caruthers, M. H., and Cech, T. R. (1993) Metal ion catalysis in the Terahymena ribozyme reaction. *Science* **361,** 85–88.

13

Isolation of Covalent Enzyme–DNA Complexes

Thomas C. Rowe, Dale Grabowski, and Ram Ganapathi

1. Introduction

Topoisomerase enzymes catalyze changes in DNA topology by a concerted breakage and reunion of the DNA phosphodiester backbone *(1,2)*. This process occurs via sequential transesterification reactions involving a tyrosyl hydroxyl group present in the active site of the enzyme. A novel aspect of the breakage–reunion reaction is the the formation of a transient covalent O^4-phosphotyrosyl bond between the topoisomerase protein and a phosphate residue present at the break site in the DNA. This covalent complex can be irreversibly trapped by treatment with strong protein denaturants, such as sodium dodecyl sulfate (SDS) or alkali. Type II topoisomerases (i.e., bacterial gyrase and topoisomerase IV, bacteriophage T4 topoisomerase II, African swine fever virus topoisomerase, and eukaryotic topoisomerase II) and the type I-5'-topoisomerases (i.e., bacterial topoisomerases I and III, archaebacterial reverse gyrase, and yeast topoisomerase III) have been shown to form a covalent phosphotyrosine linkage with the 5'-ends of the cleaved strands of DNA *(2)*. In contrast, cleavage of DNA by the type I-3'-enzymes (i.e., eukaryotic topoisomerase I, DNA topoisomerase V, and vaccinia, variola, and Shope fibroma viral topoisomerases) results in the formation of a phosphotyrosine bond involving the 3'-ends of the broken strands of DNA.

A number of topoisomerase inhibitors act by stabilizing the covalent topoisomerase–DNA intermediate resulting in elevated levels of protein-linked DNA breaks following treatment with SDS or alkali *(3)*. For type II enzymes, this includes the 4-quinolone class of bacterial topoisomerase II inhibitors as well as a variety of anticancer drugs that target eukaryotic topoisomerase II (i.e., epipodophyllotoxins, ellipticines, anthracyclines, amsacrines, and so

From: *Methods in Molecular Biology, Vol. 95: DNA Topoisomerase Protocols, Part II: Enzymology and Drugs*
Edited by N. Osheroff and M.A. Bjornsti © Humana Press Inc., Totowa, NJ

forth). There are also several inhibitors (camptothecins, Hoechst dyes) that have been shown to have a similar effect on the type I-3'-enzyme eukaryotic topoisomerase I.

DNA that is covalently complexed to topoisomerase proteins can be readily separated from free DNA using the K^+-SDS precipitation method *(4,5)*. SDS is a strong anionic protein denaturant that dissociates noncovalent protein–DNA interactions, but is unable to dissociate covalent protein–DNA interactions. In addition, SDS causes precipitation of covalent protein–DNA complexes in the presence of KCl without causing precipitation of protein-free DNA. This selective precipitation of protein-linked DNA provides a convenient method for the detection and isolation of covalent topoisomerase–DNA complexes. This assay was initially developed for studying the biochemical properties of the covalent topoisomerase–DNA complex. However, with the discovery that topoisomerase enzymes are targets of a number of important chemotherapeutic agents, this assay has been utilized to study the mechanism of drug inhibition as well as to measure the drug sensitivity of topoisomerase enzymes in drug-sensitive and drug-resistant cells *(6,7)*. This procedure is simple, reproducible, and can be adapted to a variety of experimental circumstances.

Examples of two applications of this procedure will be given. The first application will present the methodology for measuring covalent topoisomerase–DNA complexes in vitro *(4,5)*, whereas the second application will present a procedure for measuring topoisomerase–DNA complexes in drug-treated mammalian cells *(7)*.

2. Materials

2.1. In Vitro Assay

1. pBR322 or λ bacteriophage DNA from Boehringer Mannheim (Indianapolis, IN).
2. Nick translation kit for labeling the pBR322 or λ bacteriophage DNA can be obtained from Promega (Madison, WI) or Gibco/BRL (Gaithersburg, MD).
3. [α-^{32}P] dATP (3000 Ci/mmol, 10 mCi/mL) from Dupont NEN Research Products (Boston, MA).
4. Ecolume scintillation cocktail from ICN Pharmaceuticals (Costa Mesa, CA).
5. Herring sperm DNA from Boehringer Mannheim.
6. Solutions.
   a. 10X topoisomerase I cleavage buffer: 500 m*M* Tris-HCl, pH 7.5, 100 m*M* $MgCl_2$, 5 m*M* dithiothreitol, 5 m*M* EDTA, 300 μg/mL BSA.
   b. 10X topoisomerase II cleavage buffer: 500 m*M* Tris-HCl, pH 7.5, 100 m*M* $MgCl_2$, 1 m*M* ATP, 5 m*M* dithiothreitol, 5 m*M* EDTA, 300 μg/mL BSA.
   c. Stop solution: 2% SDS, 2 m*M* EDTA, 0.5 mg/mL herring sperm DNA.
   d. 0.25 *M* KCl.
   e. Wash solution: 10 m*M* Tris HCl, pH 8.0, 100 m*M* KCl, 1 m*M* EDTA, and 100 μg/mL herring sperm DNA.

2.2. Cell-Culture Assay

1. [2-^{14}C] thymidine (>50 Ci/mmol) from Dupont NEN Research Products.
2. Ecolume scintillation cocktail from ICN Pharmaceuticals.
3. Herring sperm DNA from Boehringer Mannheim.
4. Lysis solution: 1.25% SDS, 0.4 mg/mL herring sperm DNA, 5 m*M* EDTA, pH 8.0.
5. Wash solution: 10 m*M* Tris-HCl, pH 8, 0.1 mg/mL herring sperm DNA, 100 m*M* KCl, 1 m*M* EDTA.

3. Methods

3.1. In Vitro Assay for Quantitating Covalent Topoisomerase–DNA Complexes

The in vitro assay provides a convenient approach for quantitating the levels of covalent complexes formed between purified topoisomerases and labeled DNA substrates. This approach can also be used to purify labeled DNA sequences selectively that are covalently complexed to topoisomerase enzymes for subsequent biochemical analyses *(8)* (*see* **Note 1**).

1. Label pBR322 or λ bacteriophage DNA by nick translation in the presence of [α-^{32}P] dATP (3000 Ci/mmol, 10 mCi/mL) using a nick translation kit (*see* **Note 2**). After labeling, the DNA should be phenol-extracted and then ethanol-precipitated twice in the presence of 2 *M* ammonium acetate to remove proteins and unincorporated nucleotides prior to being used in the cleavage assay *(9)*.
2. Reaction conditions for forming drug-stabilized covalent topoisomerase I/II complexes with DNA are as follows:

   Topoisomerase I cleavage reaction (final volume 50 μl)

5 μL	10X topoisomerase I cleavage buffer
5 ng	[^{32}P]-nick-translated pBR322 or λ bacteriophage DNA
Variable	drug (for camptothecin, a useful range is from 10–100 μ*M*)
10 U	topoisomerase I (1 U is the amount of enzyme that catalyzes the relaxation of 0.3 μg of supercoiled plasmid DNA under standard reaction conditions)

   Bring the reaction volume to 50 μL with H_2O.

   Topoisomerase II cleavage reaction (final volume 50 μL)

5 μL	10X topoisomerase II cleavage buffer
5 ng	[^{32}P]-nick-translated pBR322 or λ bacteriophage DNA
Variable	drug (for etoposide, a useful range is from 10–100 μ*M*)
10 U	topoisomerase II (1 U is the amount of enzyme that catalyzes the unknotting of 0.3 μg of P4 DNA under standard reaction conditions)

   Bring the reaction volume to 50 μL with H_2O.
3. Incubate the cleavage reactions at 37°C for 10 min.
4. Following a 10-min incubation at 37°C, terminate the reactions by adding 100 μL of stop solution, vortex for 10 s, and then heat at 65°C for 10 min.

5. To precipitate the covalent topoisomerase–DNA complexes, add 50 µL of 0.25 *M* KCl, vortex for 10 s, and then place the sample on ice for 10 min (*see* **Note 3**).
6. Collect the precipitate by centrifugation at 10,000*g* for 10 min at 4°C.
7. Remove the supernatant by vacuum aspiration (into a radioactive waste container) and resuspend the pellet in 200 µL of wash solution (10 m*M* Tris, pH 8.0, 100 m*M* KCl, 1 m*M* EDTA, and 100 µg/mL herring sperm DNA). Then heat the sample at 65°C for 10 min with periodic mixing.
8. Cool the sample on ice for 10 min, and then centrifuge for 10 min at 10,000*g* at 4°C to pellet the topoisomerase–DNA complexes.
9. After removing the supernatant, resuspend the pellet in 200 µL of H_2O at 65°C for 5 min. Transfer the solubilized pellet to a scintillation vial containing 4 mL of Ecolume scintillation fluid, mix, and count.
10. The stimulation of covalent topoisomerase–DNA complexes by drug can be expressed as a ratio of the dpm in drug-treated vs control, nondrug-treated samples. Alternatively, the amount of DNA precipitated can be expressed as a fraction of the total input DNA.

3.2. Measurement of Covalent Enzyme–DNA Complex in Cultured Cells

The K+-SDS assay can be easily used to detect covalent complexes between protein and labeled DNA in cultured cells and has a variety of applications (*see* **Note 4**).

1. The DNA in logarithmically growing cells in culture is labeled overnight (18–24 h) with 0.04 µCi/mL of [2-^{14}C-]thymidine (53 mCi/mmol) (*see* **Note 5**).
2. The labeled cells are pelleted by centrifugation (150*g*) for 10 min, and resuspended in fresh growth medium for an additional 3 h at 37°C to chase all the radioactivity into high-mol wt DNA. A cell count after incubation in radiolabel free media is taken, and 4×10^5 cells/mL (2 mL total volume) are treated with various concentrations of a topoisomerase I inhibitor (e.g., camptothecin) or topoisomerase II inhibitor (e.g., etoposide) for 1 h at 37°C. Control cells are treated with an equal volume of diluent used to prepare a stock solution of the inhibitor. It is preferable to have at least duplicate samples for each control and treatment condition.
3. Following treatment, cells are pelleted by centrifugation at 150*g* for 5 min. The supernatant media is carefully aspirated (into a radioactive waste container), and the cell pellet lysed by the addition of 1 mL of preheated (65°C) lysis solution.
4. After incubating for 10 min at 65°C, 250 µL of 325 m*M* KCl are added to each tube to achieve a final concentration of 65 m*M* KCl. The suspension is then vigorously vortexed for 10 s to achieve reproducible fragmentation of DNA (*see* **Note 6**). Since the fragment size of DNA influences the sensitivity of the assay, reproducible shearing of nuclear DNA during handling is important.

Table 1
Comparison of Etoposide-Induced DNA Cleavable Complex Formation in Cells Sensitive and Resistant to the Toposisomerase II Inhibitor Etoposide

	DNA cleavable complex, fold-increase	
Etoposide, μ*M*[a]	Drug-sensitive human HL-60 cells	Drug-resistant human HL-60 cells
1	2.6[b] (57%)[c]	0.8[b] (100%)[c]
2.5	3.9 (18%)	1.1 (100%)
5	7.1 (7%)	1.7 (100%)
10		3.0 (85%)
40		4.3 (33%)

[a]Cells with radiolabeled DNA were treated for 1 h with indicated concentration of etoposide.

[b]DNA cleavable complex (fold increase) is the ratio of dpm in the treated vs the untreated control.

[c]Numbers in parentheses are the survival of cells in a soft-agar colony assay when treated with the indicated concentration of etoposide.

5. Transfer 1 mL of the lysate to a 1.5-mL microcentrifuge tube, and cool on ice for 10 min to allow precipitation to occur (*see* **Note 3**). The tube is then centrifuged (10,000*g*) for 10 min at 4°C in a microcentrifuge equipped with a rotor for 1.5-mL tubes.
6. Remove the supernatant, and resuspend the pellet in 1 mL of wash solution. Heat the tube at 65°C for 10 min with periodic mixing, and then place on ice for an additional 10 min to reprecipitate the covalent topoisomerase–DNA complexes. Centrifuge the tube at 10,000*g* for 10 min at 4°C to recover the precipitate.
7. Remove the supernatant, and repeat the wash procedure in **step 6** (*see* **Note 7**).
8. Resuspend the washed pellet in 500 μL H_2O preheated to 65°C, and add to 4 mL of Ecolume biodegradable liquid scintillation cocktail and count.
9. The data are normalized for [2-^{14}C-]thymidine counts in the untreated control sample or, alternatively, to the total trichloroacetic acid precipitable counts in the control. The stimulation of covalent topoisomerase–DNA complexes by drug is expressed as a ratio of dpm in drug-treated vs the untreated control cells. The results from a typical experiment using this methodology are outlined in **Table 1**.

4. Notes

1. To purify labeled DNA sequences covalently complexed with topoisomerase proteins, resuspend the pellet at the end of **step 8 (Subheading 3.1.)** of the in vitro procedure in 50 μL of 10 m*M* Tris-HCl, pH 8.0, 1 m*M* EDTA, 1 mg/mL proteinase K, and incubate for 1 h at 37°C to remove the topoisomerase protein from the DNA. If necessary, the deproteinized DNA can then be further purified by phenol extraction and ethanol precipitation *(9)* depending on the type of DNA analysis that will be done.

2. A variation of the in vitro assay can be used to distinguish 5'- vs 3'-linked covalent topoisomerase–DNA complexes *(4)*. Instead of using nick-translated DNA, a linear DNA substrate labeled at either its 3'- or 5'-end is used. When this approach is used, cleavage reactions are terminated with 100 µL of a 37°C alkaline stop solution (0.2 *M* NaOH, 2% SDS, 2 m*M* EDTA, 0.5 mg/mL herring sperm DNA), vortexed, and incubated an additional 10 min at 37°C. The alkaline stop solution not only traps the covalent topoisomerase–DNA complex, but also denatures the DNA duplex. The samples are then neutralized by adding 50 µL of a solution containing 0.25 *M* KCl, 0.4 *M* HCl, and 0.4 *M* Tris-HCl, pH 7.9, vortexed, and placed on ice for 10 min. These two steps replace **steps 4** and **5** of **Subheading 3.1.** All other steps remain the same. A covalent linkage of topoisomerase to the 5'-end of the cleaved DNA strands will cause precipitation of a DNA substrate that is labeled at its 3'-, but not 5'-end. Correspondingly, a covalent linkage of topoisomerase to the 3'-end of the cleaved DNA strands will cause precipitation of a DNA substrate that is labeled at its 5'-, but not 3'-ends.
3. As an alternative to the centrifugation method, the covalent topoisomerase–DNA precipitate formed at this step can also be isolated by filtration through glass fiber disks using a filter manifold (Millipore, Bedford, MA) *(8)*. In this case, the samples are filtered through Whatman GF/A or GF/C glass fiber filters prewetted with filter wash buffer (10 m*M* Tris-HCl, pH 8.0, 100 m*M* KCl, 1 m*M* EDTA). The filters are then washed five times with 5 mL of 4°C filter wash buffer. Following this, the filters are washed twice with 10 mL of 95% ethanol (4°C) and once with 10 mL of 70% ethanol. The filters are then dried under an infrared heat lamp before placing in 5 mL of Scintiverse II scintillation cocktail (Fisher Scientific, Pittsburg, PA) to count.
4. The cell culture assay has the following potential applications:
   a. Screening of compounds for their potential effects in stimulating topoisomerase I- or II-mediated DNA cleavable complex formation.
   b. Screening of tumor cells with sensitive and resistant forms of the topoisomerase enzyme.
   c. Evaluation of conditions or agents that affect the formation and stability of topoisomerase-mediated DNA cleavable complex.
5. The radiolabeling of DNA with [^{14}C] vs [^{3}H]thymidine is important for obtaining reproducible results in cells actively growing during the treatment with radiolabel. In our experience, the proliferation of tumor cells is markedly inhibited in culture following labeling with [^{3}H]thymidine. In contrast, the proliferation of cells in culture is not affected by labeling with [^{14}C]thymidine.
6. Vortexing of cells to shear the DNA in **step 4 (Subheading 3.2.)** is critical for obtaining consistent results. In our experience, passing the lysate repeatedly through a syringe and a 23-gage needle does not always ensure uniformity between samples. Also, the vortexing vs syringing cuts down on the amount of radioactive waste generated

7. Additional washing of the pellet in **step 7 (Subheading 3.2.)** does not significantly reduce the background signal and may result in a loss of covalent topoisomerase–DNA complexes.

Acknowledgments

The authors would like to thank Kristen Schneider for her critical reading of this manuscript. This work has been supported by the United States Public Health Service Grants CA35531, CA60158, and GM 47535.

References

1. Wang, J. C. (1994) DNA topoisomerases as targets of therapeutics: an overview. *Adv. Pharmacol.* **29,** 1–19.
2. Roca, J. (1995) The mechanisms of DNA topoisomerases. *Trends Biochem. Sci.* **20,** 156–160.
3. Chen, A. Y. and Liu, L. F. (1994) DNA topoisomerases: essential enzymes and lethal targets. *Annu. Rev. Pharmacol. Toxicol.* **34,** 191–218.
4. Liu, L. F., Rowe, T. C., Yang, L., Tewey, K. M., and Chen, G. L. (1983) Cleavage of DNA by mammalian DNA topoisomerase II. *J. Biol. Chem.* **258,** 15,365–15,370.
5. Muller, M. T. (1983) Nucleosomes contain DNA binding proteins that resist dissociation by sodium dodecyl sulfate. *Biochem. Biophys. Res. Commun.* **114,** 99–106.
6. Zwelling, L. A., Hinds, M., Chan, D., Mayes, J., Sie, K. L., Parker, E., Silberman, L., Radcliffe, A., Beran, M., and Blick, M. (1989) Characterization of an amsacrine-resistant line of human leukemia cells. Evidence for a drug-resistant form of topoisomerase II. *J. Biol. Chem.* **264,** 16,411–16,420.
7. Rowe, T. C., Chen, G. L., Hsiang, Y. H., and Liu, L. F. (1986) DNA damage by antitumor arcidines mediated by mammalian DNA topoisomerase II. *Cancer Res.* **46,** 2021–2026.
8. Trask, D. K., DiDonato, J. A., and Muller, M. T. (1984) Rapid detection and isolation of covalent DNA/protein complexes: application to topoisomerase I and II. *EMBO J.* **3,** 671–676.
9. Sambrooks, J., Fritsch, E. F., and Maniatis, T. (1989) *Molecular Cloning, A Laboratory Manual,* 2nd ed. Cold Spring Harbor Laboratory Press, Cold Spring Harbor, NY.

14

ICE Bioassay

Isolating In Vivo Complexes of Enzyme to DNA

Deepa Subramanian, Christine Sommer Furbee, and Mark T. Muller

1. Introduction

DNA topoisomerases I and II (topo I and topo II) are ubiquitous nuclear enzymes that regulate DNA topology by breaking and resealing one or both strands of a DNA duplex. Topoisomerase I makes a single-stranded break in a DNA duplex, mediates passage of the intact strand through the break, and then reseals it. Type II topoisomerases create transient breaks in both strands of a duplex, pass an intact DNA segment through that break, and then reseal the cleavage site. This ability of topoisomerases to catalyze concerted breaking and rejoining of DNA strands makes them biologically important enzymes that are involved in crucial cellular processes ***(1–3)***.

The cleavage/religation mechanisms of both topo I and II involve a reaction intermediate commonly referred to as a cleavable or covalent complex, in which DNA is transiently linked to the enzyme by a phosphotyrosine linkage. Under normal conditions, the half-life of a covalent complex is quite short, since rapid religation of the nick or scission is greatly favored ***(4,5)***. This rapid religation is probably a safeguard against inadvertent damage to the genome. Low levels of the in vivo topoisomerase I/DNA covalent intermediates can be detected by arresting the reaction using protein denaturants ***(4,6)*** (such as ionic detergents) to lyse the cells rapidly and denature the protein, thereby trapping it on DNA. Several classes of clinically important antineoplastic and antibacterial drugs have been shown to target topos I and II by shifting the equilibrium of the cleavage/religation reaction, thus prolonging the half-life of the

From: *Methods in Molecular Biology, Vol. 95: DNA Topoisomerase Protocols, Part II: Enzymology and Drugs*
Edited by N. Osheroff and M.A. Bjornsti © Humana Press Inc., Totowa, NJ

$$\text{Topo} + \text{DNA} \underset{K_{cl}}{\overset{K_{rel}}{\rightleftharpoons}} \text{Topo/DNA complex}$$

$$\text{No drug: } K_{rel} >>> K_{cl}$$

$$\text{With drug: } K_{rel} > K_{cl}$$

Fig. 1. In the absence of enzyme inhibitors, the equilibrium of the topoisomerase/DNA cleavage and religation reaction heavily favors rapid religation of DNA knicks or double-stranded scissions. Several classes of topoisomerase inhibitors lengthen the half-life of the cleavage intermediate, thereby increasing the amount of complexes trapped on cell lysis with sarkosyl.

topoisomerase/DNA reaction intermediates (**Fig. 1**). Treatment with these drugs therefore greatly enhances trapping of covalent complexes on addition of ionic detergents. By prolonging the half-life of the nicked or cleaved DNA intermediates, these drugs are believed to turn topoisomerases into DNA damaging agents *(7,8)*. Technically, these drugs are not necessarily inhibitors, but are classified as topoisomerase poisons. The clinical efficacy of several antineoplastic drugs that target topoisomerase II has been shown to correlate with their ability to stabilize the enzyme/DNA intermediate *(9–11)*. There is substantial evidence that topoisomerase-mediated DNA damage in response to these drugs is enhanced by processes related to DNA template activity (e.g., replication, transcription, and repair) *(11)*; therefore, it becomes especially important to examine in vivo drug effects in the presence of ongoing DNA templating.

Since these enzymes are important targets for novel chemotherapeutic and antibacterial drug development, assays that measure DNA damage inflicted by topoisomerases and drugs under physiological conditions provide valuable predictive information. The ICE bioassay measures genomic DNA cleavage mediated specifically by topoisomerases in vivo. Whereas general DNA damage can be analyzed by methods, such as alkaline elution *(12)*, the ICE bioassay is specific for topo I or II. The method involves a physical separation of enzyme/DNA complexes from free proteins and utilizes antibodies to detect DNA-bound topo I or II. It therefore allows quantitation of covalent topoisomerase/DNA complexes in any tissue or cells after exposure (in vitro or in vivo) to topoisomerase inhibitors (*see* **Subheading 3.**). The utility of the ICE bioassay is evident when one considers that commonly used in vitro assays for agents active against topoisomerases do not necessarily reflect the effects

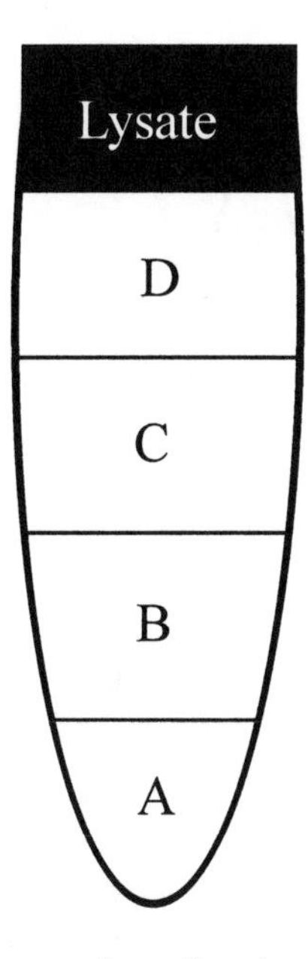

Fig. 2. Schematic representation of cesium chloride step gradient.

one would observe in the cellular environment. Chemical modification and degradation of topoisomerase inhibitors have been observed in vitro at physiological pH and in cells *(13,14)*. Drug uptake, relocation to the site of enzyme activity, and efficacy can be best evaluated using methods, such as the ICE bioassay, which measure a drug's ability to damage DNA in vivo.

To carry out the assay, cells are exposed to a topoisomerase poison under conditions where the endogenous enzyme should be actively engaging cellular DNA, followed by addition of sarkosyl to effect rapid cell lysis. Under these conditions, proteins are denatured, and electrostatic interactions between protein and DNA are disrupted. The lysates are loaded onto cesium chloride density gradients (**Fig. 2**) and centrifuged to dissociate noncovalent protein–DNA interactions *(15–17)*, and to achieve separation of free proteins and DNA-complexed protein. After fractionation of the gradients, immunoblotting using an appropriate topoisomerase antibody measures topo I or topo II in the DNA peak.

This chapter describes two basic applications of the ICE bioassay to HeLa cells. The first experiment (**Fig. 3**) details treatment of HeLa cells plus/minus the topo II inhibitor VM26 (100, μ*M*). In the presence of the drug, rapid sarkosyl lysis is able to trap the enzyme DNA covalent complex. Cesium chloride centrifugation separates the bound and free protein, and immunoblotting of the gradient fractions with an anti-topo II antibody clearly indicates that

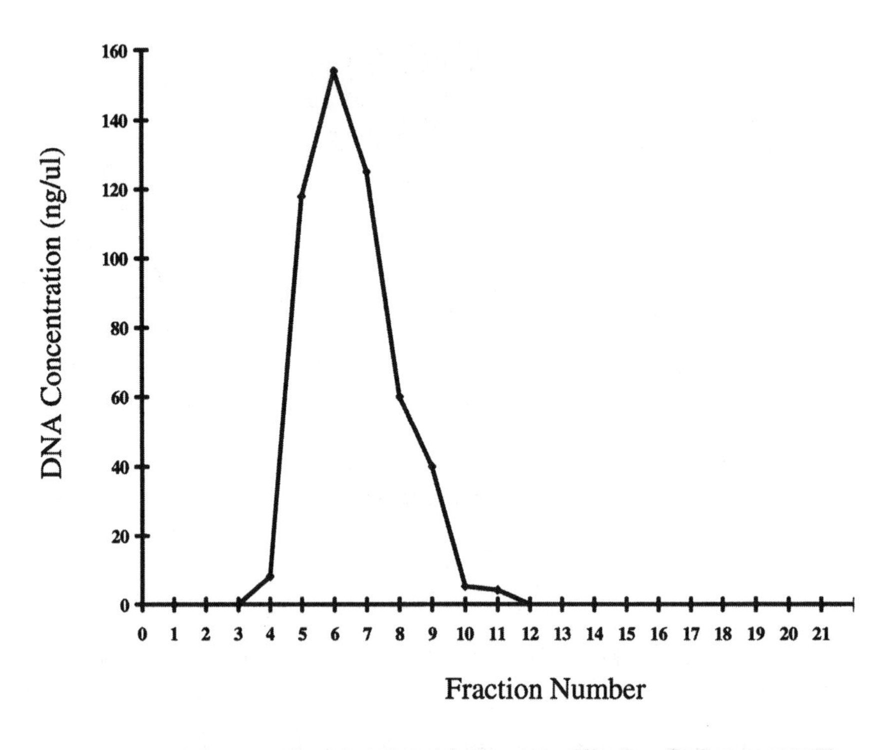

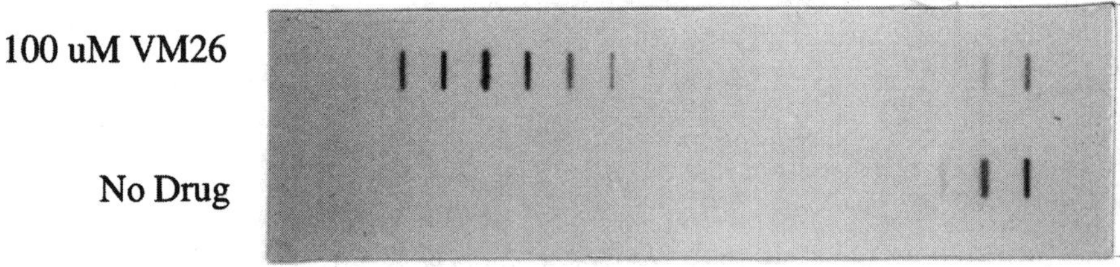

Fig. 3. Detection of VM26-induced topoisomerase II/DNA covalent complexes in HeLa cells. HeLa cells were treated with 100 μ*M* VM26 for 30 min at 37°C, followed by sarkosyl lysis and analyzed by the ICE bioassay. Top, DNA profile (measured by fluorometry) from a single representative gradient. Bottom, immunoblot of gradient fractions from untreated and VM26-treated cells. Blots were analyzed using a rabbit antibody to topo II followed by visualization using horseradish peroxidase conjugated goat antirabbit antibody and detection using chemiluminesence.

topo II is present in the DNA peak. The second experiment details application of the ICE bioassay to trap topo I covalent complexes using topotecan, a well-characterized topo inhibitor ***(18)***. Drug-induced cleavages are shown to be reversible by incubating the cells in drug-free medium after the initial exposure.

Although the ICE bioassay is clearly useful to determine the ability of drug candidates to damage DNA in vivo, this methodology can also be applied to examine topo I- or topo II-mediated DNA damage in response to physiological or genetic manipulations. The ICE bioassay has been successfully applied to tissue-culture cells, patient blood ***(18)***, and biopsy specimens, and should be adaptable to any cell or tissue source, provided that a homogeneous suspension of mostly single cells can be achieved. The method is also useful for determining whether a prospective chemotherapeutic agent is active against endogenous topoisomerase in vivo, and should increase our ability to assess its potential efficacy in a clinical setting. The basis of the ICE methodology could be adapted to identify tumors likely to be responsive or resistant to specific topoisomerase inhibitors.

2. Materials

2.1. Cell Culture and Drug Exposure

1. Cultured cells, e.g., HeLa, HL60, MCF7.
2. Other sources, such as tissue biopsies, peripheral blood, and bone marrow aspirates.
3. Medium: DMEM + 10% bovine calf serum for HeLa cells.
4. RPMI + 10% fetal bovine serum for HL60.
5. Topoisomerase poisons, e.g., camptothecin, etoposide.
6. DMSO or methanol (negative control).
7. 1X TE: 10 m*M* Tris-HCl, pH 8.0, 1 m*M* EDTA.
8. 1% Sarkosyl in 1X TE.
9. 2% Sarkosyl in 1X TE.

2.2. Centrifugation of Lysates

1. Cesium Chloride.
2. Polyallomer tubes, 14 × 89 mm (Beckman).
3. Mineral oil.
4. SW41 rotor and buckets (SW50.1 rotor can also be used)
5. Tube perforation and fractionation device.

2.3. Immunoblotting

1. Hybond C nitrocellulose membrane (Amersham).
2. 25 m*M* sodium phosphate buffer, pH 6.5: Prepare 1 *M* solutions of monobasic and dibasic sodium phosphate. Mix appropriate volumes of both solutions to obtain buffer at pH 6.5.
3. Slot-blot or dot-blot manifold (Schleicher and Schuell).
4. 1X TBST: 20 m*M* Tris-HCl, pH 7.5, 137 m*M* NaCl, and 0.1% Tween-20. Make a 10X stock of TBST and dilute as required.
5. Nonfat dried milk (BLOTTO).
6. Primary antibodies against topo I and II.
7. [^{125}I]-protein A (ICN).

8. Horse radish peroxidase-conjugated goat antirabbit or goat antimouse (Bio-Rad).
9. Chemiluminesence detection kit (Amersham).

3. Methods

3.1. Preparation of Lysates

The ICE bioassay can be used with a variety of cultured cells or other sources of cells (*see* **Note 1**). Approximately 1×10^6 to 1×10^7 cells should be used per treatment. Cells should be actively metabolizing. Cells are treated with topoisomerase poisons as follows:

1. Adherent cell lines (e.g. HeLa cells): Cells are cultured in 100-mm dishes, and a single Petri dish is used per treatment (90% confluent, approx 1×10^7 cells). The medium is removed from the cells and replaced with serum-free medium (1 mL) containing topoisomerase inhibitors (e.g., 50 μ*M* camptothecin or 100 μ*M* etoposide, final concentration) and cells are incubated at 37°C. A negative control containing no drug (solvent only) should be done in parallel. At the appropriate times, drug containing medium is removed, and cells are lysed by the direct addition of 1% sarkosyl equilibrated to 37°C (3 mL/plate). The sarkosyl is squirted on the plate two to three times to ensure complete and rapid lysis (*see* **Note 2**).
2. Suspension cells (e.g., HL60 cells): The cell suspension from one flask (approx 1×10^7 cells) is transferred to a 15-ml centrifuge tube, and cells are pelleted by low-speed centrifugation (1000*g* for 5 minutes). Cells are resuspended in 1 mL of serum-free medium, and topoisomerase inhibitors are added at the required concentration. Cells are incubated at 37°C, and at the appropriate time lysed by the addition of an equal volume of 2% sarkosyl equilibrated to 37°C (final concentration 1%). The tube is quickly inverted several times to ensure complete lysis.

3.2. Cesium Chloride Density Gradients

1. Step gradients for the SW41 buckets (polyallomer tube size 14mm X 89mm) are set up by making four layers (2 mL/layer) of cesium chloride of decreasing densities from the bottom to the top (for the SW50.1 buckets, use 1 mL/layer) (*see* **Note 3**).
2. Prepare a stock solution of cesium chloride by dissolving 120 g of cesium chloride in 70 mL of 1X TE (density = 1.86 g/cc, refractive index 1.414).
3. From the above stock solution prepare 4 different solutions according to **Table 1**. Layer 2 mL of each step in a polyallomer tube (for the SW41 rotor) carefully using a 5-mL pipet as indicated in **Fig. 2** (*see* **Note 4**)
4. Overlay the lysates on top of the gradient, top off with mineral oil, and centrifuge the tubes at 125,000*g* for 18–24 h at 20°C. Do not centrifuge at 4°C, since this will cause the cesium chloride to precipitate and may lead to rotor failure.

Table 1
For Each Gradient Use the Following Ratios of Cesium Chloride and TE

	A	B	C	D
TE	0.075 vol	0.2 vol	0.45 vol	0.55 vol
CsCl Stock	0.925 vol	0.8 vol	0.55 vol	0.45 vol

5. Fractionate the gradients (300-µL samples for SW41 gradients) from the bottom of the tubes. Ideally, a tube perforator should be used to collect drops from the bottom of the tube (*see* **Note 5**).

   Note: The DNA peak is located near the bottom of the gradient distributed over three to five fractions. In some cases, these fractions might be viscous. Covalent cleavage complexes will be located in the DNA peak. Proteins including free topoisomerases are located at the top of the gradient. Do not pipet fractions from the top of the gradients, since proteins from the top might contaminate the DNA-containing fractions.
6. Locate the DNA fractions in the gradient by measuring absorbency at 260 nm (dilute 30 µL of each fraction in 270 µL of water prior to measurement). Alternately, DNA concentrations can be accurately measured by fluorometry.

3.3. Immunodetection

1. Cut Hybond C nitrocellulose membranes to fit slot-blot or dot-blot manifolds, and soak in 25 m*M* sodium phosphate buffer, pH 6.5, for at least 15 min. Assemble the slot-blot device by placing two pieces of Whatman filter paper under the nitrocellulose membrane.
2. Dilute aliquots from each fraction (50–100 µL) with an equal volume of 25 µ*M* sodium phosphate buffer, pH 6.5, and apply the samples to the blot using a vacuum (*see* **Note 6**). Additionally, a standard curve should be set up on the same or parallel blot using purified topoisomerase (4–40 ng range).
3. Remove the membrane, and rinse briefly with 25 m*M* sodium phosphate buffer.
4. The membranes are then equilibrated in 1X TBST for 15 min with gentle agitation.
5. The membranes are blocked in 1X TBST containing 5% nonfat dried milk (BLOTTO) at room temperature for 2–3 h with agitation followed by three washes (10 min/wash) in 1X TBST.
6. Dilute the primary antibody in 30–50 mL of TBST, according to the size of the blot (the solution should cover the blot completely). Typically, antibodies are diluted to 1:1000- to 1:10,000-fold as appropriate for that specific antibody. Incubate blot in the primary antibody solution for at least 6 h at room temperature (*see* **Note 7**). The tray containing the blot should be covered with plastic wrap to prevent evaporation of the solution. The filters are then washed three times (10 min/wash) in TBST.

7. The membranes can then be processed in two ways:
   a. The membranes can be incubated in TBST containing [^{125}I]protein A at 0.4 µCi/mL of TBST (30–40 mL solution) for 2 h at room temperature followed by three final washes (10 min/wash). The membranes are dried slightly, wrapped in plastic wrap, and visualized by autoradiography.
   b. Alternatively, the blots can be visualized by chemiluminescence using kits, such as Amersham's ECL detection system. In this case, the primary antibody incubation should be restricted to 1 h. Following the three washes in TBST from **step 6**, the blots are incubated in an appropriate dilution of horse-radish peroxidase-conjugated secondary antibodies (goat antimouse or goat antirabbit depending on primary antibody) in TBST for 30 min at room temperature. The blots are washed four times in TBST (10 min/wash) followed by detection as per the kit protocol.

3.4. Quantitation of Covalent Complexes

1. Topoisomerase signals on the immunoblots may be quantified either by scanning the autoradiogram by densitometry or by analyzing the blots on a PhosphorImager. The signals are compared to the standard curve to determine the amount of topoisomerase (ng) in the DNA fractions (*see* **Note 8**).
2. DNA concentrations are determined by fluorometry or spectrophotometry.
3. Complex formation can be represented as ng of topoisomerase bound/µg DNA.

3.5. Results

Figure 3 shows a typical experiment with HeLa cells treated with the topo II inhibitor, VM26 (100 µ*M*). Topo II can be detected in the free protein fractions in both drug-treated and control experiments; however, only in the presence of VM26 is topo II detected in the DNA fractions.

Drug-induced topoisomerase–DNA covalent complexes dissociate on drug removal in vitro *(7,18)*. To determine if the ICE bioassay can detect reversibility of topoisomerase–DNA complexes, the following experiment was done. HeLa cells were treated with topotecan (a water-soluble analog of camptothecin) for 30 min and either lysed directly in sarkosyl or incubated in drug-free medium for 30 min prior to lysis. The results show that topo I–DNA covalent complexes are detected when cells are lysed immediately after drug treatment; however, complexes are reversed following a 30-min drug-free chase (**Fig. 4**).

4. Notes

1. The ICE bioassay can also be used with peripheral blood and cell suspensions prepared from tissue samples. If tissue samples are being used, fresh specimens should be obtained (frozen samples tend to give a lot of background signals in the blots and therefore should not be used). Tumor samples should be stored in saline

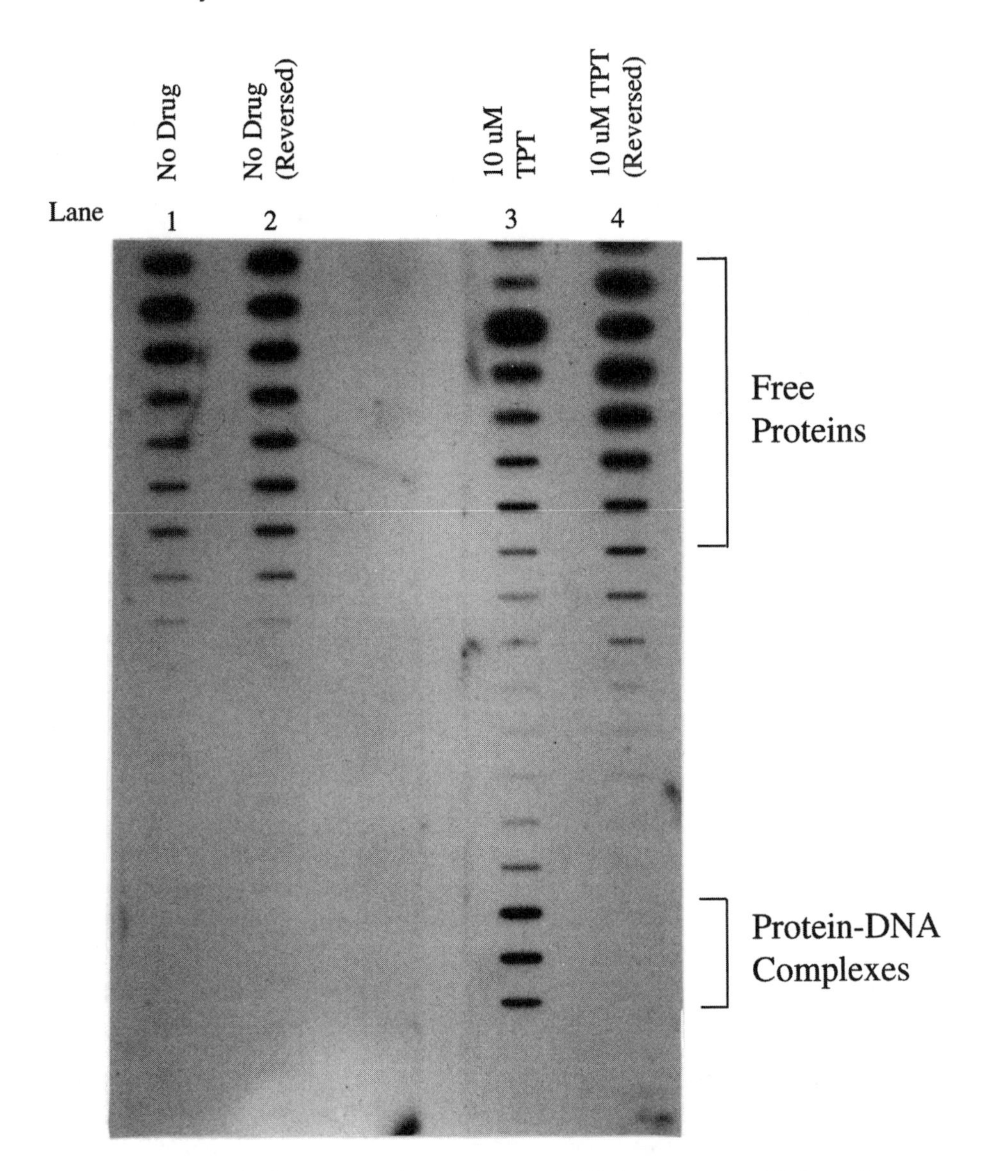

Fig. 4. Reversal of topotecan-induced topo I–DNA complexes. HeLa cells were treated with 10 μ*M* topotecan or no drug for 30 min at 37°C (two 100-mm Petri dishes/treatment). One set of plates from each treatment was then lysed with sarkosyl; the other set of plates was incubated in drug-free medium for 30 min at 37°C followed by sarkosyl lysis.

at 4°C prior to use. Once the cell suspensions are prepared, they should be treated with inhibitors as described in **Subheading 3.1.**

2. Sometimes, lysates may be very viscous. In this case, shear the lysates by drawing the sample into a syringe with an 18-gage needle two to three times or use brief sonication.
3. If fewer cells are used, smaller gradients can be prepared in SW50.1 polyallomer tubes (13 × 51 mm). Layer 1 mL of each cesium chloride step in the tubes and centrifuge the gradients in the smaller rotor.
4. The gradients must be prepared very carefully to ensure that the DNA forms a tight peak near the bottom of the gradient (usually three fractions). If gradients are not layered properly, the DNA peak may spread out and be more diffuse. This is magnified when fewer cells are used and the DNA yields are low.
5. The gradients should be carefully inspected before fractionation. The lysates may contain a large amount of debris, unlysed cells, membrane aggregates and so on. Large aggregates from the lysates may sediment near the DNA peak. In this case, a chaotrope, such as sodium thiocyanate (final concentration, 6 *M*), can be added to the lysate to aid in total lysis of the cells.
6. Viscosity in the DNA fractions may also interfere with application of the sample to the blot. To minimize this problem, reduce the sample volume or shear the DNA with an 18-gage needle and syringe prior to application.
7. Primary antibody incubations can be done for longer periods of time. Typically, blots are incubated in primary antibody overnight at room temperature. Additionally, diluted antibody solutions can be reused two to three times provided they are stored at –20°C in between experiments.
8. A negative control must always be run for each experiment. In the absence of inhibitors, there should be a low topoisomerase signal in the DNA peaks. The gradients have a built-in positive control, since topoisomerase signals should always be detected at the top of the gradient where free proteins are located. Failure to detect a signal at the top of the gradient indicates that there is a problem with the detection system.

References

1. Vosberg, H. P. (1985) DNA topoisomerases: Enzymes that control DNA conformation. *Curr. Top. Microbiol. Immunol.* **114,** 19–102.
2. Wang. J. C. (1985) DNA topoisomerases. *Annu. Rev. Biochem.* **54,** 665–697.
3. Osheroff, N. (1989) Biochemical basis for the interactions of type I and type II topoisomerases with DNA. *Pharmacol. Ther.* **41,** 223–241.
4. Trask, D. K., DiDonato, J. A., and Muller, M. T. (1984) Rapid detection and isolation of covalent DNA/protein complexes: application to topoisomerase I and II. *EMBO J.* **3,** 671–676.
5. Gromova, I. I., Kjeldsen, E., Svejstrup, J. Q., Alsner, J., Christiansen, K., and Westergaard, O. (1993) Characterization of an altered catalysis of a camptothecin-resistant eukaryotic topoisomerase I. *Nucleic Acids Res.* **21,** 593–600.

6. Morham, S. G. and Shuman, S. (1992) Covalent and noncovalent DNA binding by mutants of vaccinia DNA topoisomerase I. *J. Biol. Chem.* **267,** 15,984–15,992.
7. D'Arpa, P. and Liu, L. F. (1989) Topoisomerase-targeting antitumor drugs. *Biochem. Biophys. Acta.* **989,** 163–177.
8. Schneider, E., Hsiang, Y. H., and Liu, L. F. 1990. DNA topoisomerases as anticancer drug targets. *Pharmacology* **21,** 149–183.
9. Zwelling, L. A. (1985) DNA topoisomerase II as a target of antineoplastic drug therapy. *Cancer Metastasis Rev.* **4,** 263–276.
10. Glisson, B. S. and Ross, W. E. (1987) DNA topoisomerase II: A primer on the enzyme and its unique role as a multidrug target in cancer chemotherapy. *Pharmacol. Ther.* **32,** 89–106.
11. Liu, L. F. (1989) DNA topoisomerase poisons as antitumor drugs. *Annu. Rev. Biochem.* **58,** 351–375.
12. Covey, J. M., Jaxel, C., Kohn, K., and Pommier, Y. (1989) Protein-linked DNA strand breaks induced in mammalian cells by camptothecin, an inhibitor of topoisomerase I. *Cancer Res.* **49,** 5016–5022.
13. Hertzberg, R. P., Caranfa, M. J., Holden, K. G., Jakas, D. R., Gallagher, G., Mattern, M. R., Mong, S. M., Bartus, J. O., Johnson, R. K., and Kingsbury, W. D. (1989) Modification of the hydroxy lactone ring of camptothecin: inhibition of mammalian topoisomerase I and biological activity. *J. Med. Chem.* **32,** 715–720.
14. Burke, T. G., Mishra, W. K., Wani, M. C., and Wall, M. E. (1993) Lipid bilayer partitioning and stability of camptothecin drugs. *Biochemistry* **32,** 5352–5364.
15. Muller, M. T., Pfund, W. P., Mehta, V. B., and Trask, D. K. (1985) Eukaryotic type I topoisomerase is enriched in the nucleolus and catalytically active on ribosomal DNA. *EMBO J.* **4,** 1237–1243.
16. Trask, D. K. and Muller, M. T. (1988) Stabilization of type I topoisomerase-DNA covalent complexes by actinomycin D. *Proc. Natl. Acad. Sci. USA* **85,** 1417–1421.
17. Shaw, J. L., Blanco, J., and Mueller, G. C. (1974) A simple procedure for isolation of DNA, RNA, and protein fractions from cultured animal cells. *Anal. Biochem.* **65,** 125–131.
18. Subramanian, D., Kraut, E., Staubus, A., Young, D. C., and Muller, M. T. (1995) Analysis of topoisomerase I-DNA complexes in patients administered topotecan. *Cancer Res.* **55,** 2097–2103.

15

DNA-Unwinding Test Using Eukaryotic DNA Topoisomerase I

Linus L. Shen

1. Introduction

The DNA unwinding effect of a drug, derived from either DNA intercalation or groove binding, represents a warning signal for its possible link to side effects or toxicity. DNA unwinding results in a lengthening of the double helix. This increase in length can be detected by physicochemical (such as electric dichroism) or hydrodynamic methods (such as viscosity and sedimentation measurements) using short segments of linear DNA. These methods mostly are labor-intensive or lack sensitivity. Alternatively, the DNA-unwinding effect of a drug can be tested utilizing basic DNA topology manipulations conducted by DNA-modifying enzymes *(1)*. The assay uses an endonuclease-linearized plasmid DNA, which is then rejoined by the action of DNA ligase in the presence of a unwinding drug that causes a decrease in DNA total twist (the number of total base pairs divided by the number of base pairs making a complete double-helical turn). The end result of such a nicking–rejoining process in the presence of a DNA-unwinding agent is the formation of a relaxed underwound DNA molecule with a deficient linking number (the number of times one strand goes around the other in a duplex ring). The relaxed underwound DNA converts to supercoiled DNA on removal of the drug owing to the recovery of normal DNA twist, and this supercoiled DNA formation is thus an indication of the unwinding effect. To be described in this chapter is a simplified version of this method for determining the effect of DNA unwinding using eukaryotic topo I, which possesses both endonuclease and ligase functions, and does not require magnesium ions for activities. The usefulness of this method has been demonstrated with many DNA binding agents, including the antibacterial quinolones *(2,3)* and antitumor quinobenoxazines *(4)*.

From: *Methods in Molecular Biology, Vol. 95: DNA Topoisomerase Protocols, Part II: Enzymology and Drugs*
Edited by N. Osheroff and M.A. Bjornsti © Humana Press Inc., Totowa, NJ

The method of the unwinding test utilizing plasmid DNA and an eukaryotic type I DNA topoisomerase has been described by Pommier et al. *(5)*. As illustrated in **Fig. 1**, a native supercoiled or relaxed plasmid DNA is first incubated with the test compound prior to enzymatic relaxation by DNA topoisomerase I. An excess amount of the enzyme is added to the drug–DNA mixture to assure complete relaxation of the DNA molecule, which is under topological constrain imposed by the unwinding agent. Subsequent removal of the drug from DNA during electrophoresis or by phenol extraction causes the appearance of the supercoiled DNA band on an agarose gel. The experiments are performed in a drug concentration-dependent manner, and tested in two experimental sets employing supercoiled (set A) and relaxed (set B) DNA as the starting substrates. The samples are then run simultaneously on the same agarose gel. The strategy of using two forms of DNA substrates is for distinguishing between drug-dependent unwinding and topoisomerase I inhibition, which may lead to false-positive results (*see* **Note 1**). For a true unwinding effect, the end results of the test, indicated by topoisomer band shifting at critical unwinding concentrations should be identical between the two sets of experiments. Another advantage of the method of utilizing eukaryotic topoisomerase I is that the test may be carried out in the absence of magnesium ions, as the enzyme does not require divalent cations for activity (*see* **Note 2**).

2. Materials

2.1. Closed-Circular DNA and DNA Topoisomerase I

The closed-circular DNA can be any bacterial plasmid DNA with a proper size. ColE1 DNA has been routinely used in our laboratory and proven to be convenient owing to its ease of isolation with high yield as well as its proper size (6.6 kb) for easy detection of its topological conversions on agarose gel. It is recommended to use the SDS/lysozyme lysis procedure followed by cesium chloride density gradient purification procedure *(6)* for isolating plasmid DNA of good quality (*see* **Note 3**). The method should yield >90% supercoiled and <10% nicked DNA. The relaxed form of DNA can be prepared using calf-thymus DNA topoisomerase I (Life Technologies, Gibco/BRL, Gaithersburg, MD) using reaction conditions described by the manufacturer. Prepare DNA topoisomerase I working solution (*see* **Note 4**): 200 U of the enzyme in 80 µL of reaction buffer. Keep the enzyme solution on ice.

2.2. Drug Samples

Drug stock solutions (2 mg/mL) are prepared preferably using the reaction buffer, or using other solvents, such as DMSO, weak acid, or weak base (*see* **Note 5**). Prepare 20X drug working solutions in microfuge tubes by 1:1 serial

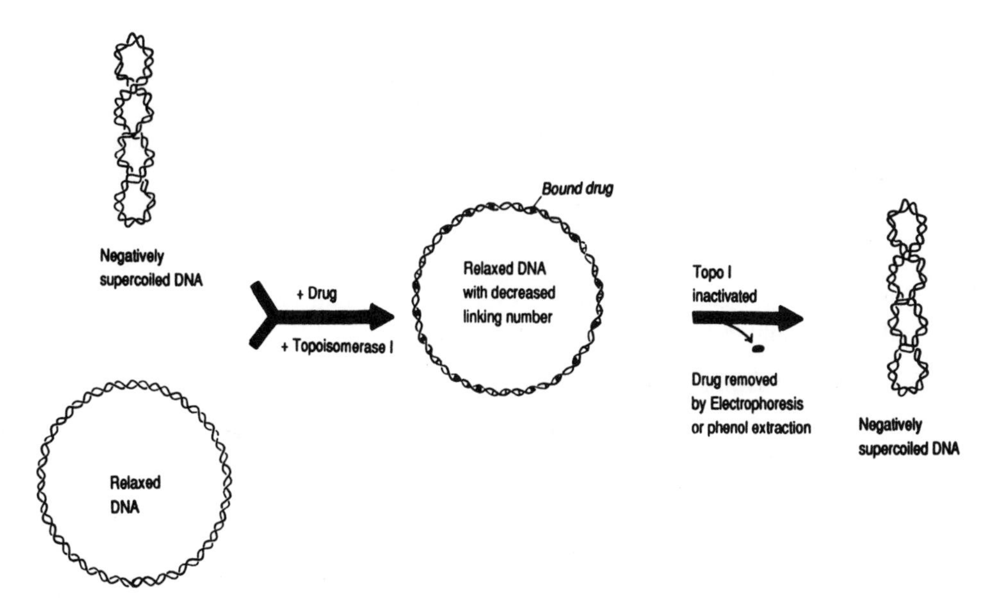

Fig. 1. Schematic presentation of the principle of DNA-unwinding test using eukaryotic DNA topoisomerase I and dual substrates.

dilutions using the same solvent. Typically, five such 1:1 dilutions from the stock, giving a total of six drug concentrations, will suffice to provide a suitable concentration range for the unwinding test. Prepare 20X positive control drug solutions, such as ethidium bromide, at 8 and 20 μg/mL.

2.3. Reaction Buffers and Agarose-Gel Electrophoresis Solution

1. Reaction buffer: 50 m*M* Tris-HCl, pH 7.5, 20 m*M* KCl, 1 m*M* Na-EDTA, 1 m*M* dithiothreitol, 30 μg/mL BSA, and $MgCl_2$ if needed (*see* **Note 3**).
2. 1X TBE buffer: 89 m*M* Tris, 89 m*M* boric acid, 2 m*M* EDTA, pH 8.3.
3. 5X gel-loading solution: 2% SDS, 14% ficoll, 0.1% bromphenol blue.

3. Methods

3.1. Unwinding Reactions

1. Label two sets of 1.5-mL capacity microfuge tubes: set A: 1–10, set B: 11–20 (*see* **Note 1**).
2. Prepare DNA solution in reaction buffer: mix 3 μg supercoiled and relaxed DNA in 180 μL buffer for set A and set B, respectively.
4. Dispense 15 μL of supercoiled and relaxed DNA to each of the 10 tubes for set A and set B, respectively.
5. Add 1 μL of 20X positive control (such as 8 and 20 μg/mL ethidium bromide) to

tubes 3 and 4 for set A, and to tubes 13 and 14 for set B. Add 1 μL 20X drug working solutions to each corresponding tubes (tube 5–10 and 15–20). For "no drug" control tubes (1, 2 for set A and 11, 12 for set B), add 1 μL of the blank solvent. Mix the contents in the tube by gentle tapping. Incubate for 10 min at room temperature.

6. Add 4 μL of topoisomerase I working solution to tubes 2–10 and 12–20. Add 4 μL reaction buffer to "no enzyme" control tubes 1 and 11. Mix with gentle tapping.
7. Incubate the reaction mixture tubes for 60 min at 37°C.
8. Optional (*see* **Note 6**): Extract the DNA with an equal volume of phenol/chloroform/isoamyl alcohol according to a standard procedure *(7)*. Ethanol-precipitate the DNA, and resuspend in 20 μL of 1X TBE.

3.2. Agarose-Gel Electrophoresis

1. Cast a horizontal 1% agarose gel in 1X TBE with 3–5 mm thickness.
2. Add 5 μL of gel loading solution to the DNA sample. Load samples to the agarose gel. Run the gel at low voltage (3 dc v/cm) for 16 h.
3. Stain the gel with 0.5 μg/mL ethidium bromide in 1X TBE for 20 min. Destain with distilled water for at least 30 min (longer destaining up to 3 h is allowed). Photograph the DNA bands on the gel visualized on a UV transilluminator using Polaroid type 665 positive/negative black-and-white instant pack film.

3.3. Analysis of Results

The DNA unwinding effect is normally expressed as the UC_{50}, the drug concentration at which the relaxed DNA band is shifted to the midpoint between the supercoiled and the relaxed DNA bands, or as the minimum unwinding concentration (MUC), the drug concentration at which the first sign of unwinding appears. Normally, a visual judgment is enough for a good estimation. For a precise quantitative determination, the relative amount of DNA bands can be determined by using laser densitometry.

3.4. Results and Examples

3.4.1. General Results

Figure 2 shows three possible outcomes of an unwinding experiment using the current method. The gel on the left in each panel shows results obtained

---

Fig. 2. *(see facing page)* Agarose gels demonstrating three possible cases for an unwinding test result. Each agent was tested in two sets of experiments using supercoiled (set A) and relaxed (set B) ColE1 plasmid DNA as the stating substrate. Panel I: the test compound, A-71974, shows negative DNA-unwinding results, and the catalytic activity of DNA topoisomerase I used in the test is not inhibited. The absence of inhibition is indicated by the complete conversion of the supercoiled DNA to the relaxed DNA even at the highest drug concentration shown in set A. Both gels (sets A and B) give identical results, and this validates the results. Panel II: the test compound, A-71460, inhibits DNA topoisomerase I at concentrations higher than 16 μg/mL.

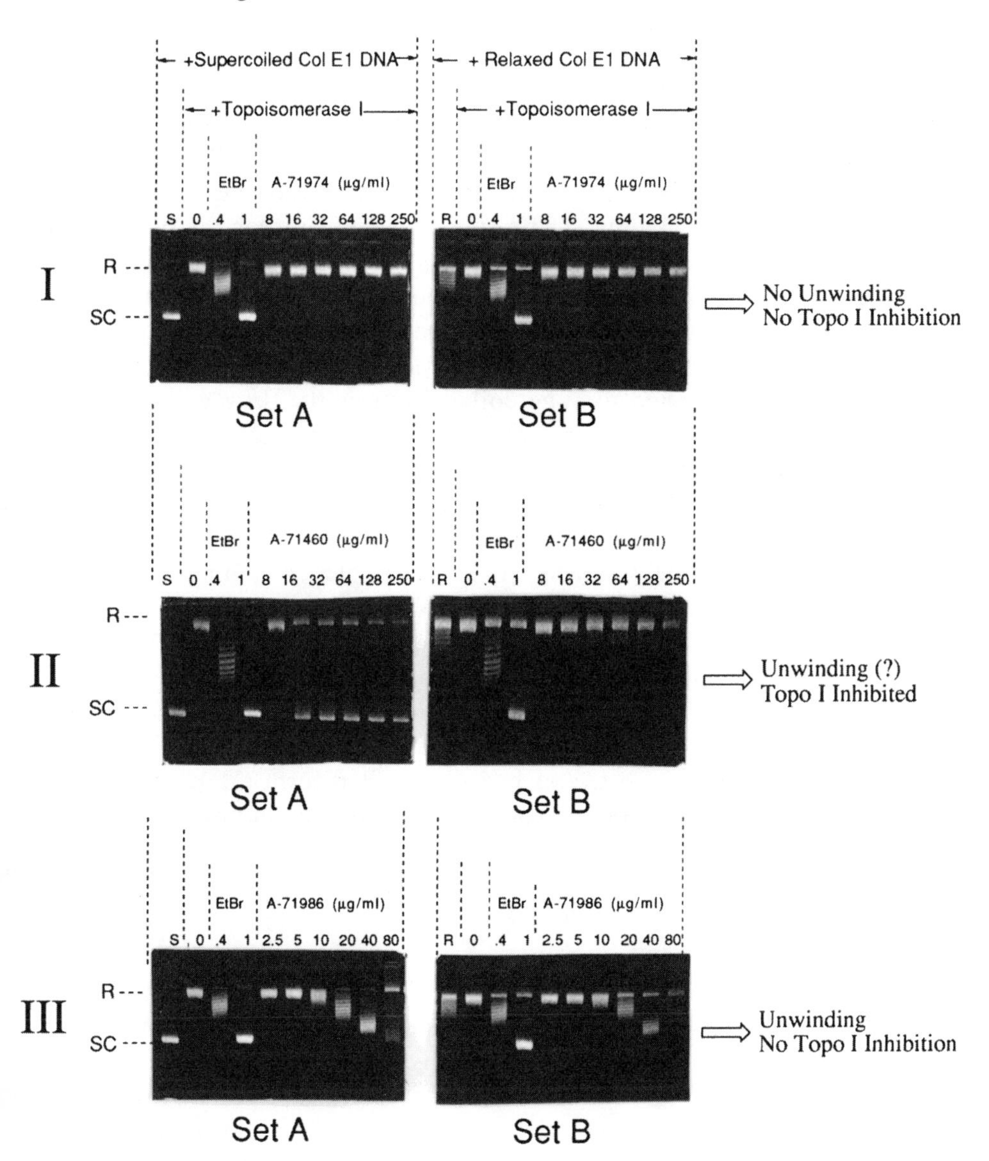

Therefore, the test results at inhibitory concentrations (>16 μg/mL) are invalid. In this case, set A and set B show different patterns. Panel III: the test compound, A-71986, shows a positive unwinding activity. Notice that both set A and set B show identical results. Ethidium bromide (EtBr) at final concentrations of 0.4 and 1 μg/mL was used in each gel and served as positive controls. Numbers on top of each lane are the final drug concentrations in μg/mL. S and R denote supercoiled DNA and relaxed DNA controls without adding DNA topoisomerase I and drug, respectively. On the left side of each panel, R and SC indicate the positions of relaxed and supercoiled DNA bands, respectively. *See text* for further explanations.

with supercoiled DNA substrate (set A), and the gel on the right shows the results when relaxed DNA is used (set B). Ethidium bromide at final concentrations of 0.4 and 1 µg/mL is used as a positive control for each group of experiments. As illustrated in **Fig. 1**, the formation of supercoiled DNA band after a sequential process of topoisomerase I treatment and drug removal is an indication of DNA unwinding. The results obtained between the two sets of experiments, either positive or negative, should be identical. Panels I and III in **Fig. 2** give examples showing such cases, and they represent a negative and a positive unwinding results, respectively. The result of a positive unwinding effect (panel III) is characterized with a unique distributive shifting of the topoisomers from relaxed to supercoiled form as the drug concentration increases. This distributive pattern in DNA band shifting can also be seen with ethidium bromide controls in all of the gels in **Fig. 2**. When results between set A and B are not identical, this indicates that the results of the unwinding test are invalid probably owing to an inhibition of topoisomerase I. Such an example is shown in panel II in which only the samples containing supercoiled DNA substrate (set A) show a dose-dependent shift from relaxed to supercoiled species as drug concentration increases. This is a predictable pattern when a drug inhibits the enzymatic activity of DNA topoisomerase I, as both starting substrates remain at their original positions at the inhibitory concentrations (>16 µg/mL for A-71460 shown in set A of panel II). In addition, the shifting from relaxed (at 8 µg/mL) to supercoiled form (at 16 µg/mL) is processive unlike the distributive pattern observed with an unwinding agent (**Fig. 2**, panel III).

3.4.2. Example 1: Magnesium-Dependent DNA Unwinding of Quinolones

Quinolone antibacterials are known to be DNA-targeted agents as demonstrated by a radioligand binding method using pure DNA ***(8)***. Using a rather complex gel electrophoretic method, norfloxacin and nalidixic acid have been shown to be capable of unwinding the DNA in the presence of magnesium ***(9)***. Utilizing the current simplified method, such unwinding effects are demonstrated in **Fig. 3**, which shows the results of three selected quinolones tested in the absence and presence of magnesium ions. Oxolinic acid, a weak gyrase inhibitor with IC_{50} (drug concentration causing 50% inhibition of the enzyme activity) of about 3 µg/mL ***(10)***, does not unwind DNA even at the highest test concentration of 400 µg/mL. Norfloxacin, a more potent gyrase inhibitor with IC_{50} equal to about 1 µg/mL ***(10)***, shows a weak magnesium-dependent unwinding effect with UC_{50} at about 50 µg/mL in the presence of magnesium ions and at about 200 µg/mL in the absence of added magnesium ions. The

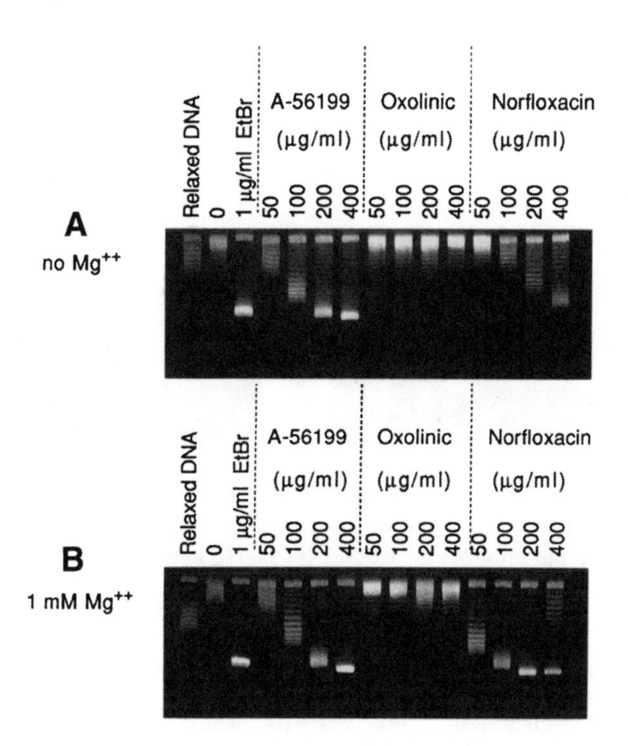

Fig. 3. Result of DNA-unwinding test of three selected quinolone analogs (A-56199, oxolinic acid, and norfloxacin) in the absence (**A**) and presence (**B**) of 1 m*M* $MgCl_2$. Notice that only relaxed DNA substrate is used in this test. Structures of these compounds can be found in **ref.** *(8)*.

unwinding effect is apparently not totally owing to the magnesium chelating to the 3-carboxyl and 4-keto groups on the quinolone ring, since the ethyl-ester derivative of norfloxacin (A-56199), which lacks a chelating effect, maintains its unwinding potency with a similar UC_{50} equal to about 100 μg/mL regardless of the presence or absence of magnesium ions, whereas the antigyrase activity drops more than two orders of magnitude with IC_{50} equal to 250 μg/mL *(8)*. The lack of correlation between the IC_{50} and UC_{50} values of these compounds does not support the notion that the antigyrase activity is derived from a DNA-unwinding effect. The magnesium-dependent unwinding effect of norfloxacin can be antagonized by polyamines (**Fig. 4**) or histones as demonstrated in previous publications *(2,3)*. The antagonizing effect is not owing to inhibition of topoisomerase I by polyamines, since the antagonizing effect was not observed using the same test with other DNA-unwinding agents, such as the anticancer quinobenoxazines *(4)*.

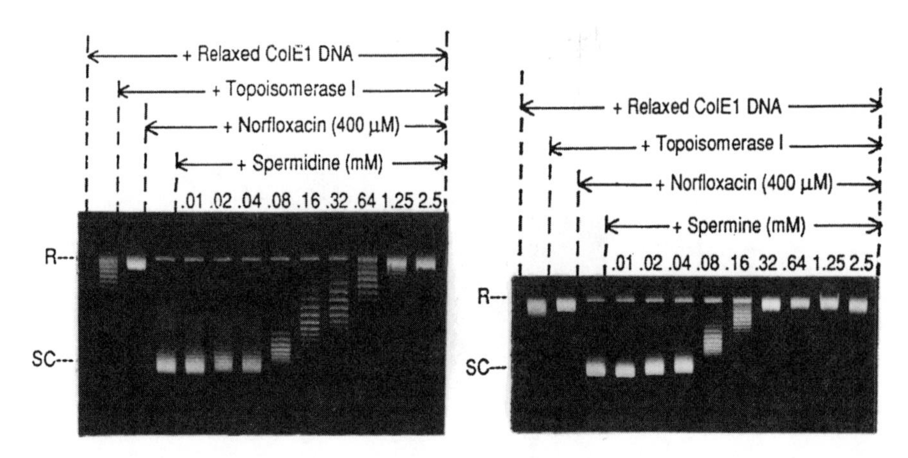

Fig. 4. The antagonizing effect of spermidine (right panel) and spermine (left panel) on norfloxacin-dependent DNA unwinding. Assays are performed with relaxed ColE1 DNA substrate in the presence of 1 m*M* $MgCl_2$. Numbers on top of each lane are the concentrations of the added polyamine.

3.4.3. Example 2: DNA-Unwinding Effect of Some Antibiotics

Chartrusin and elsamicin A are two structurally related antitumor antibiotics with antitumor potency about 10–30 times stronger for the latter compound ***(11)***. Elsamicin A was reported to inhibit a number of enzymes, including helicases ***(12)*** and human topoisomerase II ***(13)***. It also has been shown to inhibit T7 RNA polymerase ***(14)*** and to cleave directly DNA in the presence of ferrous iron and reducing agent ***(15)***. Chartreusin was also reported to inhibit human topoisomerase II ***(13)*** and *Alcaligenes* sp. topoisomerase I ***(16)***. Such a nonspecific inhibition pattern of these agents may be derived from their DNA-unwinding effect, which is shown in **Fig. 5**. The unwinding effect of elsamicin A is shown to be fivefold stronger than chartreusin, with UC_{50} values of about 0.6 and 3 µg/mL, respectively.

3.4.4. Example 3: DNA Unwinding by the Quinobenoxazines

Quinobenoxazines, a class of tetracyclic quinolone derivatives (**Fig. 6** for structures), are potent antitumor agents ***(17)*** that act as catalytic inhibitors of mammalian topoisomerase II ***(4)***. Results shown in **Fig. 7** demonstrate that quinobenoxazines possess DNA unwinding activity, which can be correlated with their antineoplastic and DNA topoisomerase II inhibitory activities (**Table 1**). The unwinding effect may be maximized in the presence of 10 m*M* magnesium ions, but contrary to the antibacterial quinolones, the effect of quinobenoxazines cannot be antagonized by polyamines (data not shown).

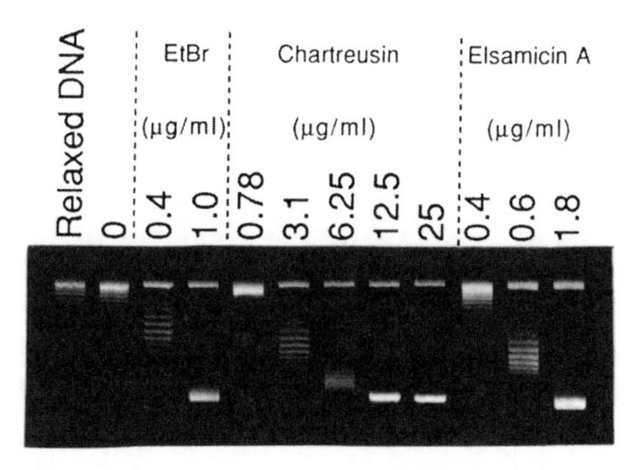

Fig. 5. Results of DNA-unwinding test of chartreusin (A) and elsamicin A (B) using relaxed DNA substrate. Ethidium bromide (EtBr) is used as a control.

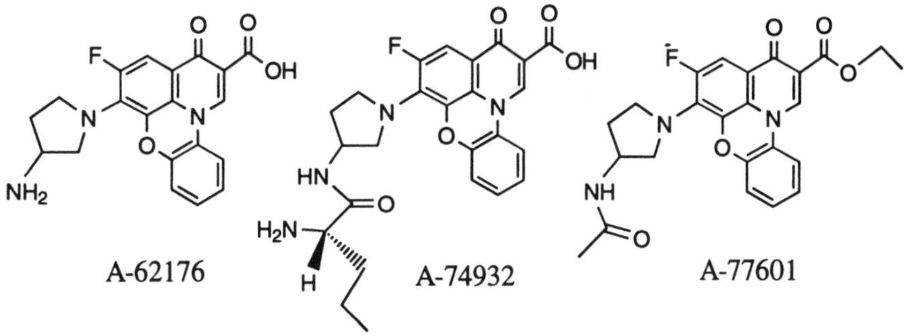

A-62176 A-74932 A-77601

Fig. 6. Chemical structures of A-62176, A-74932 and A-77601.

4. Notes

1. It should be emphasized that the unwinding assay is valid only if the topoisomerase I is not inhibited by the drug. It is therefore desirable that both relaxed and supercoiled DNA substrates are used, such as those experiments shown in **Fig. 2**, for distinguishing an unwinding effect from topoisomerase I inhibition. When either a negative or a positive unwinding effect needs to be determined, a single-substrate test is sufficient. To assess and demonstrate the positive unwinding activity of a compound, relaxed DNA should be used as the substrate. The test will correctly identify an unwinding effect, since the unwinding signal (supercoiled DNA formation) is impossible to be generated from relaxed DNA substrate if the enzyme is inhibited. By the same principle, if a negative unwinding activity needs to be demonstrated, a test using only the

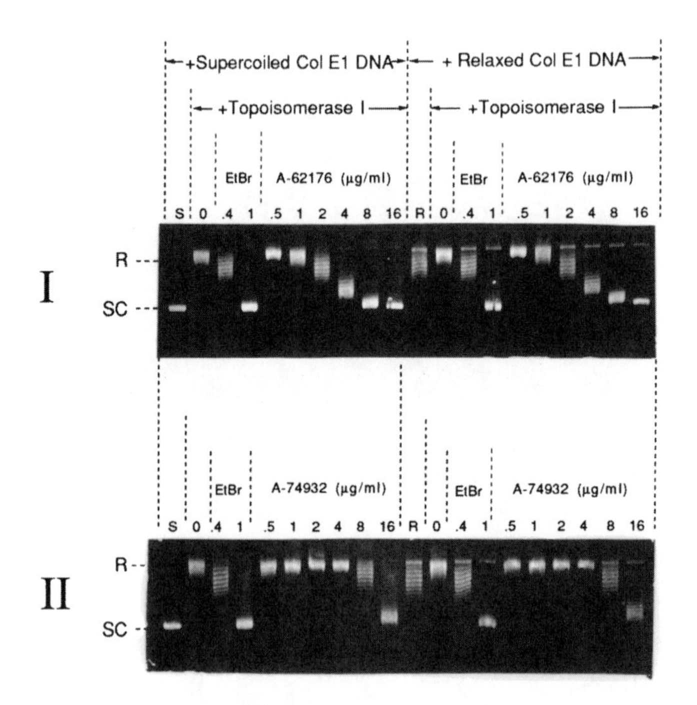

Fig. 7. Test of DNA unwinding by quinobenoxazines using both supercoiled (left side of the gel) and relaxed (right side of the gel) DNA substrates. Results of A-62176 and A-74932 are shown in panels I and II, respectively. Experiments were performed in the absence of added magnesium ions.

supercoiled DNA substrate will suffice. In this test, the negative results (showing only the relaxed DNA band) cannot be an artifact owing to topoisomerase I inhibition, since such a negative unwinding signal (relaxed DNA formation) cannot be produced from supercoiled DNA substrate if topoisomerase I is inhibited. For this reason, the positive DNA unwinding activities shown in **Figs. 3, 4,** and **5** cannot be an artifact owing to topoisomerase I inhibition.

2. The independence of eukaryotic DNA topoisomerase I activity on metal ions facilitates the use of this approach to investigate the metal ion effect on DNA unwinding, such as that demonstrated by quinolones (*see* **Subheading 3.4.2.**).
3. The plasmid DNA isolated by commercial column purification kits has been shown to be unsuitable for this type of assay.
4. In most cases, 10 relaxation units of topoisomerase I per reaction are sufficient for the test. If enzyme availability and purity are not limited; 20 or more units can be used to minimize any possible topoisomerase inhibition by the test compound.
5. The maximum concentration of DMSO allowed for the test without interfering with the test result is about 5% in the final reaction mixture. For using acid or

Table 1
Cytotoxicity, Decatenation Inhibition, and DNA Unwinding for the Quinobenoxazines

	Quinobenoxazines		
	A-62176	A-74932	A-77601
MUC			
Without added $MgCl_2$	2	8	>128
With 10 m*M* $MgCl_2$	0.5	3	128
Decatenation inhibition[b]	0.45	2.0	>100
HT29 (human colon cancer cell line), IC_{50}	0.1[a]	0.2[a]	1.4
A546 (human breast cancer cell line), IC_{50}	0.26[a]	0.6[a]	2
P-388 (leukemia cell line), IC_{50}	0.02[a]	0.04[a]	0.5

Unit = µg/mL; MUC = minimum unwinding concentration.
[a]From ref. *(17)*.
[b]Drug concentration that inhibits 50% of the conversion of catenated to decatenated k-DNA catalyzed by calf-thymus topoisomerase II.

base, suspend drug in distilled water to give a concentration near 2 mg/mL, and titrate the suspension with small aliquots of 0.1 *M* KOH or HCl until the compound is dissolved. Add enough water to make a 2 mg/mL final concentration.

6. Overnight electrophoresis normally is sufficient to remove the drug from DNA. This is true even for some strong DNA intercalators, such as ethidium bromide, adriamycin, or *m*-AMSA.

Acknowledgment

The author would like to thank Claude Lerner for his critical reading of the manuscript.

References

1. Wang, J. C. (1982) DNA topoisomerases. *Sci. Am.* **247,** 94–109.
2. Shen, L. L., Baranowski, J., Wai, T., Chu, D. T. W., and Pernet, A. G. (1989) The binding of quinolones to DNA: should we worry about it? in *International Telesymposium on Quinolones.* (Fernandes, P. B., ed.), J. R. Prous Science Publishers, Barcelona, Spain, pp. 159–170.
3. Shen, L., Baranowski, J., Chu, D. T. W., and Pernet, A. G. (1989) DNA unwinding properties of new quinolones—the antagonizing effect of polyamines. Abstract 29. *Abstracts of the ICAAC.* Houston, TX.
4. Permana, P. A., Snapka, R. M., Shen, L. L., Chu, D. T. W., Clement, J. J., and Plattner, J. J. (1994) Quinobenoxazines: a class of novel antitumor quinolones and potent mammalian DNA topoisomerase II catalytic inhibitors. *Biochemistry* **33,** 11,333–11,339.

5. Pommier, Y., Covey, J. M., Kerrigan, D., Markovits, J., and Pham, R. (1987) DNA unwinding and inhibition of mouse leukemia L1210 DNA topoisomerase I by intercalation. *Nucleic Acid Res.* **15,** 6713–6731.
6. Sambrook, J., Fritsch, E. F., and Maniatis, T. (1989) *Molecular Cloning. A Laboratory Manual,* 2nd ed., Cold Spring Harbor Laboratory Press, Cold Spring Harbor, NY, p.1.36–1.46.
7. Sambrook, J., Fritsch, E. F., and Maniatis, T. (1989) *Molecular Cloning. A Laboratory Manual,* 2nd ed., Cold Spring Harbor Laboratory Press, Cold Spring Harbor, NY, p.E.3–E.4.
8. Shen, L. L. and Pernet, A. G. (1985) Mechanism of inhibition of DNA gyrase by analogues of nalidixic acid: the target of the drugs is DNA. *Proc. Natl. Acad. Sci., USA* **82,** 307–311.
9. Tornaletti, S. and Pedrini, A. M. (1988) Studies on the interaction of 4-quinolones with DNA by DNA unwinding experiments. *Biochim. Biophys. Acta.* **949,** 279–287.
10. Shen, L. L., Mitscher, L. A., Sharma, P. N., O'Donnell, T. J., Chu, D. W. T., Cooper, C. S., Rosen, T., and Pernet, A. G. (1989) Mechanism of inhibition of DNA gyrase by quinolone antibacterials. A cooperative drug-DNA binding model. *Biochemistry* **28,** 3886–3894.
11. Konishi, M., Sugawara, K., Kofu, F., Nishiyama, Y., Tomita, K., Miyaki, T., and Kawaguchi, H. (1986) Elsamicins, new antitumor antibiotics related to chartreusin. I. Production, isolation, characterization and antitumor activity. *J. Antibiot. (Tokyo)* **39,** 784–791.
12. Bachur, N. R., Johnson, R., Yu, F., Hickey, R., Applegren, N., and Malkas, L. (1993) Antihelicase action of DNA-binding anticancer agents: relationship to guanosine-cytidine intercalator binding. *Mol. Pharmacol.* **44,** 1064–1069.
13. Lorico, A. and Long, B. H. (1993) Biochemical characterisation of elsamicin and other coumarin-related antitumour agents as potent inhibitors of human topoisomerase II. *Eur. J. Cancer* **29a,** 1985–1991.
14. Portugal, J. (1995) Abortive transcription of the T7 promoter induced by elsamicin A. *Anticancer Drug Des.* **10,** 427–38.
15. Parraga, A., Orozco, M., and Portugal, J. (1992) Experimental and modelling studies on the DNA cleavage by elsamicin A. *Eur. J. Biochem.* **208,** 227–233.
16. Yoshida, T., Habuka, N., Takeuchi, M., and Ichishima, E. (1986) Inhibition of DNA topoisomerase I from Alcaligenes Sp. by chartreusin. *Agricultural and Biol. Chem.* **50,** 515,516.
17. Chu, D. T. W., Hallas, R., Clement, J. J., Alder, J., McDonald, E., and Plattner, J. J. (1992) Synthesis and antitumor activities of quinolone antineoplastic agents. *Drugs Exp. Clin. Res.* **18,** 275–282.

16

Drug–DNA Interactions

David E. Graves

1. Introduction

The interactions of small molecules with nucleic acids have provoked considerable interest in the field of antitumor drug design over the past three decades; however, critical information linking the physical–chemical properties associated with these complexes with their biological effectiveness remains unclear. Significant progress has been made toward unraveling the structural and dynamic properties of many ligand–DNA complexes, which has provided pivotal insight into the design and development of more effective second-generation chemotherapeutic agents for the successful treatments of many types of cancer (*see* **ref. *1*** and references therein). Over a decade ago, a primary cellular target for many of these agents was identified to be topoisomerase II through formation of a ternary complex among the ligand, DNA, and enzyme (*see* reviews in **refs. *2–5***). Interestingly, DNA binding affinity did not directly correlate with antitumor activity; however, modulation of topoisomerase II activity was shown to coincide with biological activity of these agents ***(6)***. Interactions of ligands with DNA are being explored using a variety of physical and biochemical methods in an effort to determine the chemical and physical basis of novel binding phenomena, such as DNA base sequence selectivity, correlation of structure–activity relationships dictating the geometry and thermodynamics of drug–DNA complexes, the influences of substituent modifications on the drug–DNA complex, and the correlation between these chemical and physical properties with the compounds' effectiveness in eliciting topoisomerase II inhibition. The methods described herein provide several basic approaches used in examining drug–DNA interactions.

From: *Methods in Molecular Biology, Vol. 95: DNA Topoisomerase Protocols, Part II: Enzymology and Drugs*
Edited by N. Osheroff and M.A. Bjornsti © Humana Press Inc., Totowa, NJ

1.1. Mechanism of Drug Activity

The molecular mechanisms responsible for eliciting the biological actions for most antitumor agents remain unknown despite extensive research efforts. The correlation between the capacity for many of these drugs to bind DNA and resulting biological activities has been well documented. However, the exact molecular mechanism responsible for chemotherapeutic activities of these agents remains unknown. The ability to interact with DNA and subsequently form a ternary complex between the drug–DNA and enzyme is thought to be a key determinant in the inhibition of topoisomerase II by antitumor agents, such as *m*-AMSA, adriamycin, daunorubicin, ellipticine, mitoxantrone, and actinomycin D *(7–12)*. Several biophysical methods are available for examining the interactions of ligands with nucleic acids. A brief overview of several of these methods is provided below.

2. Materials

2.1. Buffer Preparation

Experimental methods used to obtain accurate binding isotherms of ligands with DNA require stable pH and ionic conditions. Generally, phosphate buffers such as sodium phosphate (0.01 *M*), are used to maintain the pH of the solution to approx 7.0. Additionally, 0.001 *M* disodium EDTA is added to the buffer to chelate metal ions and act as a bactericide. Sodium chloride can be added to the buffer solution as needed with 0.1–0.2 *M* being used to mimic physiological conditions. Doubly distilled or Milli-Q ultrapure water is used to dissolve all buffer components. Prior to use in binding studies, all buffer solutions are degassed and filtered through a 0.45-μ filter.

2.2. DNA Preparation

For determination of binding isotherms to native DNAs, large genomic DNAs from calf thymus or salmon sperm are readily available from commercial sources. The DNAs are dissolved in sodium phosphate buffer (0.01 *M*), pH 7.0, containing 0.001 *M* disodium EDTA and 0.01 *M* sodium chloride at a concentration of 2 g/mL. The DNA solution is extensively sheared by sonication for a total of 30 min at 4°C. Following sonication, the sodium concentration is raised to 200 m*M* by addition of NaCl and the solution treated with RNase at 0.5 mg/mL for 30 min at 37°C. Proteinase K at 0.5 mg/mL is added and incubated at 37° for 1 h. The DNA solution is extracted three times with an equal volume of chloroform:phenol (1:1 v/v). The DNA solution is then extracted three times with water-saturated ether, followed by extensive dialysis against 0.01 *M* sodium phosphate buffer (pH 7.0), 0.001 *M* disodium EDTA,

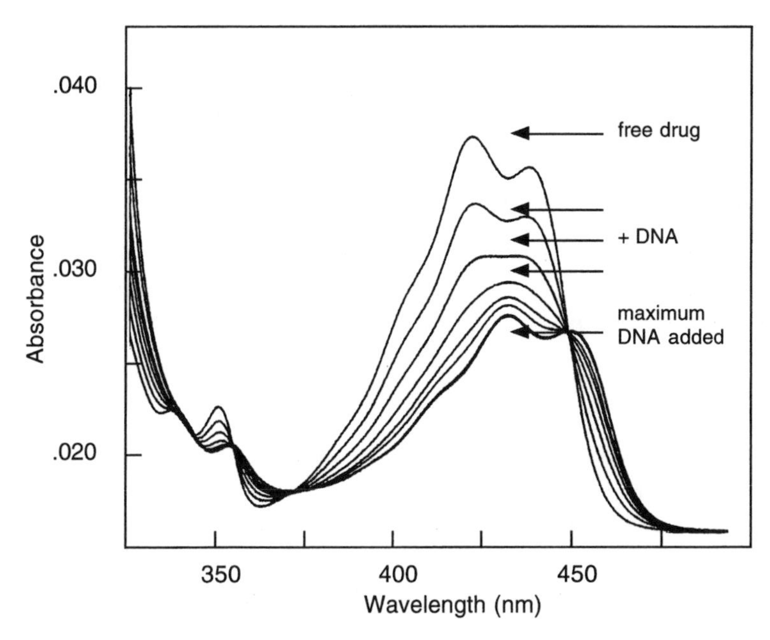

Fig. 1. Typical composite visible-absorbance spectra demonstrating the bathochromic and hypochromic shifts incurred on intercalation of the DNA binding agent, acridine-4-carboxamide to calf thymus DNA in 0.01 *M* sodium phosphate buffer, pH 7.0, 0.001 *M* disodium EDTA, 0.1 *M* NaCl. The drug concentration was maintained at 1.5 μ*M* whereas the DNA concentration varied from 0 to 100 μ*M* (basepairs). Temperature was maintained at 20°C.

0.1 *M* sodium chloride *(13)*. DNA concentrations are determined by UV absorbance at 260 nm using a molar absorptivity of 13,200 M^{-1}/cm in units of DNA bp.

3. Methods

3.1. Spectral Properties of DNA Binding Ligands

Ligands that bind DNA via intercalation are generally aromatic planar heterocycles that exhibit strong absorption bands in the visible region of the electromagnetic spectrum. Intercalation of the planar ligand chromophore between adjacent DNA basepairs induces both hypochromic and bathochromic shifts in the absorption spectrum of these drugs. This hypochromic effect can be utilized to probe the equilibrium binding properties of a drug with DNA by assuming a two-state model in which there exists two absorbing species (free and bound drug) with each state having a characteristic molar absorptivity. Composite absorbance spectra are shown in **Fig. 1** illustrating typical absorption properties of the ligand on binding. As observed in this figure, addition of DNA results in both bathochromic and hypochromic shifts on complex formation. Changes in the absorbance can be used to quantitate the association constant, K_a, and binding site size, n, from these data.

3.1.1. Absorbance Titration

The equation relating the change in absorbance to the concentration of bound drug is described as follows. In accordance with the Beer-Lambert Law, the total drug absorbance is the sum of the absorbances of the free and bound drug species:

$$A = \varepsilon_f C_f + \varepsilon_b C_b \quad (1)$$

where A is the total (observed) absorbance (assuming unit pathlength), C_f is the free drug concentration (M), and C_b is the bound drug concentration (M). The total drug concentration is the sum of the free and bound drug concentrations, $C_t = C_f + C_b$, where C_t is the total drug concentration. Solving for C_f and substituting into the first equation yields the following equations:

$$A = \varepsilon_f (C_t - C_b) + \varepsilon_b C_b \quad (2)$$

$$A = \varepsilon_f C_t - \varepsilon_f C_b + \varepsilon_b C_b \quad (3)$$

$$A = \varepsilon_f C_t + (\varepsilon_b - \varepsilon_f) C_b \quad (4)$$

Since the expected drug absorbance ($A_{expected}$) in the absence of DNA is given by $A_{expected} = \varepsilon_f C_t$, and the difference in molar absorptivities of the free and bound drug species can be represented by $\Delta\varepsilon = \varepsilon_b - \varepsilon_f$, the former equation can be simplified to $A = A_{expected} + \Delta\varepsilon C_b$. Solving for C_b yields the equation:

$$\frac{A - A_{expected}}{\Delta\varepsilon} = C_b \text{ or } \frac{\Delta A}{\Delta\varepsilon} \quad (5)$$

by which the concentration of bound drug can be determined from the change in absorbance (ΔA) ***(14,15)***.

Titration experiments are performed on a research-grade, double-beam, UV-visible spectrophotometer equipped with a constant temperature refrigerated circulating bath. In order to increase sensitivity, cylindrical cuvets of 5- to 10-cm pathlengths are used. DNA solution of known volume and concentration is added to both the reference and sample cuvets and baseline recorded for the visible region of the drug spectrum. Over the course of the titration, small amounts of a known stock drug solution are added to the DNA solution and the absorbance (A_{obs}) observed for each addition.

The amount of free drug is determined by the difference:

$$C_f = C_t - C_b \tag{6}$$

The extinction coefficient of the bound drug is determined by titration of a known concentration of drug with known amounts of DNA until no further change in absorbance of the drug is observed (reverse titration). Extrapolation to the abscissa of a plot of the drug's absorbance vs 1/[DNA] provides a close approximation of ε_b.

3.1.2. Fluorescence Titration

Fluorescence of the drug chromophore can be used to monitor DNA binding in a manner similar to that described for visible spectroscopy. Once appropriate conditions have been determined (i.e., slit width, λ_{ex}, excitation wavelength and λ_{em} [emission wavelength]), the forward titration is performed as described above. The ratio of the fluorescence intensity of the drug in the absence of DNA (I_f) and presence of DNA (I_{obs}) is used to calculate the amount of bound drug (C_b) according to the equation:

$$C_b = \frac{I_{obs} - I_f}{I_f(V - 1)} \; (C_t) \tag{7}$$

where V is the ratio of I_{obs}/I_f obtained by titrating the drug with a known quantity of DNA until no further fluorescence change is observed ***(16–18)***. C_f is determined as the difference between C_t and C_b.

3.1.3. Ethidium Displacement Assay

The fluorescence properties of ethidium and the ethidium–DNA complex provides a unique probe for determining the binding constants of drugs with nucleic acids. The intercalative complex formation by ethidium binding to DNA results in a marked enhancement of fluorescence ($V = \sim 21$), resulting in excellent sensitivity for examining drug–DNA interactions. This method, which is essentially a competition binding assay, determines the drop in ethidium fluorescence on addition of a second drug. The ethidium displacement assay provides a rapid and reproducible method for determining the binding affinity of the second drug. In order to use this method, the drug must neither absorb at the excitation or emission wavelength of ethidium (546 and 595 nm, respectively). The concentration of drug that results in a 50% drop in ethidium fluorescence (i.e., displace 50% of the ethidium) is calculated and used to determine the binding constant. If one assumes that all DNA sites are equivalent for both the target drug and ethidium and the solution is at equilibrium, then the following expressions can be used:

$$K_e = \frac{C_{\mathrm{DNA}}}{E_f} \tag{8}$$

where K_e is the equilibrium binding constant for ethidium, C_{DNA} is the concentration of DNA (basepairs), and E_f is the free ethidium concentration. Similarly, the binding constant for the added drug can be determined by the equation:

$$K_d = \frac{D_b}{D_f C_{\mathrm{DNA}}} \tag{9}$$

where D_b is the concentration of bound drug, C_{DNA} is the DNA concentration in basepairs, and D_f is the free drug concentration. From Eqs. x and y, it follows that:

$$K_d = \frac{D_b E_f K_e}{E_b D_f} \tag{10}$$

By measuring the degree of displacement of ethidium, the values of E_f and D_f can be calculated, and utilizing the DNA–ethidium association constant, K_d can be computed. From **Eq. 10**, C_{50} (concentration of drug required to displace 50% of the ethidium) should be inversely related to the drug–DNA association constant ***(19)***.

The spectroscopic methods described above provide means for obtaining accurate binding isotherms for ligand–DNA interactions. Variations in types of DNAs, temperature, ionic strengths, and pH can be used to probe biophysical properties associated with complex formation. However, these methods do have limitations based on their dependence of the spectral properties of the compounds and magnitude of binding constants. Reliable binding isotherms can be most effectively obtained at ligand concentrations near to the inverse of association constant. Extremely high or low binding constants introduce problems owing to the inability to determine accurately the bound drug concentration and free drug concentrations, respectively. In the procedures listed above, C_f is determined as the difference between C_t minus C_b, thus providing an indirect determination of this value. An alternative method for examining drug–DNA interactions would be to measure directly both C_b and C_f. This can be achieved using equilibrium dialysis.

3.1.4. Equilibrium Dialysis

Dialysis cells are available from various vendors that have precision machined cells on either side of the dialysis membrane and access ports to each side of the dialysis cell. Generally, the equilibrium dialysis experiment is set up such that DNA is contained on one side of the cell, separated by a dialysis membrane from buffer on the other side of the cell. Drug is introduced into either side of the dialysis cell through the access port. Equilibrium of the cells is achieved in a controlled-temperature environment through gentle shaking or rocking over a 24-h period. During this time, the drug is equilibrated between the two cells. Aliquots are taken from both cells and measured by UV-visible spectroscopy, fluorescence, and/or liquid scintillation counting. On the side containing the DNA, bound and free drug species will be in equilibrium, and the free drug concentration will be in equilibrium with the free drug on the other side of the dialysis membrane. Thus, C_f can be measured directly through UV-visible or fluorescence spectroscopy with reference to a standard curve prepared drug concentrations of known concentration. On the side containing the DNA, both bound and free drug concentrations are present. C_{total} (C_b plus C_f) is determined by addition of dimethylsulfoxide (DMSO) to a 1:1 (v/v) solution, which effectively dissociates all of the bound drug species from the DNA. The amount of drug is measured by UV-visible or fluorescence spectroscopy again with reference to a standard curve prepared drug concentrations of known concentration in 50% DMSO. The bound drug concentration, C_b, is determined from C_t minus C_f. Thus, in this method allows for direct measurement of C_f, but in this case, C_b must be estimated as the difference between C_t and C_f.

3.2. Determination of Equilibrium Binding Parameters

Binding isotherms obtained by the methods described above can be illustrated in a number of ways. Most common are the direct plot (r vs C_f) and scatchard plot (r/C_f vs r). Typical examples of these types of plots are shown in **Fig. 2**. Theoretical curves describe the DNA binding isotherms using the neighbor exclusion model ***(20,21)*** and a nonlinear least-squares analysis as well as the Levenberg-Marquardt algorithm of the noncooperative neighbor exclusion equation:

$$\frac{r}{C_f} = \mathrm{K}(1 - nr)\left[\frac{1 - nr}{1 - (n - 1)r}\right]^{n-1} \tag{11}$$

where r is the molar ratio of bound drug to basepair, K is the intrinsic binding constant (M^{-1}), and n is the site exclusion size (basepairs). This equation is based on a ligand binding analysis method originally developed by Scatchard, but is statistically modified to account for the occlusion of adjacent binding sites by a ligand. In this fashion, DNA binding affinities and binding site sizes were obtained for each compound of this study.

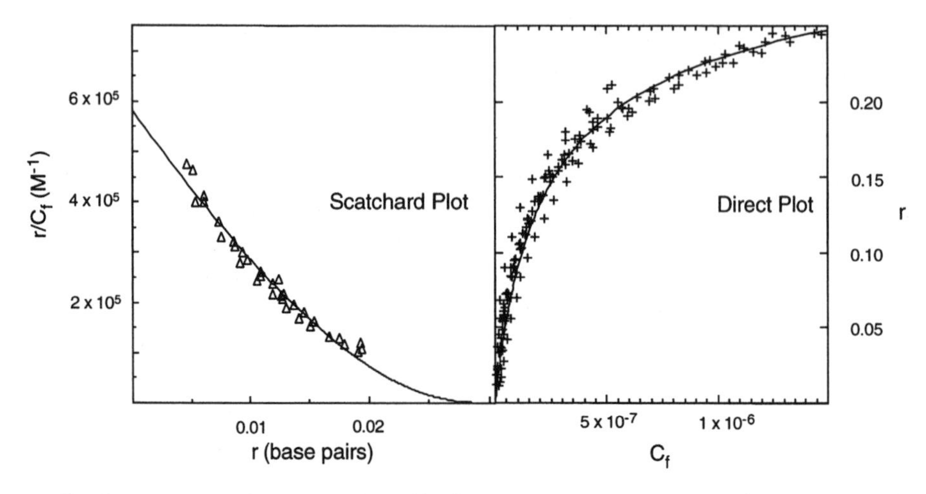

Fig. 2. Representative DNA–drug binding isotherms in the form of a scatchard plot (left panel) and a direct plot (right panel). Theoretical fits (solid lines) based on the neighbor exclusion model provide estimated values for the intrinsic binding constant (K_{obs}) and binding site size, *n* (drugs bound per basepair).

References

1. Waring, M. (1981) Inhibitors of nucleic acid synthesis, in the *Molecular Basis of Antibiotic Action*, 2nd ed. (Gale, E. F., ed.), Wiley, London, pp. 274–341.
2. Liu, L. F. (1989) DNA topoisomerase poisons as antitumor drugs. *Ann. Rev. Biochem.* **58,** 351–375.
3. Drlica, K. and Franco, R. J. (1988) Inhibitors of DNA topoisomerases. *Biochemistry* 2253–2259.
4. Froelich-Ammon, S. J. and Osheroff, N. (1995) Topoisomerase poisons: Harnessing the dark side of enzyme mechanism, *J. Biol. Chem.* **270,** 21,429–21,432.
5. Ross, W. E. (1985) DNA topoisomerases as targets for cancer therapy. *Biochem. Pharmacol.* **34,** 4191–4195.
6. Pommier, Y., Schwartz, R. E., Zwelling, L. A., and Kohn, K. (1985) Effects of DNA intercalating agents on topoisomerase II induced DNA strand cleavage in isolated mammalian cell nuclei. *Biochemistry* **24,** 6406–6410.
7. Nelson, E. M., Tewey, K. M., and Liu, L. F. (1984) Mechanism of antitumor drug action: Poisoning of mammalian DNA topoisomerase II on DNA by 4′-(9-acridinylamino)-methanesulfon-*m*-anisidide. *Proc. Natl. Acad. Sci. USA* **81,** 1361–1365.
8. Pommier, Y., Minford, J. K., Schwartz, R. E., Zwelling, L. A., and Kohn, K. W. (1985) Effects of the DNA Intercalators 4'-(9-acridinylamino)-methanesulfon-*m*-anisidide and 2-methyl-9-hydroxyellipticinium on topoisomerase II mediated DNA strand cleavage and strand passage. *Biochemistry* **24,** 6410–6416.

9. Tewey, K. M., Rowe, T. C., Yang, L., Halligan, B. D., and Liu, L. F. (1984) Adriamycin-induced DNA damage mediated by mammalian DNA topoisomerase II. *Science* **226,** 466–468.
10. Bodley, A., Liu, L. F., Israel, M., Seshadri, R., Koseki, Y., Giuliani, F. C., et al. (1989) DNA Topoisomerase II-mediated intercation of doxorubicin and daunorubicin congeners with DNA. *Cancer Res.* **49,** 5969–5978.
11. Woynarowski, J. M., Sigmund, R. D., and Beerman, T. A. (1989) DNA minor groove binding agents interfere with topoisomerase II medicated lesions induced by epipodophyllotoxin derivative VM-26 and acridine derivative *m*-AMSA in nuclei from L1210 cells. *Biochemistry* **28,** 3850–3855.
12. Sorensen, B. W., Sinding, J., Andersen, A. H., Alsner, J., Jensen, P. B., and Westergaard, O. (1992), Mode of action of topoisomerase II-targeting agents at a specific DNA sequence. *J. Mol. Biol.* **228,** 778–786.
13. Chaires, J. B., Dattagupta, N., and Crothers, D. M. (1982) Studies on interaction of anthracycline antibiotics and dexoyribonucleic acid: equilibrium binding studies on interaction of daunomycin with deoxyribonucleic acid. *Biochemistry* **21,** 3933–3940.
14. Chaires, J. B. (1985) Thermodynamics of the daunomycin–DNA interactions: ionic strength dependence of the enthalpy and entropy. *Biopolymers* **24,** 403–415.
15. Waring, M. J. (1965) Complex formation between ethidium bromide and nucleic acids. *J. Mol. Biol.* **13,** 269–282.
16. Blake, A. and Peacocke, A. R. (1968) The interaction of aminoacridines with nucleic acids. *Biopolymers* **6,** 1225–1253.
17. LePecq, J. B. and Paoletti, C. (1967) A fluorescent complex between ethidium bromide and nucleic acids. *J. Mol. Biol.* **27,** 87–106.
18. Graves, D. E. and Krugh, T. R. (1983) Single-cell partition analysis—a direct fluorescence technique for examining ligand–macromolecule interactions. *Anal. Biochem.* **134,** 73–81.
19. Cain, B. F., Baguley, B. C., and Denny, W. A. (1978) Potential antitumor agents. Deoxyribonucleic acid polyinteralating agents. *J. Med. Chem.* **21,** 658–668.
20. McGhee, J. D. and von Hippel, P. H. (1974) Theoretical aspects of DNA–protein interactions: cooperative and non-cooperative binding of large ligands to a one-dimesional homogenous lattice. *J. Mol. Biol.* **86,** 469–489.
21. Crothers, D. M. (1968) Calculation of binding isotherms for heterogeneous polymers. *Biopolymers* **6,** 575–584.

17

Quinolone Interactions with DNA and DNA Gyrase

Linus L. Shen

1. Introduction

One of the most powerful methods for investigating the molecular mechanism of drug action is the use of radioligand binding technique. The method allows investigators to pinpoint the direct target of the drug, and the nature of the interaction in terms of binding affinity, specificity, and cooperativity as well as molar binding ratios. These parameters are crucial for deducing a mechanistic model for the drug's action and thereby assist in the process of drug design by furthering our understanding of the established structure–activity relationship of that class of compounds. The method has been best exemplified by the investigations of mode of action of quinolone antibacterials that target the bacteria-specific type II DNA topoisomerase, DNA gyrase (reviewed in **refs. *1–5***).

Quinolones are a group of low-mol-wt, synthetic and extremely potent "poison"-type topoisomerase inhibitors that act by stabilizing the enzyme–DNA complex, termed the cleavable complex. The consequence of the stabilization is at least threefold:

1. Inhibition of the turnover of this essential enzyme.
2. Stalling of replication fork or transcription complex movement.
3. Induction of irreversible DNA strand breaks, which transform the enzyme into a poison leading to cell death.

Quinolone resistance mutations have been mainly mapped to the *gyrA* locus, leading to a general belief that the target of the drug is the A-subunit of DNA gyrase.

From: *Methods in Molecular Biology, Vol. 95: DNA Topoisomerase Protocols, Part II: Enzymology and Drugs*
Edited by N. Osheroff and M.A. Bjornsti © Humana Press Inc., Totowa, NJ

Detailed understanding of the mechanistic action of quinolones largely came from radioligand binding studies. Binding studies of quinolones to DNA and to gyrase revealed that quinolones do not bind to the enzyme, but instead bind to pure DNA *(6)*. Subsequent studies with various DNA substrates and DNA–enzyme complexes led to the proposal of a cooperative drug binding model in which the drug molecules bind to a single-stranded DNA pocket created by the enzyme DNA gyrase *(7–9)*. Analysis of binding affinity and cooperativity concluded that quinolones, with their simple molecular structure and few functional binding groups, acquire high binding affinity via self-association of the drug molecules. Such a cooperative binding process enhances the binding strength and also provides flexible adjustment in fitting into the receptor conformation of the binding pocket. This radioligand binding technique also has provided a helpful tool for the elucidation of the chiral discrimination phenomenon of ofloxacin stereoisomers *(10)*. The results provided further insight into the mechanism of inhibition showing that DNA binding affinity and cooperativity are necessary, but not sufficient in determining the drug potency. The number of drug molecules capable of being assembled in the pocket (the maximum molar binding ratio) and the positioning of the C-7 substituents of the drug are crucial factors for potency.

The purpose of this chapter is to provide detailed procedures of a convenient ultrafiltration technique for radioligand binding study, which has been used extensively in the mechanistic studies of quinolones as described above. The basic principle of the membrane ultrafiltration method for radioligand binding to proteins has been described by Paulus *(11)* and Markus et al. *(12)*. Our current applications of this ultrafiltration method are facilitated by the use of the Amicon Centrifree device for measuring the binding of quinolones to DNA or DNA gyrase as described previously *(6,7)*. This method is suitable even with low-affinity ligands, since the procedures allow the maintenance of a ligand-receptor binding equilibrium throughout the entire process. Alternative procedures using spin-column filtration techniques, and the use of fluorescence detection methods replacing radioactivity counting is also provided.

2. Materials

2.1. Binding Ligands

For simplicity and better sensitivity, it is more desirable to use radioactivity for ligand concentration determination, although other spectroscopic methods, such as fluorescence determination (**Fig. 1**), can also serve the purpose (*see* **Note 1**). When using radiolabeled drugs, an SA > 500 mCi/mmol is required for studies of high-affinity ligands, since experiments at low drug concentrations (<1 μ*M*) may be required. When working at high drug concentrations,

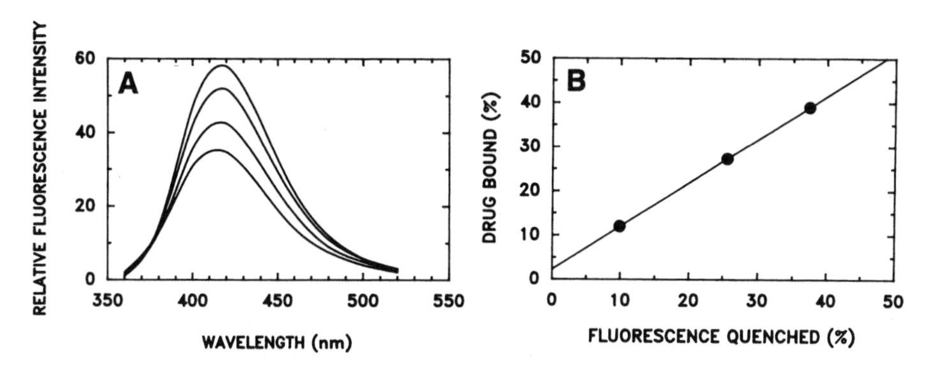

Fig. 1. The quenching of the intrinsic fluorescence of norfloxacin by interacting with single-stranded DNA. **(A)** Fluorescence emission spectra of norfloxacin (1.2 μ*M*) excited at 340 nm. Curves from top to bottom correspond to the spectra of norfloxacin solution (1.5 mL) added with 0, 10, 30, and 60 μL of thermally denatured calf thymus DNA (3.65 mg/mL) to give nucleotide/drug ratios of 0, 6, 18, and 36, respectively. The addition of DNA caused no more than 4% of volume change. **(B)** Correlation of the percentage of drug bound and the percentage of fluorescence quenched. The percentage of drug bound is determined by a ultrafiltration technique using [^{3}H]norfloxacin. Both parameters shown have been corrected by the dilution factors. Reprinted from **ref.** ***22***.

spiking of the working solution with cold unlabeled drug is necessary (*see* **Note 2**). [^{3}H]norfloxacin, the binding ligand used as an example in this chapter, can be custom-synthesized by a radioligand provider, such as New England Nuclear (Boston, MA), as described ***(6)***. The radiolabeled quinolone stock is stored in 50% ethanol and is stable for about 1 yr (*see* **Note 3**). The concentration of norfloxacin solution can be determined using the molar extinction coefficient of 36,000 at 270 nm. Unlabeled norfloxacin may be purchased from Sigma Chemical Co. (St. Louis, MO).

2.2. Enzyme

Subunits A and B of DNA gyrase can be purified separately from the *Escherichia coli*-overproducing strains N4186 and MK47 ***(13)*** or JMtacA and JMtacB strains ***(14)***, respectively. Purified subunits A and B can be stored in 50 m*M* Tris-HCl, pH 7.5, containing 50% glycerol, 1 m*M* EDTA, and 1 m*M* dithiothreitol at –70°C. DNA gyrase holoenzyme is reconstituted from equimolar amounts of subunit A and B preparations, and then stored at –25°C for up to 2 wk before use. Gyrase activity can be measured using a standard supercoiling assay ***(6,15)***. A supercoiling unit is defined as the minimum amount of reconstituted gyrase that will maximally supercoil 0.2 μg of relaxed plasmid in

1 h at 30°C. An SA value of 10^6 U/mg of the holoenzyme *(13)* can be used to estimate the molar concentration of active gyrase present. This is done to compensate for any inactive subunits or incomplete reconstitution.

2.3. DNA

ColE1, pBR322, and pUC9 plasmids have been used satisfactorily for the binding studies of quinolones to DNA of various topological forms. To assure good quality, they should be isolated using SDS/lysozyme lysis procedure followed by cesium chloride density gradient purification *(16)*. DNA linearization by restriction endonuclease digestion (*see* **Note 4**) and the preparation of relaxed DNA using calf thymus DNA topoisomerase I are performed at reaction conditions suggested by the manufacturer (Life Technologies, Gibco BRL, Gaithersburg, MD). After the reaction, the DNA preparations are extracted by a standard phenol-chloroform extraction procedure *(17)* followed by an ethanol precipitation step. The concentration of DNA can be determined spectrophotometrically assuming an absorbance of 1 at 260 nm for a 50 µg/mL solution of duplex DNA.

2.4. Ultrafiltration Device

Membrane ultrafiltration method is recommended for all legitimate binding studies (*see* **Note 5**). The disposable Centrifree micropartition device (AMICON product #4104 with mol-wt cutoff of 30 kDa) has been shown to be more convenient for studying ligand binding to macromolecular receptors (*see* **Note 6**) than alternative devices, such as the Paulus pressure-cell (11). A Sorvall RC5B superspeed centrifuge equipped with a HS-4 swinging bucket rotor with #00389 adapters can be used for the centrifugation step.

2.5. Binding Buffer

This consists of 50 m*M* HEPES, pH 7.4, 20 m*M* KCl, 5 m*M* $MgCl_2$, 1 m*M* EDTA, and 1 m*M* dithiothreitol.

3. Methods

3.1. Binding Reactions

1. Preparation of binding reaction mixtures: Binding reaction mixtures containing approx 5 pmol of DNA and/or DNA gyrase (*see* **Note 7**) in 200 µL the binding buffer are preincubated for 15 min at room temperature.
2. An equal volume (200 µL) of [^{3}H]norfloxacin solution in binding buffer at a designated concentration is added and incubated for another 60 min. A control sample for determining initial ligand concentration is prepared in the same way, except that no receptor is added.

3.2. Membrane Ultrafiltration Procedures

1. The reaction mixtures are transferred to a Centrifree Micropartition device and centrifuged for 30 min at 3000*g* at room temperature to separate the bound and free ligands. Most of the solution passes through the membrane in about 10 min.
2. Determining the radioactivity in the filtrates (*see* **Note 8**): an aliquot (e.g., 200 µL) of the filtrate that collected in the reservoir at the lower end of the device is removed, and its radioactivity is determined by scintillation counting. The amount of ligand bound (pmol) to the receptor and the molar binding ratio (R) are calculated according to the following equations:

$$\text{Amount of ligand bound (pmol)} = (\text{CPM}_o - \text{CPM}) \times (400/200) \times (\text{CF}) \quad (1)$$

$$\text{Molar binding ratio } (R) = (\text{pmol of ligand bound})/(\text{pmol of receptor}) \quad (2)$$

where CPM and CPM_0 represent the radioactivities of free and initial ligand in 200-µL filtrates with and without the receptor, respectively. (CF) denotes the conversion factor from CPM to pmol of the drug (*see* **Note 2**). (400/200) is for volume correction when 200 µL filtrate out of a total of 400 µL is counted.

3.3. Data Analysis

Binding data may be presented by a Scatchard plot or a Klotz plot ***(18)***. In this chapter, only the Klotz plot, where molar binding ratio (R) is plotted against the logarithm of free ligand concentration, is used in analyzing the binding data. A semiempirical approach can be employed to obtain quantitatively the binding parameters ***(19)*** by using a nonlinear least-square computer program, such as RS1 (Bolt, Beranek and Newman Software Products Corporation, Cambridge, MA) or any equation curve-fitting software, to fit the binding data with the Hill's equation:

$$R = R_m/[1 + (K_d/D)^n] \quad (3)$$

where R, R_m, K_d, and D represent the molar binding ratio, maximum molar binding ratio, apparent dissociation constant, and free ligand concentration, respectively, and n is the Hill constant. When $n = 1$, the binding is purely noncooperative, and for larger n values, the binding is said to be increasingly cooperative. The R_m value is chosen manually as the R value at the apparent saturation plateau position, but other parameters are fitted by the computer program.

3.4. The Alternative Spin-Column Binding Method

An alternative method to separate bound from free drug in a radioligand binding assay is the use of the spin-column technique ***(20)***. Since the equilibrium condition of the binding is being perturbed during the separation process, dissociation of bound ligand is inevitable with this approach. Owing to its simplicity, however, this method has been widely used to determine the binding of high-affinity ligands to receptors (*see* **Note 5**). We have used this method to study the binding of [^{3}H]norfloxacin to DNA–DNA gyrase complex *(7,10)*.

1. Prepare 30 µL reaction mixtures: 50 m*M* Tris-HCl, pH 7.5, 1 m*M* dithiothreitol, 6 m*M* $MgCl_2$, 20 m*M* KCl, ± 4 pmol ColE1 DNA, ± 4 pmol reconstituted active DNA gyrase. Incubate for 30 min at room temperature.
2. Add 30 µL radiolabeled quinolone, such as [^{3}H]norfloxacin, of varying concentration, and incubate for another 60 min.
3. Pipet 1 mL buffer-equilibrated Sephadex G-50 fine into a 1-mL plastic syringe equipped with a porous polyethylene disk at the lower end. Place the syringe in a test tube, and spin in a swinging bucket centrifuge at 100*g* for 2 min to remove excess buffer.
4. Apply 50 µL of buffer to rehydrate the top of the resin bed. This ensures consistently low controls and higher recovery of protein and DNA under the conditions used.
5. Apply 50 µL binding sample to the spin column, and centrifuge for 2 min. The effluent is diluted with water to 2.5 mL and the absorbance at 260 nm may be used to estimate the recovery of DNA (routinely >85%). The sample is then counted to determine the amount of ligand [^{3}H]norfloxacin that comes through the column with the macromolecule. Controls normally show <0.15% of the added counts eluted from the column in the absence of DNA or gyrase.

3.5. Results—Parameters for Elucidating Mode of Action

3.5.1. Binding Affinity and Cooperativity

The bindings of norfloxacin to supercoiled DNA, using radiolabeled drug and using the fluorescence property of unlabeled drug, are shown in **Fig. 2A** and **B** in Klotz plots. Both experiments show that the drug has a dissociation constant (K_d) of about 0.9 µ*M*, a surprisingly high binding affinity for a drug with a simple chemical structure and few functional binding groups *(8,9)*. The binding cooperativity, expressed as the Hill constant (n), is also shown in the figure. The high value of the Hill constant ($n = 4$) implies that the binding is highly cooperative to the binding site on the supercoiled DNA. The high cooperativity means that the binding of the first drug molecule to the receptor site facilitates the binding of the second and subsequent drug molecules by either a conformational change of the receptor (an allosteric effect) or by a drug self-association mechanism *(9)*.

3.5.2. Binding Specificity

Studies on the binding of quinolones to various forms of pure DNA *(8)* and to gyrase–DNA complexes *(7)* provide important binding specificity informa-

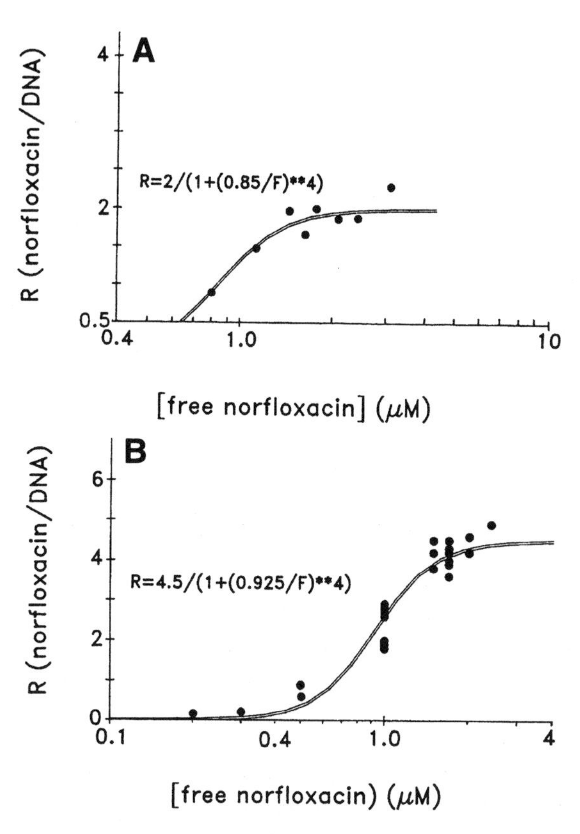

Fig. 2. Binding of norfloxacin to supercoiled DNA analyzed by Klotz plots. **(A)** [^{3}H]norfloxacin binding to supercoiled pBR322 DNA. **(B)** Unlabeled norfloxacin binding to supercoiled ColE1 DNA using the fluorescence concentration determination method. Curve is drawn according to the Hill's equation shown. Double asterisk (**) in the equation represents the exponentiation operator. Reprinted from **ref.** ***8***.

tion leading to a better understanding of the mode of drug–DNA interaction and its role in DNA gyrase inhibition. **Figure 3** shows the binding of [^{3}H]norfloxacin to ColE1 DNA of different structural forms in a drug concentration range near its supercoiling inhibition constant (K_i = 1.8 μ*M*). The results demonstrate that the binding is greater with the heat-denatured single-stranded DNA than the double-stranded DNA of either linear or covalently closed relaxed form. Further studies with native double-stranded DNA species of varying percentages of GC contents showed that there is no correlation between the amount of drug bound and the percentage of GC content of the native DNA *(8)*. These results suggest that when DNA strands are intact, the drug binding is limited, and no base preference can be demonstrated. However, this is not the case for single-stranded DNA. The binding preference to single-

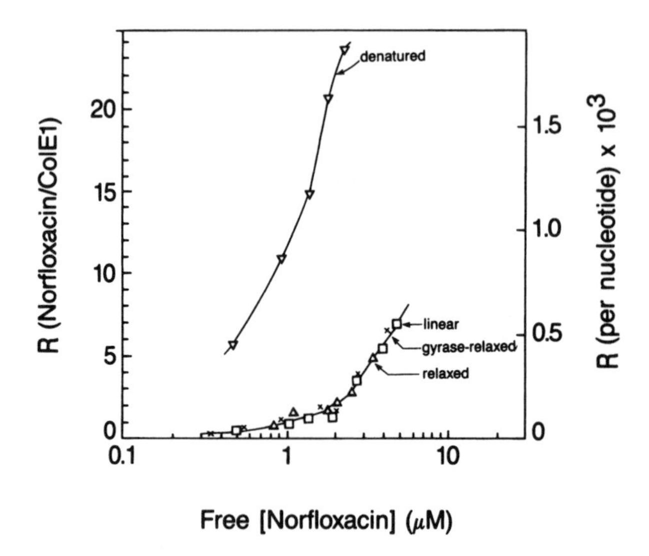

Fig. 3. Binding of [^{3}H]norfloxacin to ColE1 DNA of different structural forms at a concentration range near its supercoiling inhibition constant (1.8 μ*M*). Denatured DNA is prepared by heating at 98°C for 20 min and then rapid cooling to 25°C. The thermal denaturation process gave DNA a 25% increase in absorbance at 260 nm. The gyrase-DNA complex is prepared by incubating 3.5 pmol of relaxed ColE1 DNA with 10 pmol of gyrase holoenzyme in the absence of ATP in 100 μL of binding buffer for 60 min before adding [^{3}H]norfloxacin. Amount of DNA used in each experiment is constant at 3.5 pmol (data obtained from **ref. *6***).

stranded DNA has been studied using synthetic DNA homopolymers. **Figure 4A** shows base selectivity on the drug binding to single-stranded polydeoxyribonucleotides, demonstrating a preference for poly(dG), poly(dA), poly(dT), and poly(dC) in decreasing order, whereas binding to double-stranded poly(dA)-poly(dT) is virtually undetectable. Binding to poly(dG) is distinctively greater than to the other three polydeoxyribonucleotides. One unique structural feature of the guanine base is that it has two common hydrogen bond donors, but the other bases have only one. These results thus suggest that hydrogen bonds may be involved with the drug binding, possibly between the hydrogen bond donor groups on the base and the 4-keto and/or the 3-carboxyl group on the quinolone ring. A direct and more legitimate comparison of base binding specificity has been obtained from the binding studies of [^{3}H]norfloxacin to poly(dI) and poly(dG). The only structural difference between guanosine and inosine is that the former has an extra amino group at the C-2 position of the purine ring (**Fig. 4A** inset), which is an important hydrogen bond donor when pairing complementary strands. As shown in

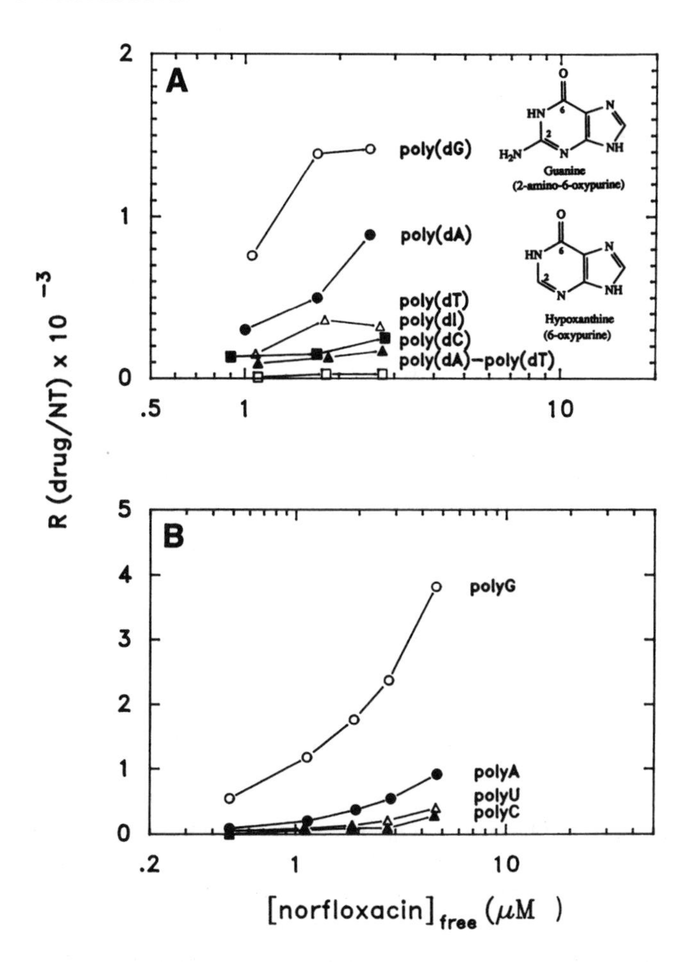

Fig. 4. DNA base binding preference of norfloxacin as demonstrated by [^{3}H]norfloxacin binding to synthetic polydeoxyribonucleotides **(A)** and to polyribonucleotides **(B)**. Data reprinted from **ref.** *8*. Note that hypoxanthine constitutes the base moiety of the inosine molecule.

Fig. 4A, the existence of such an extra hydrogen-bond donor renders a fivefold increase in binding for poly(dG) from poly (dI). The result again suggests that the drug binds to unpaired DNA bases via hydrogen bonding. Similar results supporting this conclusion were also obtained with synthetic polyribonucleotides (**Fig. 4B**).

In conclusion, the affinity and specificity of quinolone binding to pure DNA may be viewed at two levels. First, the binding specificity is controlled by the topological structure of DNA, i.e., the drug binds preferentially to single-stranded DNA or supercoil, which contains single-stranded region rather than

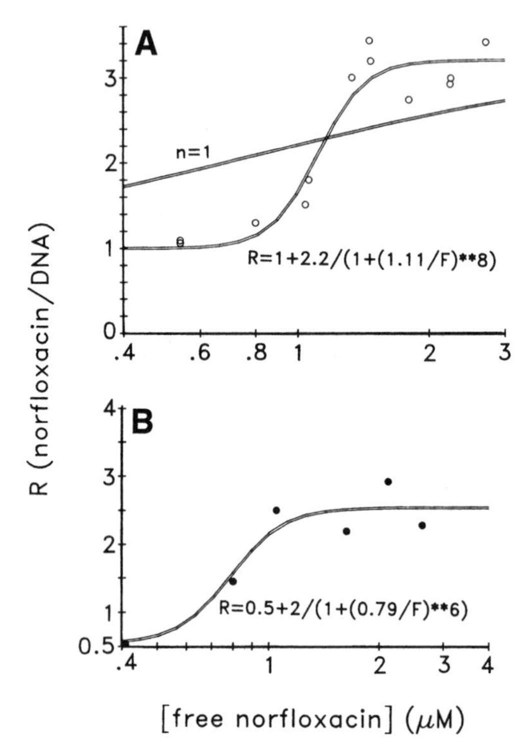

Fig. 5. Bindings of [^{3}H]norfloxacin to complexes formed between linear ColE1 DNA and DNA gyrase in the presence **(A)** and absence **(B)** of ATP (1 m*M*) presented in the form of Klotz plots. Complexes were formed by mixing equimolar amounts of DNA gyrase and DNA. Curve is drawn according to the Hill's equation shown. Double asterisk (**) in the equation represents the exponentiation operator. Also shown in panel (A) is the curve drawn with $n = 1$ with the other parameters remaining the same as for the other curve. Reprinted from **ref.** ***8***.

to double-stranded DNA *(6)*. The second level of binding specificity can only be seen after the DNA strands are separated, and the drug prefers to bind to guanine than the other three bases.

Drug binding studies with DNA–DNA gyrase complex reveal another level of drug binding specificity which contributes to the specificity of DNA gyrase inhibition. **Figure 3** shows that the amount of the drug binding to the complexes formed between relaxed DNA substrate and DNA gyrase (in the absence of ATP to prevent supercoiling) is low, indistinguishable from the amount bound to DNA alone. When linearized DNA is used in place of the covalently closed relaxed DNA substrate, the binding is greatly enhanced (**Fig. 5**) with a binding pattern similar to that obtained with supercoiled DNA alone shown in **Fig. 2**. The induction of a binding site (in the absence of ATP) is presumably owing to the absence of topological constrain resulted from the liberation of DNA ends. Similar results have been obtained when a nonhydrolyzable ATP analog, which induces a conformational change in the complex necessary for supercoiling, is used in the binding studies with the relaxed DNA–DNA gyrase complex (data not shown). Therefore, this level of binding specificity is seemingly created and controlled by the enzyme through an induction of a single-stranded binding pocket, and this binding mechanism renders the specificity at the enzyme inhibition level.

All the above binding results have led to the proposal of an inhibition model that quinolones bind cooperatively to a single-stranded DNA pocket created by DNA gyrase possibly through hydrogen bonds during or before the strand-breaking step, thus trapping the covalent enzyme–DNA intermediate *(9)*.

3.5.3. Binding Reversibility

The radioligand filter binding method described above may be used to test the reversibility of drug binding to DNA. Using this approach, the binding of quinolones to DNA has been demonstrated to be noncovalent and readily reversible by a simple washing step with the binding buffer. It has been also shown that quinolone binding to DNA may be prevented or reversed by addition of histones to the DNA, which results in the formation of insoluble complexes *(21)*.

4. Notes

1. The ligand concentration in the filtrate may be determined by other physicochemical means, such UV absorption or fluorescence measurements. Quinolones, for example, possess a strongly fluorescent chromophore *(22)*, whereas DNA molecules are essentially nonfluorescent. These fluorescence properties can be capitalized to investigate the interaction of the drug with DNA. As shown in **Fig. 1A**, norfloxacin's intrinsic fluorescence intensity is quenched on addition of single-stranded DNA, a preferable binding receptor for quinolones. The percentage of quenching correlates with the percentage of drug bound to DNA determined by the radioligand binding method (**Fig. 1B**), indicating a near-complete fluorescence quenching of the DNA-bound drug. This phenomenon demonstrates that fluorescence intensity measurement cannot be used to detect the results of equilibrium dialysis experiments, since the DNA-bound drug molecules remain silent and undetectable. The technique, however, may be employed for the determination of free ligand concentration in our presently described ultrafiltration binding method. In these experiments, the same ultrafiltration procedures are used, except that ligand concentrations are determined by measuring the drug's intrinsic fluorescence intensity instead of its radioactivity. We have used this approach to determine the binding of various quinolones to DNA when radiolabeled drugs were not available *(8)*. One example of using this fluorescence method is shown in **Fig. 2B**.

2. At higher drug concentrations, the radiolabeled drug solution should be spiked with cold drug to avoid wasting high radioactivity in the binding mixture. This dilution process changes the specific radioactivity of the drug and the CF value (in the **Eq. 1**) by the same proportion.
3. Older radiolabeled quinolone stock may be repurified by reversed-phase HPLC. Typically 100–150 nmol norfloxacin, stored in ethanol:H_2O (1:1), are brought to dryness and redissolved in 50 µL H_2O containing 0.1% trifluoroacetic acid. The sample is loaded onto a Waters µBondapak C18 column equilibrated with the same mobile phase and eluted with a 0–80% gradient of acetonitrile over a period of 40 min. Elution of the drug may be followed at 330 nm. Under these conditions the [^{3}H]norfloxacin has a retention time of approx 19 min, identical to that observed with the unlabeled drug.
4. *Eco*RI is commonly used for preparing linearized ColE1 or pRB322 DNA, which has a single restriction site on the plasmids for the restriction endonuclease.
5. The alternative spin-column binding method, described in **Subheading 3.4.**, can only be used for demonstrating the existence of high-affinity binding. However, a negative result generated by this alternative method does not prove the absence of a binding. This is extremely important in performing legitimate binding studies, since misleading conclusions may thus be derived.
6. The binding of radioligands to small peptides, DNA fragments, or oligonucleotides may be determined using Amicon YM-2 Centrifree micropartition devices, which have membranes with a mol-wt cutoff of 1 kDa.
7. When DNA gyrase is used in the binding studies, the enzyme stock should be dialyzed against the binding buffer immediately before the binding experiment. The storage buffer contains 50% glycerol, which may interfere with the binding especially with the direct membrane-counting method (*see* **Note 8**). The dialysis may be achieved by two cycles of concentration/dilution procedures in an Amicon Centricon-30 device at 4°C. The buffer-exchange procedure normally takes <2 h.
8. If a thin, uniform membrane can be found on a Centrifree device lot, the amount of bound drug may be determined directly from the radioactivity retained on the membrane after subtracting the background counts. This direct measurement of bound ligand on the filter is beneficial, since a smaller amount (up to fivefold less) of receptors can be used. Prior to using this direct membrane-counting method, however, it is recommended that the membrane lots should be tested. The background radioactivity (CPM contributed by free ligand occupied on the membrane) should be less than that equivalent to 20 µL of the testing solution and the standard error of the background should be ± 5%.

Acknowledgment

The author would like to thank Claude Lerner for his critical reading of the manuscript.

References

1. Cozzarelli, N. R. (1980) DNA gyrase and the supercoiling of DNA. *Science* **207,** 953–960.
2. Gellert, M. (1981) DNA topoisomerases. *Ann. Rev. Biochem.* **50,** 879–910.
3. Wang, J. C. (1982) DNA topoisomerases. *Sci. Am.* **247,** 94–109.
4. Sutcliffe, J. A., Gootz, T. D., and Barrett, J. F. (1989) Biochemical characteristics and physiological significance of major DNA topoisomerases. *Antimicrob. Agents Chemother.* **33,** 2027–2033.
5. Reece, R. J. and Maxwell, A. (1991) DNA gyrase: structure and function. *Crit. Rev. Biochem. Mol. Biol.* **26,** 335–375.
6. Shen, L. L. and Pernet, A. G. (1985) Mechanism of inhibition of DNA gyrase by analogues of nalidixic acid: the target of the drugs is DNA. *Proc. Natl. Acad. Sci. USA* **82,** 307–311.
7. Shen, L. L., Kohlbrenner, W. E., Weigl, D., and Baranowski, J. (1989) Mechanism of quinolone inhibition of DNA gyrase. Appearance of unique norfloxacin binding sites in enzyme-DNA complexes. *J. Biol. Chem.* **264,** 2973–2978.
8. Shen, L. L., Baranowski, J., and Pernet, A. G. (1989) Mechanism of inhibition of DNA gyrase by quinolone antibacterials: Specificity and cooperativity of drug binding to DNA. *Biochemistry* **28,** 3879–3885.
9. Shen, L. L., Mitscher, L. A., Sharma, P. N., O'Donnell, T. J., Chu, D. W. T., Cooper, C. S., et al. (1989) Mechanism of inhibition of DNA gyrase by quinolone antibacterials. A cooperative drug–DNA binding model. *Biochemistry* **28,** 3886–3894.
10. Morrissey, I., Hoshino, K., Sato, K., Yoshida, A., Hayakawa, I., Bures, M. G., et al. (1996) Mechanism of the differential activities of ofloxacin enantiomers. *Antimicrob. Agents Chemother.* **40**, 1775–1784.
11. Paulus, H. (1969) A rapid and sensitive method for measuring the binding of radioactive ligands to proteins. *Anal. Biochem.* **32,** 91–100.
12. Markus, G., DePasquale, J. L., and Wissler, F. C. (1978) Quantitative determination of the binding of ε-aminocaproic acid to native plasminogen. *J. Biol. Chem.* **253,** 727–732.
13. Mizuuchi, K., Mizuuchi, M., O'Dea, M. H., and Gellert, M. (1984) Cloning and simplified purification of Escherichia coli DNA gyrase A and B proteins. *J. Biol. Chem.* **259,** 9199–9201.
14. Hallett, P., Grimshaw, A. J., Wigley, D. B., and Maxwell, A. (1990) Cloning of the DNA gyrase genes under tac promoter control: overproduction of the gyrase A and B proteins. *Gene* **93,** 139–142.
15. Otter, R. and Cozzarelli, N. R. (1983) *E. coli* DNA gyrase. *Methods Enzymol.* **100B,** 171–180.
16. Sambrook, J., Fritsch, E. F., and Maniatis, T. (1989) *Molecular Cloning. A Laboratory Mannual*, 2nd ed., vol. 3, Cold Spring Harbor Laboratory Press, Cold Spring Harbor, NY, pp. 1.36–1.46.

17. Sambrook, J., Fritsch, E. F., and Maniatis, T. (1989) *Molecular Cloning. A Laboratory Mannual*, 2nd ed., vol. 3, Cold Spring Harbor Laboratory Press, Cold Spring Harbor, NY, p. E3.
18. Klotz, I. M. (1974) Protein interactions with small molecules. *Accounts Chem. Res.* **7,** 162–168.
19. Cantor, C. R. and Schimmel, P. R. (1980) *Biophysical Chemistry*, part III, W. H. Freeman and Company, New York, pp. 863, 864.
20. Penefsky, H. S. (1977) Reversible binding of Pi by beef heart mitochondrial adenosine triphosphatase. *J. Biol. Chem.* **252,** 2891–2899.
21. Shen, L. L., Baranowski, J., Wai, T., Chu, D. T. W., and Pernet, A. G. (1989d) The binding of quinolones to DNA: should we worry about it? in *International Telesymposium on Quinolones* (Fernandes, P. B., ed), J. R. Prous Science Publishers, Barcelona, Spain, pp. 159–170.
22. Shen, L. L. (1989) A reply: "Do quinolones bind to DNA?"—yes. *Biochem. Pharmacol.* **38,** 2042–2044.

18

Bactericidal Assays for Fluoroquinolones

Thomas D. Gootz

1. Introduction

Fluoroquinolone antibacterials rapidly kill susceptible bacteria at clinically relevant drug concentrations; all evidence indicates that these drugs exert their lethality on *Escherichia coli* by inhibiting the normal function of the DNA gyrase (a bacterial topoisomerase II) in this organism. Mutations around a limited region of the DNA gyrase A subunit (residues Ala67–Gln106) have been shown to confer resistance to these agents in *E. coli* *(1)*. This has also been documented in many other organisms, indicating that DNA gyrase is a universal target for these drugs. This relationship has been somewhat complicated recently by the observation that in *Staphylococcus aureus*, mutations in another type II enzyme, topoisomerase IV, also confer resistance to fluoroquinolones *(2)*.

Many of the mechanisms of action and bactericidal studies performed with fluoroquinolones have established a direct relationship between growth inhibition and in vitro inhibition of purified DNA gyrase activity. This is illustrated for several quinolones in **Fig. 1**, which shows the relationship between DNA gyrase inhibition and the minimal inhibitory concentration (MIC) against intact *E. coli* cells *(3)*. Here and elsewhere, the MIC is defined as the lowest concentration of drug that inhibits visible growth of an organism in test medium. In this analysis, there is a direct relationship between the MIC and the minimum concentration of drug that promotes formation of a cleavable complex with gyrase and DNA in vitro. There is a closer correlation between cleavable complex formation and MIC than between inhibition of gyrase supercoiling activity and MIC. This supports the notion that the lethal event induced by quinolones in the bacterial cell is not inhibition of gyrase catalytic

From: *Methods in Molecular Biology, Vol. 95: DNA Topoisomerase Protocols, Part II: Enzymology and Drugs*
Edited by N. Osheroff and M.A. Bjornsti © Humana Press Inc., Totowa, NJ

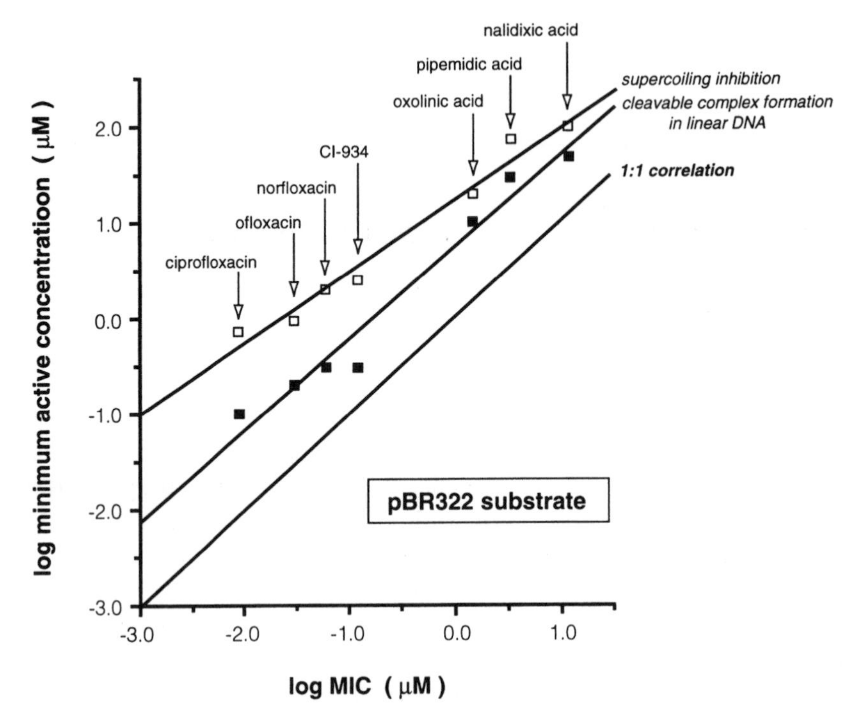

Fig. 1. Correlations between inhibition of the supercoiling reaction and formation of the cleavable complex with fluoroquinolones using pBR322 and *E. coli* DNA gyrase in vitro (redrawn from **ref. *3*** with permission).

activity, but rather formation of a cleavable complex among drug, enzyme, and DNA. This complex disrupts normal DNA replication, resulting in a lethal process described by "the poison hypothesis," which envisions a blockage of the polymerases required for normal DNA and cellular replication following exposure to fluoroquinolones *(4)*. It is well known that fluoroquinolones cause rapid inhibition of DNA synthesis in cells and that this property is expressed at the bactericidal concentration for each drug. Induction of a DNA repair system, referred to as the SOS response, also follows exposure of bacterial cells to fluoroquinolones *(5)*. It has also been observed that maximum cell killing and induction of RecA synthesis occur at similar quinolone concentrations. Higher drug levels inhibit protein synthesis, as well, which can actually antagonize both cell killing and RecA induction. This paradoxical relationship between drug concentration and killing is well described in the literature and has prompted several investigators to study the optimum bactericidal concentration (OBC) for quinolones *(6)*. The OBC for a given organism and drug is

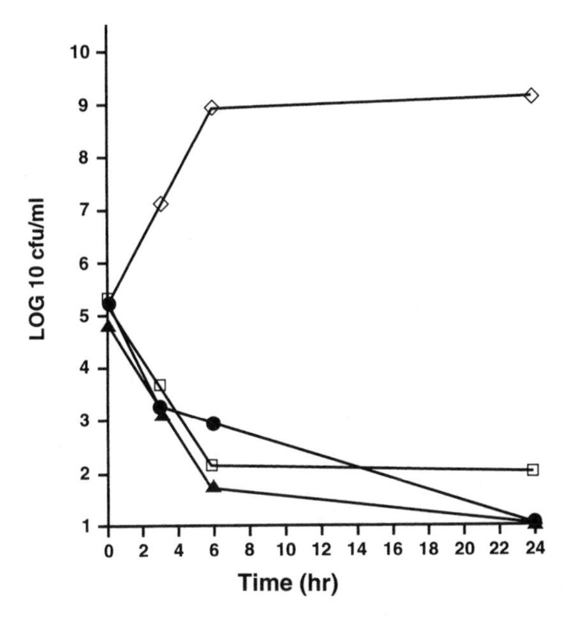

Fig. 2. Killing kinetics of *E. coli* 51A1052 at the MIC for trovafloxacin, ciprofloxacin, or sparfloxacin. Tests performed in cation-supplemented Mueller-Hinton broth starting with cells in stationary phase. ◇ 51A1052 control, ● trovafloxacin (0.0625 µg/mL), ▲ ciprofloxacin (0.0625 µg/mL), and □ sparfloxacin (0.0625 µg/mL). (Graph reprinted from **ref.** *7* with permission).

determined across a concentration range of four orders of magnitude and is yet another way (beyond the MIC) to assess the intrinsic potency of a given agent.

Although fluoroquinolones have low OBCs for highly susceptible organisms, such as *E. coli*, OBCs for species, such as *S. aureus* and *Pseudomonas aeruginosa*, are up to 10-fold higher. This is consistent with the observation that killing rates for quinolones against such species are often slower. This property tends to correlate with the poorer clinical cure rates often observed against these pathogens, particularly in immunocompromised patients. Thus, use of standardized bactericidal tests can be an important method for predicting the clinical utility of new fluoroquinolones *(5)*.

Examination of the kinetics of bacterial killing by fluoroquinolones *(7)* reveals that the initial loss of cell viability is characteristically rapid, reaching a plateau after 6–8 h of drug exposure (**Fig. 2**). The remaining cells do not constitute a resistant subpopulation, however, since on regrowth, they are equally susceptible to the fluoroquinolone. The viable count method used in **Fig. 2** is useful for measuring the rates of cell killing by fluoroquinolones against bacterial strains with different susceptibilities (MICs) to individual compounds. This method is also useful for studying the effects of different media and growth incubation conditions on the bactericidal activity of fluoroquinolones and other novel topoisomerase II inhibitor compounds.

Since a number of factors can affect the results of bactericidal assays with fluoroquinolones, careful standardization of these tests is required if the results are to be of value. Bactericidal tests are often useful for characterizing the lethal activity of new compounds compared with that observed using clinically available fluoroquinolones, such as ciprofloxacin. Only by including control compounds tested by standardized methods can the killing effect of new chemical entities be accurately assessed. Such tests are also helpful for preliminary studies with new structural analogs whose mechanism of action may not be solely based on inhibition of topoisomerase(s).

The National Committee for Clinical Laboratory Standards (NCCLS) has outlined exact protocols for conducting MIC, MBC (Universal bactericidal concentration), and timed kill studies for antimicrobials *(8,9)*. These methods are strictly adhered to in the U.S. and many other parts of the world. Use of standardized media, incubation conditions, quality control strains, and standard drugs allows bactericidal tests with new drugs to be compared with results published for clinically available fluoroquinolones. The methodology in this review comes largely from NCCLS reports and will be adapted for tests with fluoroquinolones where appropriate.

As mentioned above, there are two general types of bactericidal assays that are useful for testing fluoroquinolones. The first is the MIC/MBC test, which can be formatted for use with large numbers of organisms and drugs. Results from these tests performed in broth medium should illustrate that MBC and MIC values are similar, as would be expected for a lethal inhibitor that forms a cleavable complex with topoisomerase II and DNA. The second type of bactericidal test is the timed kill curve, which owing to its more laborious nature is best suited for tests involving a limited number of bacterial isolates. As mentioned, the killing curve can provide clinically relevant information regarding the speed and degree to which cells are killed by a compound. Both methods are described here in detail.

2. Materials

2.1. Test Medium

Although in principle any type of liquid broth medium could be used in bactericidal tests, NCCLS studies have determined that highly reproducible results are obtained using cation-supplemented Mueller-Hinton broth. Since the levels of cations, such as magnesium, can influence the antibacterial activity of fluoroquinolones, the NCCLS has recommended that commercial media be supplemented with 20–25 mg Ca^{2+}/L and 10–12.5 mg Mg^{2+}/L.

To prepare cation stocks, 3.68 g of $CaCl_2 \cdot 2H_2O$ are dissolved in 100 mL deionized water (10 mg Ca^{2+}/mL) and 8.36 g of $MgCl_2 \cdot 6H_2O$ in 100 mL deionized water (10 mg Mg^{2+}/mL).

Stock solutions should be sterilized by membrane filtration and stored at 4°C. Mueller-Hinton broth is prepared according to the manufacturer's instructions (Difco, St. Louis, MO), autoclaved, and stored overnight at 4°C before cations are added. Some commercial preparations of Mueller-Hinton broth come supplemented with cations, and it is important not to add additional amounts to these preparations.

Cation-supplemented Mueller-Hinton broth is appropriate for use when testing *Staphylococcus* spp., members of *Enterobacteriaceae*, and *P. aeruginosa*. When testing fastidious organisms, such as *Haemophilus* spp., *Neisseria* spp., or streptococci, the medium must be supplemented to support their growth *(8)*. Since extremes of pH decrease the antibacterial activity of fluoroquinolones, the pH of each batch of medium should be confirmed to be between 7.2 and 7.4 at 25°C. Media can be stored for up to several weeks at 4°C.

2.2. Use of Quality-Control Strains

The most appropriate way to generate susceptibility results that can be compared with those of other investigators requires the addition of control strains. **Table 1** lists several quality-control strains available through the American Type Culture Collection (Rockville, MD) (ATCC). These have been included in numerous publications and the acceptable MIC ranges for ciprofloxacin are shown, as published by the NCCLS.

When testing new analogs in the MIC/MBC test, the MIC values of ciprofloxacin for any of the ATCC control strains must fall within the limits given in **Table 1** in order for the test results to be valid (*see* **Note 1**).

Ciprofloxacin can be obtained from Miles Inc., Pharmaceutical Division, 400 Morgan Lane, West Haven, CT 06516-4175. The address for the American Type Culture Collection is: 12301 Parklawn Drive, Rockville, MD 20852-1776 (1-800-359-7370). Current tables for interpretive data can be obtained from: National Committee for Clinical Laboratory Standards (NCCLS), 771 East Lancaster Avenue, Villanova, PA 19085 (610-525-2435).

2.3. Preparation of 0.5 McFarland Turbidity Standard

To standardize the inoculum density for a susceptibility test, a $BaSO_4$ turbidity standard equal to a 0.5 McFarland standard is used.

Table 1
ATCC Control Strains

Strain	Acceptable MIC range, mg/mL, for ciprofloxacin
E. coli ATCC 25922[a]	0.004 – 0.015
S. aureus ATCC 29213[a]	0.12 – 0.5
P. aeruginosa ATCC 27853[a]	0.25 – 1.0
Haemophilus influenzae ATCC 49247[b]	0.004 – 0.03

[a]Tested in cation-supplemented Mueller-Hinton.
[b]Tested in *Haemophilus* test medium.

1. A 0.5-mL sample of 0.048M $BaCl_2$ (1.175% w/v $BaCl_2 \cdot 2H_2O$) is added to 99.5 mL of 0.18M (0.36 *N*) H_2SO_4 (1% V/V) with stirring.
2. The correct density of the turbidity standard is verified using a 1-cm light path and matched cuvets. The absorbance at 625 nm should be 0.08–0.10 for the 0.5 McFarland standard.
3. The barium sulfate suspension is aliquoted and tightly sealed in glass tubes of the same size used to prepared the MIC inoculum. Each tube should be vortexed before use.

3. Methods

3.1. The MIC Test

The broth MIC test done in preparation for the MBC procedure can be carried out by one of two methods. Microdilution methods conducted with microtiter trays combined with automated dilution systems are valuable when large numbers of organisms are studied (*see* **Notes 2** and **3**). However, since a smaller absolute number of cells is present in the final inoculum in each well, such tests often give less reproducible MBC values ***(9,10)***. Therefore, it is preferred that macrodilution tests be employed for determining MBCs (*see* **Note 4**).

1. Macrodilution MIC tests performed by NCCLS methods use sterile 13 × 100 – mm borosilicate glass tubes and a final 2ml volume of test medium. Drug, obtained as diagnostic powder of known potency, is dissolved in broth to make a 4X working stock. This stock is filter-sterilized and serially diluted 1:2 through ten tubes (1 mL added to 1 mL of drug-free broth in the first tube). An eleventh tube of drug-free broth is included as a control. After serial dilution through ten tubes, 1 mL is discarded from the tenth tube. The series of tubes now contain 1 mL of a 2X drug solution set. A useful final test range for fluoroquinolones is 8 µg/mL to 0.015 µg/mL.
2. The bacterial inoculum is prepared by resuspending several colonies from overnight growth on an agar plate into sterile saline to equal the turbidity of the 0.5 McFarland standard (approx 1×10^8 organisms/mL).

3. Within 15 min of preparation, the adjusted inoculum suspension is diluted 1 : 100 in sterile broth (1×10^6 organisms/mL), and 1.0 mL is added to each test tube to give a final inoculum of 5×10^5 organisms/mL. The inoculum should be released beneath the surface of the antimicrobial-containing broth in order to avoid adherence of the inoculum to the side of the tube. Mix by flushing the pipet tip two to three times. Tubes should not be shaken further. This scheme gives 2 mL of inoculated broth in each tube with a 1X final concentration of drug through the 10-tube series.
4. All tubes are then placed in an air incubator at 35°C for 20 h. The NCCLS guidelines should be consulted for incubation conditions when testing fastidious organisms, such as *H. influenzae* or *Neisseria* spp. ***(11)***.
5. The MIC is defined as the lowest concentration of drug that inhibits visible growth (clear tube) of the test bacterium compared to the drug-free control. Test results are acceptable only if MIC values for the ciprofloxacin control for the ATCC strain are within the limits given in **Table 1**.

3.2. MBC Determination

After reading the MIC test, all clear tubes should be subcultured to determine the minimal concentration of fluoroquinolone that produces a bactericidal effect (MBC). The most accurate MBC methods include determining the viable cell count in all clear MIC tubes, and comparing them with the number of viable cells present in the initial inoculum ***(10)***. The lowest drug concentration to produce a 99.9% kill (3-log reduction in initial MIC inoculum) is designated as the MBC. This direct cell count method is highly accurate, but the less laborious method described below will also provide an acceptable MBC determination. This method is useful if the collection of organisms to be tested for the MBC is extensive.

1. Immediately after the tubes are examined at 20 h to determine the MIC, they should be gently mixed by hand and returned to the incubator for another 4 h. After this time, the tubes are gently mixed again, and all tubes without visible growth are subcultured (*see* **Note 5**).
2. Remove 0.01 mL from each tube to be sampled, and spread the volume on one quadrant of a drug-free agar plate. Incubate the plate overnight at 35°C.
3. Following overnight incubation, all plates are examined for growth. The lowest concentration of drug that reduced the number of cells in the plated sample to 10 CFU or less is termed the MBC. This represents a 99.9% reduction in the initial inoculum. Since fluoroquinolones are highly bactericidal at the MIC, the MBC value is usually equal to, or no more than fourfold higher than the observed MIC.

3.3. The Killing Curve Test

The MIC and MBC tests determine the potency of a fluoroquinolone against bacteria. Since these are simple broth dilution procedures, they can easily be used to assess the activity of numerous compounds against a large collection of clinically relevant bacteria.

The killing curve test is performed to assess the rate and extent to which a fluoroquinolone kills a particular pathogen of interest. This test is conducted at drug concentrations equal to and multiples above the MIC. The kill curve test for fluoroquinolones is performed using cation-supplemented Mueller-Hinton broth employing an initial inoculum of 5×10^5 cells/mL.

1. The standardized inoculum is prepared as in **step 2** of the MIC test described in **Subheading 3.1.** Flasks plugged with cotton or tubes with loose-fitting caps are used. Although the absolute volume in the test vessel is not critical, the final starting inoculum should be 5×10^5 CFU/mL. Representative sample times are 0, 2, 4, 6, 8, and 24 h.
2. The inoculated tube or flask is then mixed gently, and a 0.1-mL sample is removed. This should be serially diluted 10-fold into tubes containing sterile saline (*see* **Note 6**).
3. 0.1-mL samples from each saline tube are plated in duplicate on the surface of a drug-free agar plate. All plates are then incubated at 35°C for 20–24 h, and the number of viable colonies for each time-point sampled is determined. The data are plotted as the mean viable count $\log_{10}$ CFU/mL vs time as shown in **Fig. 2**.

4. Notes

1. The antimicrobial chosen for quality control need not be ciprofloxacin. Any marketed antibiotic for which NCCLS quality-control strain MIC ranges are published is suitable. Such quality-control data for the ATCC strains should be obtained directly from the NCCLS in **ref.** ***8***.
2. For research laboratories that study structurally novel topoisomerase inhibitors, standardized evaluation of the bactericidal activity of such agents is critical. MIC and MBC determinations should be carried out with a sufficient number of bacterial isolates to give an accurate assessment of the compound's antibacterial activity. These tests should not be carried out exclusively with laboratory strains of bacteria, since these are frequently more susceptible to new antibacterials than are recently isolated clinical strains.
3. In establishing a collection of clinical bacterial isolates, the source and date of original isolation should be recorded for use in subsequent referencing of the collection. Clinical collections should be updated periodically, as susceptibility trends change over time.
4. Controvercy exists over the use of a microdilution test for determining the bactericidal activity of antimicrobials. This centers around the fact that a smaller absolute cell number is present in each microtiter well (0.1 mL $\times 5.0 \times 10^5$ cells/mL $= 5.0 \times 10^4$ cells) than in each tube of a macrodilution test (2 mL $\times 5.0 \times 10^5$ cells/mL $= 1.0 \times 10^6$ cells). This difference in absolute cell number is important when the test organism expresses an enzyme that can inactivate the

antimicrobial being evaluated. Such inactivation is not known to confer resistance to fluoroquinolones. This being the case, microdilution MBC procedures may give comparable results to macrodilution MBC tests with these agents.

5. In **step 1** of the MBC procedure, tubes are examined at 20 h, gently shaken, and returned for incubation for four additional hours prior to subculture. This is to ensure that any surviving cells present on the inner tube surface above the surface of broth are fully exposed to drug *(12)*.
6. In all killing curve studies, it is important to minimize drug carryover from the test vessel to the drug-free subculture plate. Although 10-fold dilutions of culture samples normally minimize any effect of drug carryover, cells sampled directly from the test vessel should be washed free of drug via centrifugation and resuspension prior to plating.

References

1. Hooper, D. C. and Wolfson, J. S. (1993) Mechanisms of bacterial resistance to quinolones, in *Quinolone Antimicrobial Agents*, 2nd ed. (Hooper, D. C. and Wolfson, J. S., eds.), American Society for Microbiology, Washington, DC, pp. 97–118.
2. Ferrero, L., Cameron, B., Manse, B., Lagneaux, D., Crouzet, J., Famechon, A., et al. (1994) Cloning and primary structure of *Staphylococcus aureus* DNA topoisomerase IV: a primary target of fluoroquinolones. *Mol. Microbiol.* **13,** 641–653.
3. Crumplin, G. C. (1990) Molecular effects of 4-quinolones upon DNA gyrase: DNA systems, in *The 4-Quinolones* (Crumplin, G. C., ed.), Springer-Verlag, pp. 54–68.
4. Kreuzer, K. and Cozzarelli, N. R. (1979) *Escherichia coli* mutants thermosensitive for DNA gyrase A: effects on DNA replication, transcription, and bacteriophage growth. *J. Bacteriol.* **140**, 424–435.
5. Phillips, I., Culebras, E. Noreno, F., and Baquero, F. (1987) Induction of the SOS response by new 4-quinolones. *J. Antimicrob. Chemother.* **20,** 631–638.
6. Hooper, D. C. and Wolfson, J. S. (1993) Mechanisms of quinolone action and bacterial killing, in *Quinolone Antimicrobial Agents*, 2nd ed. (Hooper, D. C. and Wolfson, J. S., eds.), American Society for Microbiology, pp. 53–75.
7. Gootz, T. D., Brighty, K. E., Anderson, M. R., Schmeider, B. J., Haskell, S. L., Sutcliffe, J. A., et al. (1994) *In vitro* activity of trovafloxacin, a novel 7-(3-azabicyclo[3.1.0]hexyl) naphthyridone antimicrobial. *Diagn. Microbiol. Infect. Dis.* **19,** 235–243.
8. National Committee for Clinical Laboratory Standards (1995) Methods for dilution antimicrobial susceptibility tests for bacteria that grow aerobically. Document M7-A3, vol. 15, no. 25. National Committee for Clinical Laboratory Standards, Villanova, PA.
9. National Committee for Clinical Laboratory Standards (1992) Methods for determining the bactericidal activity of antimicrobials. Document M26-T, National Committee for Clinical Laboratory Standards, Villanova, PA.

10. Amsterdam, D. (1991) Susceptibility testing of antimicrobials in liquid media, in *Antibiotics in Laboratory Medicine*, 3rd ed. (Lorian, V., ed.), Williams and Wilkins, Baltimore, MD, pp. 53–100.
11. National Committee for Clinical Laboratory Standards (1995) Methods for susceptibility testing fastidious and special problem bacteria. Document M7-A3, vol. 15, no. 14, National Committee for Clinical Laboratory Standards, Villanova, PA.
12. Taylor, P. C., Schoenknecht, F. D., Sherris, J. C., and Linner, E. C. (1983) Determination of minimum bactericidal concentrations of oxacillin for *Staphylococcus aureus*: technical factors. *Antimicrob. Agents Chemother.* **23,** 142–150.

19

Drug Toxicity in *E. coli* Cells Expressing Human Topoisomerase I

Scott T. Taylor and Rolf Menzel

1. Introduction

DNA Topoisomerases (topos) are ubiquitous enzymes that catalyze the breakage and rejoining of the DNA phosphodiester backbone, which together with an intervening strand passage event, allow this these enzymes to alter DNA topology ***(1,2)***. Intermediates in the strand passage reaction involve either single- or double-stranded breaks defining type I and type II enzymes, respectively. The bacterium *Escherichia coli* possesses two type I enzymes. One, encoded by the *topA* gene, is responsible for the major DNA-relaxing activity of the cell and is essential in an otherwise wild-type *E. coli* ***(3)***. *E. coli* and human topo I (htopoI) both catalyze the relaxation of negatively supercoiled DNA. However, they differ in the details of the reactions they catalyze and share no amino acid sequence homology. *E. coli* topoI (etopoI) demonstrates a preference for binding at the junction of double- and single-stranded regions, and proceeds using a single 5'-phosphodiester intermediate ***(4–6)***. The etopoI enzyme is very efficient at relaxing highly negatively supercoiled DNA and shows progressively decreasing activity as the substrate DNA becomes more relaxed ***(4,5)***. The htopoI enzyme relaxes both negatively and positively supercoiled DNA to completion ***(7)***. htopoI shows a preference for binding double-stranded DNA and proceeds by making a single covalent 3'-phosphodiester intermediate to a tyrosine ***(7–9)***. The htopoI enzyme is the target of the antitumor drug camptothecin (CPT), which traps the covalent phosphotyrosine intermediate of the strand passage reaction ***(10)***. The bacterium *E. coli* and its etopoI enzyme ***(11)*** are resistant to CPT.

From: *Methods in Molecular Biology, Vol. 95: DNA Topoisomerase Protocols, Part II: Enzymology and Drugs*
Edited by N. Osheroff and M.A. Bjornsti © Humana Press Inc., Totowa, NJ

Plasmid-borne copies of both the yeast *(12)* and htopoI *(13)* coding sequences have been shown to possess the capacity to complement a conditional defect in the bacterial *topA* gene, and it was suggested that *E. coli* might be useful as a model system for the study of the human enzyme. Subsequent to this initial suggestion, these researchers have pursued the development of yeast model systems for the study of the eukaryotic topo I *(13,14)*.

We obtained from James Wang the plasmid ptachTOP1 *(13)* and undertook development of an *E. coli* system for the study of hTOP1 *(15)*. We were able to confirm that the plasmid ptachTOP1 is able to provide DNA-relaxing activity; it was able to complement a deletion of *topA* and extend the temperature-sensitive range of a *gyr*Bts mutant from 42 to 37°C. We also noted that during subsequent propagation, transformants displayed considerable instability affecting both the ptachTOP plasmid and the host strain's genome. We found it necessary to reclone the segment of the plasmid containing the promoter (ptac) and hTOP1 structural gene into a new plasmid, which provided a low-copy-number origin of replication (pSC101) and a copy of the *ptac*-repressing *lac*I^Q gene. This new plasmid, pMStop1 (**Fig. 1**), is stably inherited and can be induced (0.1–1.0 m*M* IPTG) to show phenotypes associated with DNA relaxation. We were also able to demonstrate that extracts produced from strains harboring pMStop1 display the EDTA-resistant DNA-relaxing activity characteristic of eukaryotic TopoI *(1)*, if and only if induced with IPTG. Furthermore, we found that this EDTA-resistant DNA-relaxing activity is inhibited by CPT. CPT inhibits htopoI by trapping an enzyme–DNA intermediate in the so-called cleavable complex *(10,16)*, and in vivo activities, such as DNA replication, process the cleavable complex into extensive DNA damage, which results in cell death *(17)*. Since bacteria harboring pMStopI grown in the presence of IPTG contains a CPT^S DNA-relaxing activity, we reasoned that such cells should be sensitive to the effects of the CPT. Our initial experiments failed to show any CPT^S for cells harboring pMStopI grown in the presence of 1.0 m*M* IPTG.

There could be many reasons that hTOPI-expressing bacteria failed to demonstrate CPT^S. Our investigations showed that permeability was the problem. We tried several mutations in an effort to demonstrate IPTG-inducible CPT^S. We found the highest level of sensitivity when we introduced the *imp*4312 mutation. The *imp* mutant was originally isolated by selecting for maltodextran uptake, and was subsequently shown to have dramatically increased sensitivity to several drugs and chemical agents *(18)*. The combination of the *imp*4312 mutation, the plasmid pMStopI, and growth with IPTG are all necessary for a CPT^S *E. coli*.

The pathway leading from the formation of the CPT-induced cleavable complex to cell death involves DNA damage *(17)*. This has been shown in

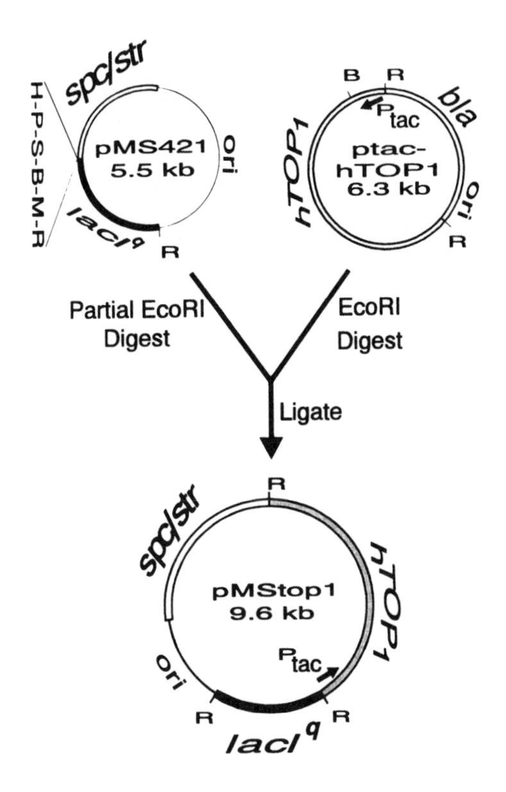

(R)EcoRI (H)HinDIII (P)PstI (S)SalI (B)BamHI (M)SmaI

Fig. 1. Creation of pMStop.

Saccharomyces cerevisiae through the elevated CPT^S seen in a *rad*52 mutant (DNA-damage-repair-deficient) and through the induction of a DNA-damage-inducible promoter following exposure to CPT ***(14,19)***. We were able to make similar observations in *E. coli* strains harboring pMStop1. An *E. coli* strain with a defect in the *recA* gene shows a 50-fold enhanced sensitivity to CPT. Measuring increases in β-galactosidase activity directed by a *sulA-lac* gene fusion is an established method for monitoring DNA damage in *E. coli* ***(20)***. We are able to show the induction of a *sulA-lac* gene fusion by CPT in an appropriately (*imp*4312) hyperpermeable strain harboring the plasmid pMStop1, if and only if IPTG was used to induce the expression of hTOP1. The utility of this *E. coli* system to detect compounds with CPT-like activity and to examine the biochemical genetics of CPT^R is self-evident.

The expression of *hTOP1* in *E. coli* with phenotypic consequences is remarkable. Several reports in the literature suggest that serine phosphorylation is required for activity ***(21,22)***. To our knowledge, this type of phosphorylation does not exist in *E. coli*. Either phosphorylation of htopoI is not absolutely required, or *E. coli* is able to carry out the appropriate phosphorylation. Either eventuality is interesting, and the matter could be resolved by further investigation.

The appearance of a DNA-damage signal following exposure of an *E. coli* strain expressing *hTOPI* to CPT is noteworthy. CPT traps a covalent complex between DNA and htopo I ***(10,16)***. In vitro this complex may be induced to resolve into single-stranded DNA breaks by the addition of a protein denaturant ***(10)***. The cell is able to process this complex into a lethal event involving extensive DNA damage ***(17)***. The molecular details of this process are not known, although the S-phase specificity of CPT toxicity demonstrates that DNA replication is involved ***(23)***. The fact that CPT-induced lethality and DNA damage can be seen in an *E. coli* strain expressing hTOP1 demonstrates that processes required to convert the cleavable complex into a lethal event are resident in *E. coli* and could be the subject of analysis in this genetically defined system.

2. Materials

2.1. Standard Microbiology

1. The strain ATCC 69736 (*thr, his trp* :: MuC^+, Δ*lac*, *malB*, *galK*, *rpsL*, *leu* :: Tn*10*, sulA-lac, *imp4132* / pMStop1).
2. Luria broth agar (1.5%) Petri dishes with 25 μg/mL spectinomycin.
3. Luria broth liquid media with and without 25 μg/mL spectinomycin.
4. Luria broth agar (1.5%).
5. Sterile glycerol.
6. Also *see* **refs. *24,25***.

2.2. β-Galactosidase Assay Reagents

1. Stock 50 mg/mL X-gal (5-bromo-4-chloro-3-indolyl-β-D-galactopyranoside) in *N*',*N*-dimethylformamide.
2. Stock isopropyl-β-D-thiogalactoside (IPTG) 100 m*M* in sterile H_2O.
3. Chlorophenol red-β-D-galactopyranoside (CPRG) 50 mg/mL in sterile H_2O.
4. Z-buffer; 0.06 *M* Na_2HPO_4, 0.04 *M* NaH_2PO_4, 0.01 *M* KCl, 0.001 *M* $MgSO_4$, 0.5 *M* β-mercaptoethanol (β-galactosidase buffer; **ref. *25***).
5. An automated pipeting station, such as the Denley Wellpro Liquid Handling System (Research Triangle, NC).
6. Microplate reader with a temperature-controlled chamber (Bio-Tek Model EL340; Winooski, VT).

3. Methods

3.1. Maintaining a "Working Frozen"

The *imp* mutation is somewhat problematic, but can easily be worked with providing certain precautions are taken. It is imperative that the culture is not kept on agar (either on a plate or on a slant) for long periods of time (more than a day or two). It is convenient to store this culture in 20% glycerol at –20°C and use small aliquots of the frozen culture to inoculate liquid broth prior to an experiment. Such a "working" frozen glycerol stock can be used for up to a year; with its lifetime being limited by the amount of material used (50–100 µL) for each inoculation. This procedure is preferable to going back to the –80°C permanent for each new experiment. Since this "working" frozen will be used over an extended period of time, it is important that its behavior be characterized prior to initiating a series of experiments.

1. Place a small amount of ATCC 69736 from a –80°C permanent culture on Luria broth agar with 25 µg/mL spectinomycin, and isolate single colonies by streaking. Pick three or four colonies to separate 1-mL culture tubes of liquid Luria broth with 25 µg/mL spectinomycin, and grow overnight at 37°C with gyratory shaking.
2. Dilute each culture into 50 mL of fresh liquid Luria broth, and allow 5 h of growth at 37°C, again with gyratory shaking. Take a 10-to 20-mL aliquot, and make it 20% glycerol using a sterile glycerol solution. Freeze 2- to 5-mL aliquots of the glycerol culture on dry ice powder (or dry ice-ethanol), and then store at –20°C. (**Note**: simply freezing by placing in the –20°C will allow the cells to settle prior to freezing and will result in poor inoculation at a later date.) With the remaining cells from each culture, verify the behavior of the isolate.

3.2. Liquid Assay Procedure

1. With 50–100 µL of frozen solid from the "working frozen" (from –20°C working frozen; **Subheading 3.1.**), inoculate a 1-mL culture of liquid Luria broth containing 25 µg/mL spectinomycin and grow overnight at 37°C with gyratory shaking.
2. Dilute the culture into 50 mL of fresh liquid Luria broth, and allow 5 h of growth at 37°C, again with gyratory shaking.
3. This culture (A_{650} = 0.6) is used an inoculum (1:100) for fresh Luria broth liquid with (1.0 m*M*; from 100 m*M* stock) and without IPTG; 190-µL aliquots (± IPTG) are added to duplicate 10-µL samples of drug (in 100% DMSO) distributed into the wells of a flat-bottom polystyrene microtiter plate. The microtiter plate is incubated overnight (12–16 h) at 30°C without shaking.
4. Following this growth and drug exposure, final cell A_{650} values are determined using a microtiter reader.
5. These cells are then made permeable by transferring separate 150-µL cultures from each well of the incubation plate to 30 µL of $CHCl_3$ present in the

corresponding well of a polypropylene microtiter plate using an automated pipeting device. The cultures are made permeable by repeated pipeting up and down (in place, six times).

6. At this time, a β-galactosidase assay is initiated by using an automated pipeting device to transfer carefully (so as to not disturb the $CHCl_3$ in the bottom of the microtiter plate well) 20 μL of the $CHCl_3$-treated culture to 180 μL of prewarmed (28°C) Z-buffer with 1 mg/mL CPRG present in each of the wells of a clear, flat-bottom polystyrene microtiter plate. The assay's progress is monitored by immediately transferring the plate to a microtiter reader (with a 28°C temperature-controlled chamber) and following the change in A_{595}.
7. Final β-galactosidase specific activities (normalized to a cell A_{650} of 1.0 to adjust for differences in growth) are calculated from a kinetic assay (with nine time-points) and expressed as A_{595}/min/cell A_{650}. The induction of increased β-galactosidase activity resulting from the increased expression of the *sulA-lac* fusion is used as a measure of DNA damage.

Figure 2 shows the results of such an experiment conducted with camptothecin (hTOP-specific DNA damage) and nalidixic acid (a general nonhTOP DNA-damaging agent.) The measured β-galactosidase levels corresponding to the indicated drug concentration for camptothecin (top panel) and nalidixic acid (bottom panel) are given. Camptothecin only causes a DNA-damage signal when IPTG is used to induce hTOP expression.

3.3. Agar Assay Procedure

1. Prepare an inoculum culture as in **steps 1** and **2** of **Subheading 3.2.** above.
2. Prepare "plus IPTG" agar plates by melting 400 mL of Luria broth with 1.5% agar and allowing it to cool to 50°C in temperature-controlled water bath (higher temperature will reduce cell viability). To this add 20 mL of the inoculum culture, 4 mL of stock 100 m*M* IPTG, and 3.2 mL of stock 50 mg/mL Xgal in dimethylformamide. Mix thoroughly, and pour 40-mL aliquots of the mixture rapidly (so that cells will cool quickly) into 150 × 15 mm plates. Let the agar solidify at room temperature, and then transfer to 4°C for 1/2 h.
3. Prepare "minus IPTG" plate by melting 400 mL of Luria broth with 1.5% agar and allowing it to cool to 50°C. To this add 10 mL of the inoculum culture and 3.2 mL of stock 50 mg/mL Xgal in dimethylformamide. Mix thoroughly and pour 40-mL aliquots of the mixture rapidly into a 150 × 15 mm plate. Let the agar solidify.
4. In the cooled and solidified plates, 4-mm cylindrical holes (wells) are cored to the bottom of the Petri dish using a small probe attached to a vacuum line. Create an array of holes (with a center to center spacing of 10–20 mm) appropriate to accommodate the samples to be tested.
5. Samples (10 μL of the drug in DMSO) may then be added to the wells created in the agar. Test plates should be transferred to a 30°C incubator and may be scored

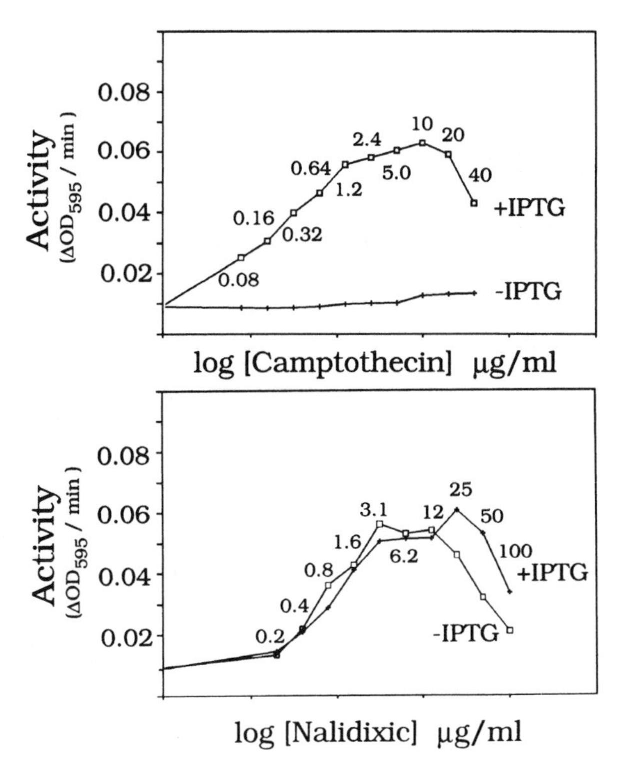

Fig. 2. DNA damage as measured by a β-galatosidase activity produced by a sulA-lac fusion. Induction of hTOP1 (+IPTG) is required for camptothecin to produce a signal.

after 14–18 h of growth. (Growth at 37°C reduces the sensitivity of the test.) Although plates can be scored immediately after growth, a better response can be seen if the test plates are transferred to a 4°C cold room and allowed to sit for 12–48 h following the growth period. Ten microliters of a 4 μg/mL camptothecin control should give a dark-blue color, whereas lower concentrations show a weaker light-blue response. The limit of detection for camptothecin is typically in the range of 10 μL of a 0.25–0.5 μg/mL solution in DMSO (2.5–5 ng total). The DMSO-alone control will give a very faint pale-blue ring in the immediate vicinity of the agar well. Any material showing a response stronger than the solvent-alone control well should be considered SOS-positive (it produces a DNA-damage signal) Specificity for hTOP-dependent DNA damage is confirmed by demonstrating the IPTG dependence of the SOS response.

An example of this test is shown in **Fig. 3** where aliquots from serial dilutions of a series of camptothecin analogs were added to the columns of sample wells (top, highest, to bottom) in replicate to both the plus IPTG (A+) and

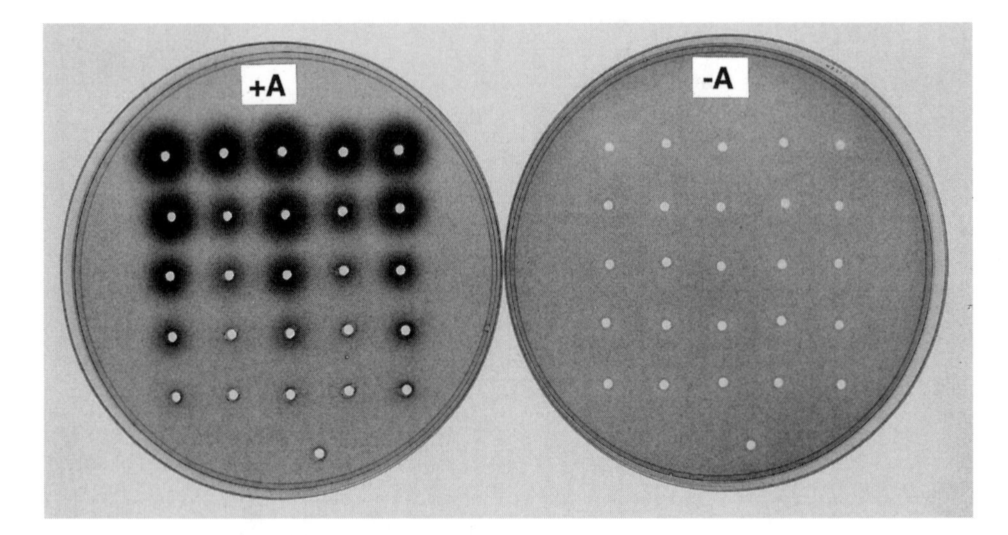

Fig. 3. DNA damage visualized by zones of X-gal hydrolysis.

minus IPTG (A–) plates. There is a single DMSO-alone control in the isolated well at the bottom of the plate.

4. Notes

1. Both the agar (**Subheading 3.3.**) and the microtiter methods (**Subheading 3.2.**) give comparable results. The microtiter method is more readily quantified. The agar method proves more useful when the samples contain complex mixtures.
2. The liquid microtiter plate assay (**Subheading 3.2.**) can be preformed with regular culture tubes (and volumes) with the more conventional *o*-nitrophenyl-β-D-galactoside substrate *(25)*. The use of chlorophenolred-β-D-galactopyranoside simply increases the sensitivity (and therefore the speed) of the procedure *(26)*. The adaptation of the procedure to a microtiter plate and automated pipeting is also a matter of convenience and has been described in more detail elsewhere *(27)*. Calculation of reaction rates improves the accuracy of the assay over that of an end-point reaction.
3. The X-gal blue color in the agar plates (**Fig. 3** and **Subheading 3.3.**) is stable, and plates may be stored after color development for several weeks at 4°C.
4. In both procedures (**Subheadings 3.2.** and **3.3.**), nonhTOP1-dependent DNA-damaging agents often give a weaker response (less β-galactosidase induction) in the plus IPTG plates. This is because the induced expression of hTOP1 causes the cells to grow more poorly, and less robust cells apparently give a less robust SOS signal. This is also the reason that the inoculum (**Subheading 3.3.**) of the plus IPTG agar plates is twice that of the minus IPTG agar plates.

References

1. Wang, J. C. (1985) DNA topoisomerases. *Annu. Rev Biochem.* **54,** 665–697.
2. Vosberg, P.-H. (1985) DNA topoisomerases: enzymes that control DNA conformation. *Curr. Top. Microbiol. Immunol.* **114,** 19–102.
3. DiNardo, S., Voelkel, K. A., Sternglanz, R., Reynolds, A. E., and Wright, A. (1982) Escherichia coli DNA topoisomerase I mutants have compensatory mutations in DNA gyrase genes. *Cell* **31,** 43–51.
4. Wang, J. C. (1971) Interactions between DNA and an *Escherichia coli* protein ω. *J. Mol. Biol.* **55,** 532–533.
5. Kirkegaard, K. and Wang J. C. (1985) Bacterial DNA topoisomerase I can relax positively supercoiled DNA containing a single-stranded loop. *J. Mol. Biol.* **185,** 625–637.
6. Depew, R. E., Liu, L., and Wang J. C. (1978) Interactions between DNA and *Escherichia coli* protein ω. *J. Biol. Chem.* **253,** 511–518.
7. Baase, W., and Wang, J. C. (1974) A ω protein from *Drosphilia melanogaster. Biochemistry* **13,** 4299–4303.
8. Been, M. D. and Champoux, J. J. (1984) Breakage of single-stranded DNA by eukaryotic type I topoisomerase occurs only at regions with the potential for base-pairing. *J. Mol. Biol.* **180,** 515–531.
9. Champoux, J. J. (1977) Strand breakage by the DNA untwisting enzyme results in covalent attachment of the enzyme to DNA. *Proc. Natl. Acad. Sci. USA* **74,** 3800–3804.
10. Hsiang, Y.-H., Hertzberg, R. P., Hecht, S., and Liu, L. (1985) Camptothecin induces protein-linked DNA breaks via mammalian DNA topoisomerase I. *J. Biol. Chem.* **260,** 14,873–14,878.
11. Drlica, K. and Franco, R. J. (1988) Inhibitors of DNA topoisomerases. *Biochemistry* **27,** 2253–2259.
12. Bjornsti, M.-A. and Wang, J. C. (1987) Expression of yeast DNA topoisomerase I can complement a conditional-lethal DNA topoisomerase I mutation in *Escherichia coli. Proc. Natl. Acad. Sci. USA* **84,** 8971–8975.
13. Bjornsti, M.-A., Benedetti, P., Viglainti G. A., and Wang, J. C. (1989) Expression of human DNA topoisomerase I in yeast cells lacking yeast DNA topoisomerase I: restoration of sensitivity to the antitumor drug camptothecin. *Cancer Res.* **49,** 6318–6323.
14. Nitiss, J. and Wang J. C. (1988) DNA topoisomerase-targeting drugs can be studied in yeast. *Proc. Natl. Acad. Sci. USA* **85,** 7501–7505.
15. Taylor, S. T. and Menzel, R. (1995) The creation of a camptothecin-sensitive *Echerichia coli* based on the expression of the human topoisomerase I. *Gene* **167,** 69–74.
16. Kjeldsen, E., Bonven, B. J., Andoh, T., Ishii, K., Okada, K., Boland, L., et al. (1988) Characterization of a camptothecin-resistant human topoisomerase I. *J. Biol. Chem.* **263,** 3912–3916.

17. Liu, L. (1989) DNA topisomerase poisons as antitumor drugs. *Annu. Rev. Biochem.* **58,** 351–375.
18. Sampson, B. A., Misra, R. M., and Benson, S. A. (1989) Identification of a new gene of *Escherichia coli* K-12 involved in outer membrane permeability. *Genetics* **122,** 491–501.
19. Eng, W.-W., Faucette, L., Johnson, R., and Sternglanz, R. (1989) Evidence that DNA topoisomerase I is necessary for the cytotoxic effects of camptothecin. *Mol. Pharm.* **34,** 755–760.
20. Kenyon, C. J., Brent, R., Ptashne, M., and Walker G. C. (1982) Regulation of damage inducible genes in *E. coli*. *J. Mol. Biol.* **160,** 445–457.
21. Durban, E., Mills, J. S., Roll, D., and Busch, H. (1983) Phosphorylation of purified Novikoff hepatoma topoisomerase I. *Biochem. Biophys. Res. Commun.* **111,** 897–905.
22. Pommier, Y., Kerrigan, D., Hartman, K., and Glaser, R. I. (1990) Phosphorylation of Mammalian DNA Topoisomerase and Activation by Protein Kinase C. *J. Biol. Chem.* **265,** 9418–9422.
23. Zhang, H., D'Arpa, and Liu, L. F. (1990) A model for tumor cell killing by topoisomerase poisons. *Cancer Cells* **2,** 23–27.
24. Silhavy, T. J., Berman, M. L., and Enquist, L. W. (1984) *Experiments with Gene Fusions*. Cold Spring Harbor Labortory Press, Cold Spring Harbor, NY.
25. Miller, J. H. (1972) *Experiments in Molecular Genetics*. Cold Spring Harbor Labortory Press, Cold Spring Harbor, NY.
26. Eustice, D. C., Feldman, P. A., Colberg-Poley, A. M., Buckery, R. M., and Neubauer, R. H. (1991) A sensitive method for the detection of β-galactosidase in transfected mammalian cells. *BioTechniques* **11,** 739–742.
27. Menzel, R. (1989) A microtiter plate-based system for the semiautomated growth and assay of bacterial cells for beta-galactosidase activity. *Anal. Biochem.* **181,** 40–50.

20

Drug-Induced Cytotoxicity in Tissue Culture

Cynthia E. Herzog and Leonard A. Zwelling

1. Introduction

Topoisomerase II is a requisite enzyme that cleaves both strands of double-stranded DNA, allowing for passage of a second DNA duplex, resulting in the unwinding of chromosomal DNA. Topo II is covalently bound to the 5'-end of the cleaved DNA in a reversible reaction. However, treatment with numerous chemotherapeutic agents, including the epipodophyllotoxins, amsacrine and Adriamycin, stabilize the topo II–DNA complex with the DNA in the cleaved state (reviewed *1,2*). The mechanisms by which complex stabilization and DNA strand breaks lead to cell death have not been elucidated, but in general, the magnitude of drug-induced DNA strand-break production correlates with the magnitude of drug-induced cytotoxicity *(3,4)*. However, there are exceptions to this general rule *(5)*.

The cytotoxicity of topo II-reactive drugs can be measured in tissue culture by numerous methods. Four methods will be given:

1. Growth curve analysis.
2. MTT assay *(6)*.
3. Soft agar cloning *(3)*.
4. Colony formation assays.

Since DNA–complex formation usually reflects topo-mediated cytotoxicity, we will also describe an SDS/KCl assay that measures protein associated with DNA in whole cells and serves as an indirect measure of drug-induced topo-mediated cytotoxicity *(3)*.

The particular method chosen to evaluate cytotoxicity is influenced by several factors. Each of these four methods can be used with adherent cells.

From: *Methods in Molecular Biology, Vol. 95: DNA Topoisomerase Protocols, Part II: Enzymology and Drugs*
Edited by N. Osheroff and M.A. Bjornsti © Humana Press Inc., Totowa, NJ

However, drug-induced cytotoxicity in suspension cells can only be quantified using growth curves or soft agar cloning. Growth curves and the MTT assay are quick assays and useful in the preliminary evaluation of new cell lines or new drugs. The information obtained in these assays can then be used to direct the drug dosages used in soft agar cloning or colony formation assays, which are both more protracted and more precise measures of cell self-renewal capacity. Soft agar cloning evaluates the ability of cells to overcome contact inhibition and to grow in an anchorage-independent manner. Since this reflects how tumors generally grow in vivo, this is the most accurate assay to evaluate what would be expected in vivo. However, not all cell lines can be induced to grow in soft agar. In such cell lines, the colony formation assay is a good alternative.

Before any of these cytotoxicity assays can be used properly, it is first necessary to determine some growth characteristics of the cells of interest in the absence of drug treatment. When tissue-culture cells are initially plated or suspension cells passaged, there is an initial lag in the growth rate, followed by a period of exponential growth and finally a plateau phase. Since many of the drugs of interest are cell-cycle-specific, it is essential that cytotoxicity assays be performed while cells are in the exponential phase of growth. Treatment of adherent cells should therefore be done when they are 40–60% confluent to assure exponential growth. Another important parameter is the plating efficiency.

$$\text{Plating efficiency} = \frac{\text{number of colonies formed}}{\text{number of cells seeded}} \times 100 \qquad (1)$$

When each colony arises from a single cell, this is also called the cloning efficiency.

Although the frequency of drug-induced, topo II-mediated DNA strand breaks has generally been shown to correlate with the degree of cytotoxicity, there are exceptions *(5)*. These breaks are reversible on removal of the drug, yet cell death still occurs *(5)*. Thus, DNA–protein complex formation and DNA strand breakage appear to be necessary, but in themselves not sufficient for cytotoxicity to occur. Many drugs have more than one mechanism of action, so cytotoxicity does not necessarily indicate a topoisomerase-mediated process. Measurement of drug-induced DNA–protein complexes is an indication of topoisomerase-mediated drug effects. Performance of the SDS/KCl assay in whole cells is a rapid method for evaluating DNA–protein interactions *(3,7–9)*.

2. Materials

2.1. MTT Assay

1. 0.25% Trypsin, 0.2% EDTA
2. Phosphate-buffered saline (PBS): 150 m*M* NaCl, 0.7 m*M* KH_2PO_4, 4.3 m*M* K_2HPO_4, pH 7.4.
3. 3-(4,5-Dimethylthiazol-2-y)-2,5-diphenyltetrazolium bromide (MTT) in PBS. Filter through 0.45-µm cellulose nitrate to remove undissolved particles.
4. Dimethyl sulfoxide (DMSO).
5. 96-Well, tissue-culture dishes.
6. Multichannel pipet.
7. Microtiter plate reader (Dynatech Laboratories, Chantilly, VA).

2.2. Growth Curve

1. Trypan blue.
2. Hemocytometer.
3. PBS: 150 m*M* NaCl, 0.7 m*M* KH_2PO_4, 4.3 m*M* K_2HPO_4, pH 7.4.

2.3. Soft Agar Cloning

1. Noble agar (Difco Laboratories, Detroit, MI).
2. 0.1 *M* EDTA, pH 7.4.
3. 12 × 75 mm plastic tubes (Falcon 2054, Becton Dickinson Labware, Lincoln Park, NJ).
4. Dark-field counter.

2.4. Colony Formation Assay

1. Thirty-five-millimeter dishes are needed.
2. 0.4% crystal violet.

2.5. SDS/KCl Assay

1. Glass filters (Whatman GF/F).
2. Filter manifold.
3. 10% trichloroacetic acid (TCA).
4. Ethanol.
5. 1 *N* HCl.
6. 0.4 *N* NaOH.
7. 325 m*M* KCl.
8. Lysis solution: 1.25% SDS, 5 m*M* EDTA, pH 8.0, 0.4 mg/mL denatured Herring sperm DNA.
9. Wash solution: 10 m*M* Tris, pH 8.0, 100 m*M* KCl, 1 m*M* EDTA, 0.1 mg/mL herring sperm DNA.

10. 65°C water bath.
11. Scintillation counter.
12. [^{3}H]thymidine.
13. [^{14}C]leucine.
14. PBS: 15 m*M* NaCl, 5 m*M* KH_2PO_4, pH 7.4.

3. Methods

3.1. MTT Assay

1. Detach cells using trypsin/EDTA (*see* **Note 1**). Resuspend cells in the appropriate amount of medium such that 100 µL contains the number of cells desired/well (*see* **Note 2**).
2. Add 100 µL cells/well, using a multichannel pipet. Allow cells to attach overnight (*see* **Note 3**). (Outside wells can be used as controls: a blank with nothing in it and background wells containing only culture medium.)
3. Suction off medium. Add 200 µL culture medium containing the appropriate drug or vehicle.
4. Drug should be diluted into culture medium to give a drug concentration that is expected to kill all of the cells. (Enough of the initial dilution should be prepared to aliquot 200 µL/well and to do subsequent dilutions.)
5. Do serial two- or threefold dilutions into medium.
6. As a control for the potential cytotoxic effects of the vehicle in which the drug is dissolved, cells should also be treated with vehicle alone at the concentration of vehicle equal to that in the highest drug dose.
7. Add 200 µL of each drug dilution (or vehicle control) to each of triplicate wells containing the attached cells.
8. Incubate at 37°C for 3 d (*see* **Note 4**).
9. Add 40 µL 0.25% MTT to each well except blank. Incubate at 37°C for 1–4 h. (In most cases, 1 h is adequate, but slower-growing cells may need longer incubation.)
10. Suction off medium containing MTT. Add 100 µL DMSO to lyse cells, and shake plates to mix thoroughly.
11. Read the absorbency at 570 nm on a microtiter plate reader (*see* **Note 5**).

$$\% \text{ Growth inhibition} = 100 \times [1 - (\text{OD}_{570} \text{ of treated cells} - \text{background} / \text{OD}_{570} \text{ of vehicle treated cells} - \text{background})] \quad (2)$$

3.2. Growth Curve

1. Treat exponentially growing cells at the desired drug concentration and time. As a control also have vehicle-treated cells.
2. At the end of drug treatment, wash cells twice with PBS.
3. Place cells in fresh medium at a concentration that would allow untreated cells to remain in exponential growth over several days.

Table 1
Agar Medium

# of Tubes	Agar, mg	ddH_2O, mL	Prewarmed medium, mL
10	50	1.25	34
50	225	5.62	160
100	450	11.25	320

4. At the desired time-points, remove an aliquot of cells for counting, and add trypan blue (*see* **Note 6**).
5. Using a hemocytometer, count the total number of live cells (*see* **Note 7**).
6. A plot of the number of cells vs time should be sigmoidal.

3.3. Soft Agar Cloning

1. Wash agar with 0.1 *M* EDTA, pH 7.4. Stir for 6–8 h. Let settle overnight. Decant EDTA, and repeat process three to four times. Rinse with H_2O. Stir for 6–8 h. Let settle overnight. Decant H_2O, and repeat process two to three times. After decanting H_2O for the last time, stir agar at 37°C until dry. Dried agar can be stored indefinitely in air-tight bottles.
2. Mix agar and H_2O according to **Table 1** (*see* **Note 8**). Autoclave for 15 min.
3. While agar is being autoclaved, warm medium to 42°C.
4. Label cloning tubes (triplicate tubes of four dilutions for each drug concentration and time-point—*see* **Step 8** *below*).
5. When agar has cooled enough to handle, add to prewarmed medium as indicated in **Table 1**, and mix thoroughly. Do not shake. Add 3 mL of agar medium to each cloning tubes. Place tubes in 37°C incubator.
6. Treat exponentially growing cells at the desired drug concentration and time.
7. At the end of drug treatment, wash cells twice with PBS.
8. Label four dilution tubes (A, B, C, D) for each concentration and time-point.
9. Place 4.5 mL medium in each dilution tube, except A tube.
10. Count washed drug-treated cells, and then dilute to 1×10^5 cells/mL. This is the A tube.
11. Add 0.5 mL from A tube to B tube (1:10 dilution, now have 1×10^4 cells/mL).
12. Add 0.5 mL from B tube to C tube (1:10 dilution, now have 1×10^3 cells/mL).
13. Add 0.5 mL from C tube to D tube (1:10 dilution, now have 1×10^2 cells/mL).
14. Add 1 mL from each dilution to each of triplicate cloning tubes containing 3 mL of medium. Cap and invert three times to disperse cells throughout tube.
15. Place tube in slushy ice for 15 min, and then place at room temperature for 45 min before placing in 37°C incubator.
16. Count colonies after 14–21 d. Count colonies when they have grown to a size that is clearly visible using the dark-field counter. Count the colonies in the lowest dilution that gives a readable number of colonies (*see* **Note 9**).

$$\text{Cloning efficiency} = \frac{\text{number of colonies in samples}}{\text{predicted number of colonies}} \quad (3)$$

$$\text{for D tubes, this would be } \frac{\text{number of colonies}}{3 \times 100}$$

$$\text{Survival fraction} = \frac{\text{cloning efficiency of drug-treated cells}}{\text{cloning efficiency of vehicle-treated cells}} \quad (4)$$

3.4. Colony Formation Assay

1. Treat exponentially growing cells at the desired drug concentration and time (*see* **Note 10**). As a control, also have untreated cells.
2. At the end of drug treatment, wash cells twice with PBS.
3. Treat cells with trypsin/EDTA to detach.
4. Using 35-mm dish, plate in triplicate with 200–300 cells/dish in 3 mL medium (*see* **Note 11**).
5. After 9–12 d stain colonies with 0.4% crystal violet and count (*see* **Note 12**). Colonies of ≥25 cells should be counted (*see* **Note 13**).

$$\text{Survival fraction} = \frac{\text{number of colonies formed by drug-treated cells}}{\text{number of colonies formed by untreated control cells}} \quad (5)$$

3.5. SDS/KCl Assay

1. Label cells for about 24 h with 0.6 µCi/mL [^{3}H]thymidine and 0.2 µCi/mL [^{14}C]leucine (*see* **Note 14**).
2. Chase with fresh medium for 2 h.
3. Detach cells with trypsin/EDTA and resuspend in medium.
4. Plate 4×10^5 cells/well using six-well plates, 5 mL/well. (You will need TCA and untreated controls, plus four drug concentrations).
5. Incubate for 1 h at 37°C.
6. At this point, place water and wash solutions at 65°C to prewarm.
7. After 1 h, pipet vigorously, and transfer to 15-mL conical tubes on ice. Wash well with 2–3 mL cold PBS.
8. Centrifuge at 1500*g*, at 4°C for 5 min. Resuspend in 5 mL PBS. Centrifuge again, then resuspend in 1 mL PBS, and transfer to microfuge tube (except for TCA control, *see* **Subheading 3.6.**).
9. Centrifuge at high speed in microfuge for 5 min.
10. Resuspend in 1 mL lysis solution. Pass lysate through 22-gage needle four times, being careful not to get too foamy.
11. Incubate for 10 min at 65°C.
12. Add 250 µL 325 m*M* KCl. Vortex vigorously. Place on ice for 10 min.
13. Centrifuge at high speed in microfuge for 15 min.
14. Resuspend in 1 mL wash solution. Incubate for 10 min at 65°C.
15. Incubate on ice for 10 min. Centrifuge at high speed for 10 min.

16. Repeat **steps 13** and **14.**
17. Resuspend in 1 mL dH_2O. Transfer to scintillation vial containing 3 mL hot H_2O. Rinse tube with 1 mL hot H_2O.
18. Add 10 mL Scinti-Verse II, and shake. Count (*see* **Note 15**).

3.6. TCA Precipitation

1. Add 1 mL 10% TCA to 1 mL labeled cells.
2. Pass onto glass filters. Rinse filter with 5 mL 10% TCA and then 70% ethanol.
3. Place filter in vial, and add 0.4 mL 1 *N* HCl. Incubate for 1 h at 65°C.
4. Add 2.5 mL 0.4 *N* NaOH, and incubate for 1 h.
5. Add 2.1 mL hot H_2O, and 10 mL Scinti-Verse II. Count.
6. Results are expressed as [^{3}H]DNA / [^{14}C]protein (*see* **Note 16**).

4. Notes

4.1. MTT Assay

1. Cell should be disaggregated to form a single-cell suspension, since clumps of cells prevent accurate plating of the same number of cells per well.
2. The seeding density of cells will need to be determined ahead of time. The seeding density should be such that an adequate number of cells are present to give an accurate reading for quantitation, but the cells do not become confluent during the 4 d of growth.
3. If cells are loosely adherent and tend to detach with manipulation, gelatin can be used to increase the adhesiveness of cells. Add 0.5% gelatin to each well to coat the bottom. Allow to sit for 30–60 min. Suction off gelatin. Plates are now ready for use.
4. Continuous exposure will allow for determination of relative sensitivity of different cell lines. If the MTT assay is being used to find the appropriate dose for use in a cloning assay in which cells will be treated with drug for 1–2 h, then cells should be treated for the appropriate amount of time, washed twice with PBS, and then placed in fresh medium without drug for 3 d.
5. MTT is converted to formazan in the mitochondria of metabolically active cells. The formazan is then detected as a blue color. This assay is therefore an indirect measure of the number of metabolically active cells.

4.2. Growth Curve

6. Although growth curves can be useful with adherent cells, this assay is easiest to perform with suspension cells, where serial aliquots can be taken from the same flask and counted over several days.
7. Trypan blue is excluded by cells with intact membranes, thereby allowing for differentiation between live (clear) and dead (blue) cells.

4.3. Soft Agar Cloning

8. This method has been used in our laboratory predominantly with HL-60 and L1210 leukemia cells, and uses a final agar concentration of 0.1% *(3)*. The

optimal agar concentration for other cell lines may be different and will have to be determined. Soft agar cloning in plates ***(10)*** and capillary tubes has also been described ***(11)***.

9. The plot of log survival fraction vs drug concentration should yield a straight line. The slope of the line indicates the effectiveness of chemotherapy. The flatter the line, the more resistant the cells, whereas a step line indicates sensitivity.

4.4. Colony Formation Assay

10. Although cells that are 40–60% confluent should be in exponential growth, use of cells in the plateau phase may reduce or abolish drug effect.
11. Various-sized plates or plating dilutions can be used to accommodate cells of different sizes and growth rates.
12. Slower-growing cells may require longer incubation times, whereas more rapidly growing cells will require shorter incubation times.
13. The plot of log survival fraction vs drug concentration should yield a straight line. The slope of the line indicates the effectiveness of chemotherapy. The flatter the line, the more resistant the cells, but a step line indicates sensitivity.

4.5. SDS/KCl Assay

14. Since topoisomerases form covalent bonds with DNA, treatment with SDS does not disassociate DNA from the protein. When protein is precipitated with KCl, the covalently linked DNA is also precipitated. By labeling both DNA with [^{3}H]thymidine and protein with [^{14}C]leucine, the [^{14}C]labeled protein serves as an internal control for cell number.
15. Both topo I and topo II form covalent bonds with DNA. Therefore, this assay does not differentiate between these two enzymes.
16. Theoretically the [^{14}C] counts should be the same for both the TCA and the SDS/KCl precipitated protein. By dividing the [^{3}H] count for each sample by the [^{14}C] count, one corrects for differences in the number of cells aliquoted or the efficiency of precipitation.

References

1. Osheroff, N., Zechiedrich, E. L., and Gale, K. C. (1991) Catalytic function of DNA topoisomerase II. *BioEssays* **13,** 269–275.
2. Liu, L. F. (1989) DNA topoisomerase poisons as antitumor drugs. *Annu. Rev. Biochem.* **58,** 351–375.
3. Zwelling, L. A., Hinds, M., Chan, D., Mayes, J., Sie, K. L., Parker, E., et al. (1989) Characterization of an amsacrine-resistant line of human leukemia cells. *J. Biol. Chem.* **264,** 16,411–16,420.
4. Davies, S. M., Robson, C. N., Davies, S. L., and Hickson, I. D. (1988) Nuclear topoisomerase II levels correlate with the sensitivity of mammalian cells to intercalating agents and epipodophyllotoxins. *J. Biol. Chem.* **263,** 17,724–17,729.
5. Gewirtz, D. A. (1991) Does bulk damage to DNA explain the cytostatic and cytotoxic effects of topoisomerase II inhibitors? *Biochem. Pharmacol.* **42,** 2253–2258.

6. Alley, M. C., Scudiero, D. A., Monks, A., Hursey, M. L., Czerwinski, M. J., Fine, D. L., et al. (1988) Feasibility of drug screening with panels of human tumor cell lines using a microculture tetrazolium assay. *Cancer Res.* **48,** 589–601.
7. Liu, L. F., Rowe, T. C., Yang, L., Tewey, K. M., and Chen, G. L. (1983) Cleavage of DNA by mammalian DNA topoisomerase II. *J. Biol. Chem.* **258,** 15,365–15,370.
8. Trask, D. K. and Muller, M. T. (1983) Biochemical characterization of topoisomerase I purified from avian erythrocytes. *Nucleic Acids Res.* **11,** 2779–2800.
9. Denstman, S. C., Ervin, S. J., and Casero, R. A., Jr. (1987) Comparison of the effects of treatment with the polyamine analogue N^1,N^8-*bis*(ethyl)spermidine (BESpd) of difluoromethylornithine (DFMO) on the topoisomerase II mediated formation of 4'-(9-acridinylamino) methanesulfon-*M*-anisidide (*m*-AMSA) induced cleavable complex in the human lung carcinoma line NCI H157. *Biochem. Biophys. Res. Commun.* **149,** 194–202.
10. Bepler, G., Carney, D. N., Gazdar, A. F., and Minna, J. D. *In vitro* growth inhibition of human small cell lung cancer by physalaemin. *Cancer Res.* **47,** 2371–2375.
11. Ali-Osman, F. and Beltz, P. A. (1988) Optimization and characterization of the capillary human tumor clonogenic cell assay. *Cancer Res.* **48,** 715–724.

21

Analysis of Drug Uptake-Flow Cytometry

Grace Zoorob and Thomas Burke

1. Introduction

Flow cytometry is a laser-based technique used for the quantitation of specific intracellular and extracellular properties of cells, bacteria, or other biological particles. It is unique in its ability to perform simultaneous multiparameter analysis and to sort single cell types from heterogeneous mixtures. Its application initiated the third phase of method development in modern cell biology, following the introduction of microspectrophotometry and autoradiography *(1)*. In comparison with other methodologies, flow cytometry offers very rapid measurements with hundreds or even thousands of cells being analyzed in 1 s with high accuracy and reproducibility.

A schematic diagram of a typical flow cytometer is shown in **Fig. 1**. The optical layout has an orthogonal configuration with the three main axes of the sample flow, laser beam, and optical axis of fluorescence detection existing at right angles to each other. Basically, a laser source emits light of a certain wavelength, which is focused on the cells' sample flow by means of one or two lenses. The stream of cells is uniformly illuminated at the analysis point (**Fig. 1**). Surrounding the analysis point, a set of lenses collects the emerging laser light and focuses it onto photomultiplier tubes (detectors). Each photomultiplier tube has a specific light filter in front of it to ensure that it only detects light of the color transmitted through its own filter. The intensity of this light (side scatter signal) is frequently related to the fluoresecence characteristics of the particles flowing through the illuminating light path. A photodiode detector, which sits directly in line with the illuminating laser beam, measures the forward scatter signal and can be related to the size and refractive index of the cells in the flowing stream. By applying different voltages to the deflecting

From: *Methods in Molecular Biology, Vol. 95: DNA Topoisomerase Protocols, Part II: Enzymology and Drugs*
Edited by N. Osheroff and M.A. Bjornsti © Humana Press Inc., Totowa, NJ

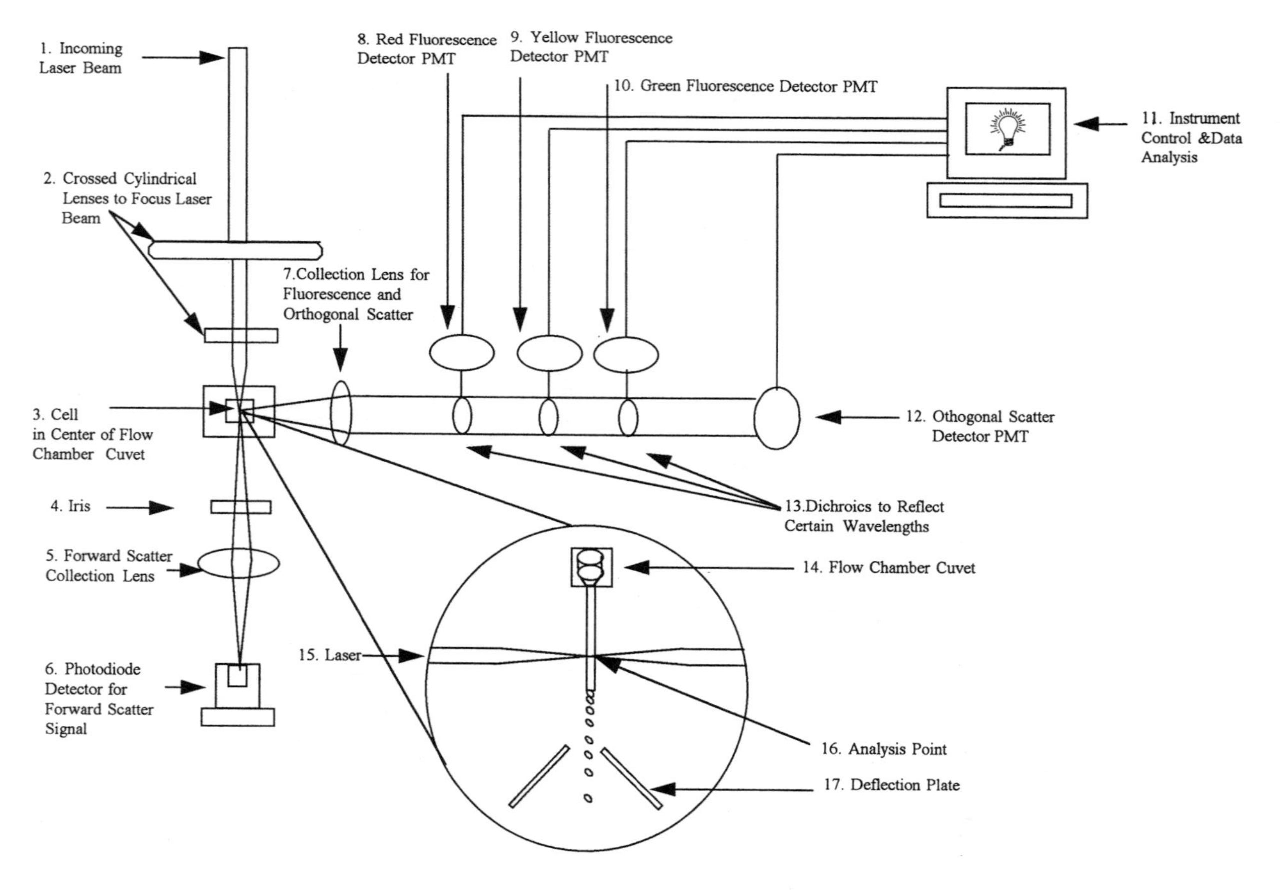

Fig. 1. Schematic diagram of a typical flow cytometer.

plates, droplets become charged and are deflected into the sorting tubes. Flow cytometry is an attractive approach for fluorescent drug uptake studies, since it measures the fluorescence signal of drugs from a single cell with high accuracy ***(2–10)***. Even the intracellular accumulation of drugs with relatively weak fluorescence (e.g., mitoxantrone) can be studied by flow cytometry using high-power laser excitation and high-gain photomultiplier detection systems ***(11,12)***. In addition, studies with nonfluorescent drugs have been conducted after preparing fluorescent analogs where the point of covalent attachment of the fluorescent label is such that significant biological activity is maintained ***(13–21)***.

1.1. Topoisomerase I and II Inhibitors

DNA topoisomerses I and II (topo I and topo II) are enzymes responsible for controlling, maintaining, and modifying structures of DNA during replication and translation of genetic material. A number of topo-active anticancer drugs have been developed, many of which are intinsically fluorescent, which allows their cellular interactions to be characterized using flow cytometric technologies. For example, topo II inhibitors, such as doxorubicin, daunorubicin, and idarubicin from the anthracycline family, as well as menogaril and mitoxantrone are all intrinsically fluorescent, which allows their cellular interactions to be studied using flow cytometry. The agents listed above interfere with topo II activity and have been used in the treatment of solid tumors, including carcinomas, soft tissue sarcomas, and leukemia.

Camptothecin (CPT) and its analogs are agents that inhibit DNA and RNA synthesis in tumor cells by targeting topo I. Because of their manageable toxicity and encouraging activity against solid tumors, several camptothecin analogs, such as 9-aminocamptothecin, topotecan, and irinotecan, have been approved for use or are currently undergoing clinical trials. CPT is a strongly fluorescent molecule and many of its analogs are fluorescent as well. The strong fluorescent emissions of the CPTs prove to be of great advantage in characterizing their cellular interactions as described below.

1.2. Uptake Studies of CPT and Its Analogs

This chapter describes a flow cytometric assay for the intracellular accumulation of the highly fluorescent camptothecin drugs with red blood cells (RBCs). The drug uptake studies are designed to show some of the features attainable by the flow cytometric technique.

CPT and its related analogs each contain an α-hydroxy-δ-lactone ring moiety that hydrolyzes rapidly under physiological conditions, i.e., at pH 7.0 or above, to yield the far less active carboxylate form of the drug ***(22,23)*** (**Fig. 2**). The fluorescent nucleus of both lactone and carboxylate forms of the drug are

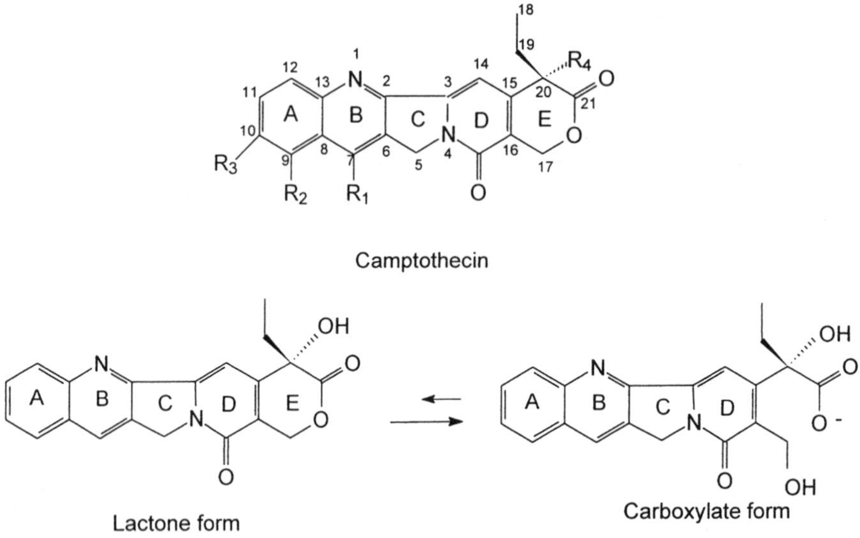

Camptothecin

Lactone form

Carboxylate form

Fig. 2. Structure of camptothecin, lactone, and carboxylate forms.

the same. Because of the dynamic nature of CPT in solution, a rapid noninterrupted flow cytometric method was developed *(24,25)* to study the uptake of CPT by red blood cells. The measurements are accomplished within 5 min after drug addition to cell suspensions (*see* **Notes 1** and **2**).

The uptake of CPT by RBCs has been studied in both PBS and PBS containing 40 mg/mL human serum albumin (HSA) (**Fig. 3**). The histogram in **Fig. 3A** represents the fluorescence intensity of RBCs only (background intensity) at a mean fluorescence of 52. The uptake of 10 μ*M* CPT in RBC is shown in **Fig. 3B** with a mean fluorescence at 250. The observed decrease in the cellular accumulation of 10 μ*M* CPT in the presence of 40 mg/mL HSA (mean fluorescence at 162) is presented in **Fig. 3C**. The results are suggestive of CPT associating with the extracellular protein (HSA), a process that reduces the accumulation of the drug in the RBCs.

In another set of experiments, the uptake of both lactone and carboxylate forms of CPT by RBCs is studied (**Fig. 4**). As pointed out before, both forms of the drug have the same fluorophore, and direct comparison of data sets is possible in order to evaluate relative changes in drug accumulation. The carboxylate form of CPT is negatively charged at neutral pH and is therefore less membrane-permeable relative to its lactone form *(26)*; therefore, the lower fluorescence intensity and the decrease in the slope of **Fig. 4A** for the carboxylate-treated cells are readily interpretable. In addition, cellular accumulation of the lactone form of CPT and a lactone lipophilic prodrug encapsulated in liposome

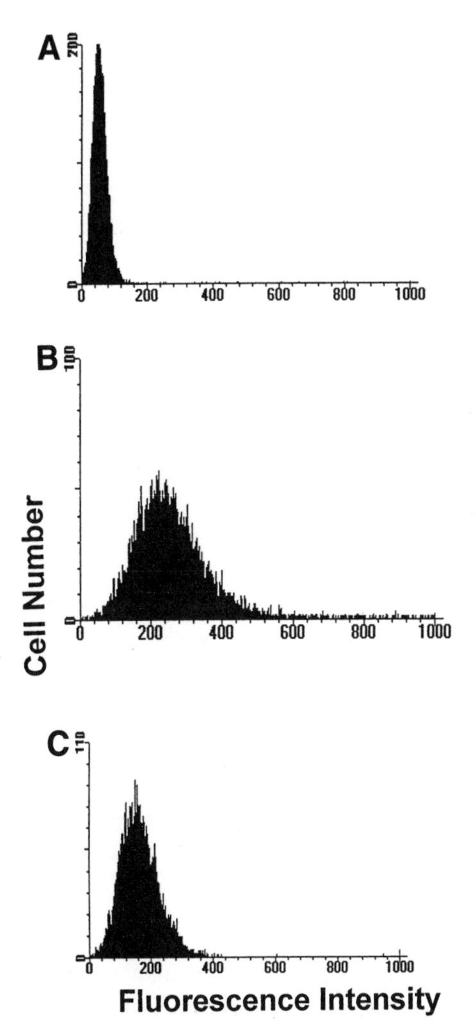

Fig. 3. Histograms of the fluorescence intensity of (**A**) RBCs only, (**B**) RBCs exposed to 10 μ*M* CPT, and (**C**) RBCs exposed to 10 μ*M* CPT in the presence of 40 mg/mL HSA.

is shown in **Fig. 4B**. The prodrug is more lipophilic than CPT and presumably interacts more extensively with RBCs yielding higher fluorescence intensities than CPT alone as seen by the dramatic increase in the slope of **Fig. 4B**.

To rule out the effects of liposomes in the previous experiment on mean fluorescence intensity, the accumulation of two lipophilic prodrugs in RBC suspensions is evaluated. Each drug has exactly the same fluorescent chromophore, and therefore, comparison of fluorescence intensity levels is a valid

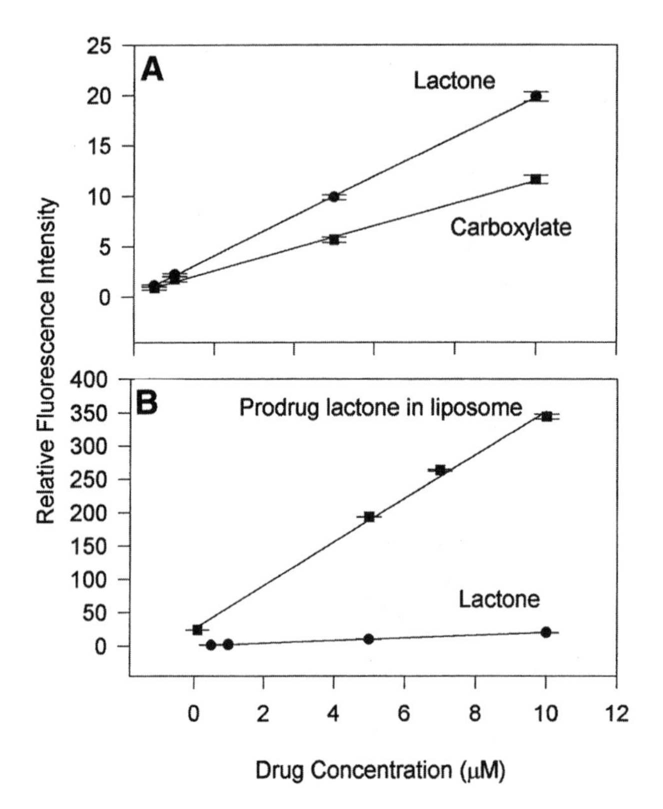

Fig. 4. Uptake of CPT analog by RBCs. **(A)** Uptake of the lactone and carboxylate forms of CPT. **(B)** Uptake of lactone forms of CPT and a lipophillic prodrug delivered as a liposomal suspension.

method for monitoring relative changes in drug accumulation within the RBCs. Prodrug 2 is more lipophilic than 1. **Figure 5** shows that the more lipophilic the drug, the better it accumulates in RBCs, hence the increase in the slopes of the cells treated with the lipophilic prodrugs relative to the control cells treated with CPT only.

The uptake of CPT and daunomycin by HL-60 cells was also studied by flow cytometry. HL-60 is a promyelocytic cell line, and the accumulation of different drugs in the cells can be monitored over time (**Fig. 6**). The mean fluorescence intensity of CPT in HL-60 increases in the first 5 min and then reaches a plateau indicating a saturation in CPT levels in the cells. On the other hand, daunomycin uptake is strongly influenced by the ability of the agent to bind DNA, resulting in more continual accumulation of the drug in the cell. CPT, which does not bind DNA with high affinity, likely equilibrates with

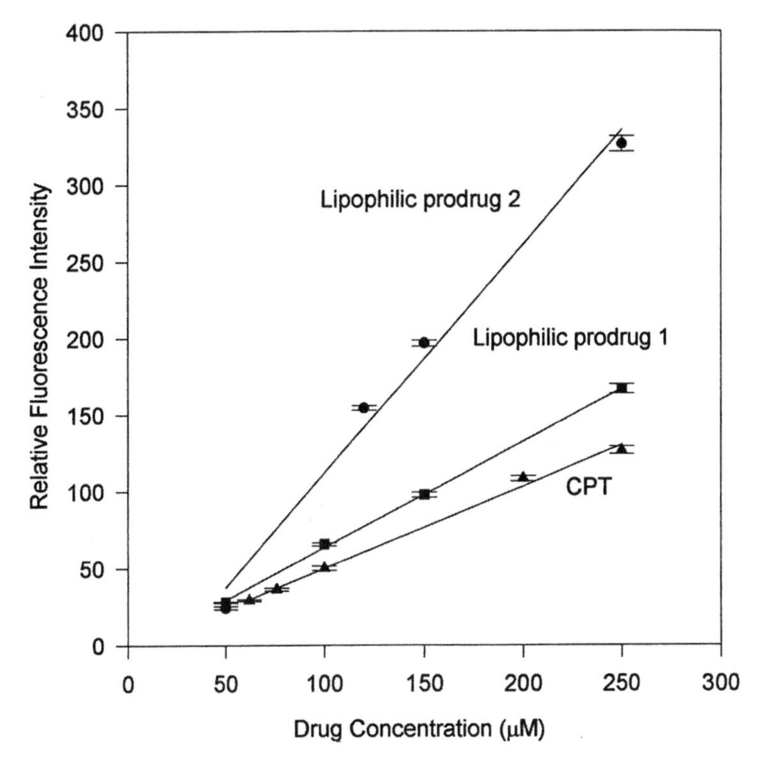

Fig. 5. Accumulation of CPT and some of lipophilic prodrugs in RBCs.

membrane components and reaches a plateau soon after drug addition to cell suspension.

1.3. Characterization of Drug Accumulation in Multidrug Resistant Cell Lines

Multidrug resistance (MDR) to cancer chemotherapeutic agents, including antibiotics, anthracyclines, and alkaloids, may be the result of the overexpression of proteins (mainly P-glycoprotein or PGP) that alter net drug cellular accumulation. MDR poses a major problem in the treatment of many cancers; therefore, the understanding of MDR and overcoming it has been a much pursued area of cancer research. Monitoring drug retention and efflux from cells can be accomplished by flow cytometry. In the past few years, several protocols have been reported in the literature covering a large number fluorescent drugs and cells lines studied by flow cytometry. The intrinsically high fluorescent anthracyclines (doxorubicin and daunorubicin) are among those being actively studied. Various flow cytometric methods, whether

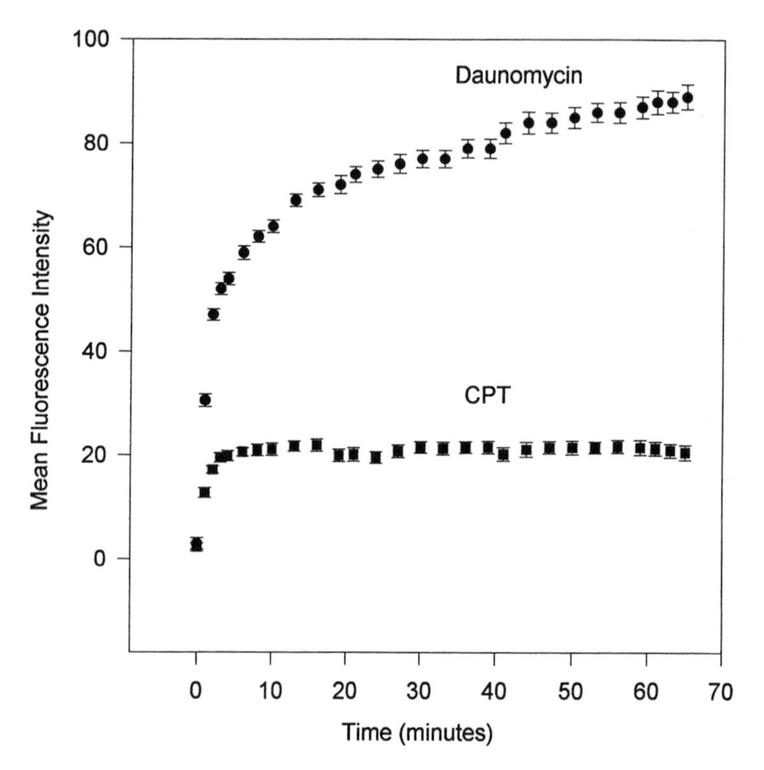

Fig. 6. Uptake of daunomycin and CPT by HL-60 cells.

time-interrupted *(2,3)* or noninterrupted *(4)* have been developed for monitoring anthracycline uptake into wide-type and PGP-positive MDR cells *(5–7)*. Other investigations concerning cell cycle-related uptake *(5)* and drug uptake into multicell spheroids *(6)* for anthracyclines have been reported. In addition to doxorubicin and daunorubicin, a fluorescence dye rhodamine 123 is also widely used as a substrate of PGP to analyze MDR phenotype by flow cytometry *(7–9)*. Recently, Doral and coworkers reported using flow cytometry for monitoring the transport, uptake, efflux, and compartmentalization of fluorescent drugs in MDR cell samples *(10)*.

2. Materials

1. A FACStar Plus (Becton-Dickinson, San Jose, CA) flow cytometer equipped with a 5-W water-cooled multiline argon ion laser operating at 350 nm is used. A band-pass filter of 445–485 is employed. Fluorescence signal intensity of 1000 events is measured at an analysis rate of 400–500 cells/s.
2. The 20(S)-camptothecin (CPT) was generously provided by the laboratories of Monroe Wall and Mansukh Wani (Research Triangle Institute, Research Triangle

Park, NC). Lipophilic CPT analogs are prepared in our laboratories. Daunomycin was purchased from Sigma Chemical Companies, St. Louis, MO. Stock solutions of the drugs are prepared in A.C.S. spectrophotometric-grade dimethyl sulfoxide (DMSO) (Aldrich, Milwaukee, WI) at a concentration of 2×10^{-3} *M* and are stored in the dark at –20°C until use.

3. Outdated packed RBCs are used (*see* **Note 3**). The HL-60 human promyelocytic leukemia cell line was kindly provided by Elaine Jacobson, Nutritional Sciences, University of Kentucky, Lexington, KY. The cells are maintained in RPMI 1640 medium (Gibco) supplemented with 10% fetal bovine serum, 50 µg/mL penicillin, 50 µg/mL streptomycin, 100 µg/mL neomycin (PSN antibiotics mixture, Gibco), 2 m*M* L-glutamine at 37°C in a humidified atmosphere containing 5% CO_2. Cell viability is assessed by the trypan blue exclusion assay. Only exponentially growing, viable cells are used in the experiments.
4. Phosphate-buffered saline (PBS): 8 m*M* dibasic sodium phosphate (Na_2HPO_4), 1 m*M* potassium phosphate monobasic (KH_2PO_4), 137 m*M* sodium chloride (NaCl), and 3 m*M* potassium chloride (KCl) at pH 7.4 are used (*see* **Note 4**). Additional PBS solutions are prepared at pH 3.0 and 10.0. The pH is adjusted using concentrated solutions of hydrochloric acid (HCl) and potassium hydroxide (KOH).
5. Centrifuge tubes (15 mL each) and pipets.
6. Polystyrene Falcon brand test tubes (12 × 75 mm) for sample introduction to the flow cytometer.
7. Bench-top centrifuge.
8. Hemocytometer.
9. Standard-Brite fluorescence beads (Coulter, Hialeah, FL).
10. Zeiss inverted microscope (Carl Zeiss, Thornwood, NY).

3. Methods

3.1. Preparation of RBCs

Suspensions containing plasma-free RBCs are prepared in PBS buffer in the following manner. Outdated RBCs (10 mL) are centrifuged at 4°C at 1000*g* for 15 min, and the supernatant is discarded. The RBCs are then carefully resuspended in an equal volume of ice-cold PBS and centrifuged at 1000*g* again. This step is repeated four times. The washed RBCs (4.5 mL) are then resuspended in PBS buffer (5.5 mL). The RBC suspension is then diluted for drug uptake studies, and the final RBC concentration is determined using a hemocytometer to be about 2×10^6 cells/ mL. RBC preparations are used the day of isolation and are kept at 4°C until immediately prior to use.

3.2. Flow Cytometric Analysis

1. Use a long band-pass filter of 445–485 nm with the flow cytometer to separate fluorescence emission from the scattered light for the drug. Another band-pass filter of 305–390 nm is used to measure the right-angle scattered light.

2. Set a reference channel (fluorescence intensity) using fluorescence standard beads. Record the output power of the laser, high voltage and gain of the PMT, and the fluorescence channel. The same reference channel should be used daily so that data generated from different days can be compared.
3. Prepare RBC suspensions as in the previous section (density of 2×10^6 cells/mL). Put in Falcon Brand test tubes, and sample by the flow cytometer for 1 min to obtain a baseline value prior to the addition of the drug stock solution. The fluorescence intensity recorded is the background signal of cells. Control the cell flow rate at 400–500 cells/s.
4. Pause the instrument and quickly add drug stock solution to a final desired concentration. Mix thoroughly and measure again until the fluorescence intensity reaches a plateau (2–5 min following drug addition to the RBC suspension). Measure the fluorescence intensity at the plateau for 3 min.
5. The mean fluorescence intensity at the plateau, corrected for background signal before adding the drug, is used to depict the intracellular accumulation of drug. The plot of fluorescence intensity vs drug concentration is linear, indicating a passive diffusion mechanism in the concentration range tested.
6. Measure the fluorescence intensity of RBC alone, and then measure it for 10 μ*M* CPT in RBC suspension. Add HSA (40 mg/mL) to 10 μ*M* CPT suspended in RBC, and measure the fluorescence. The effects of the addition of HSA on drug uptake are seen in **Fig. 3**.
7. Repeat the experiment as before by following **steps 3–5** using pH-adjusted drug stock solution to study the effects of the uptake of the lactone and carboxylate forms of CPT by RBCs. Repeat the experiment as before, but use the lactone form of a lipophilic CPT analog encapsulated in liposomes. The plot of fluorescence intensity vs drug concentration is seen in **Fig. 4**.
8. Measure the fluorescence intensities of RBCs exposed to CPT, lipophilic prodrug 1, and lipophilic prodrug 2 as before. Plot to examine the effects of altered drug lipophilicities on their uptake by RBCs (**Fig. 5**).
9. HL-60 cells in exponential phase are washed twice with PBS and resuspended in PBS at a density of 1×10^6 cells/mL. Each suspension is subjected to flow cytometry for 1 min to obtain a baseline value prior to the addition of drug stock solution. Both CPT and daunomycin display strong intrinsic fluorescence emissions, and cellular fluorescence intensities owing to drug exposure are monitored as before at times up to 65 min (*see* **Notes 5** and **6**). The uptake of CPT and daunomycin is seen in **Fig. 6**.

4. Notes

1. By preparing a CPT carboxylate solution (1 part 2 m*M* CPT lactone in DMSO, 1 part PBS buffer at pH 10.0), the interaction of the carboxylate form of camptothecin with RBCs can be monitored. Add either lactone or carboxylate stock solutions to cell suspensions.
2. It should be noted that flow cytometric measurements, like other steady-state fluorescence spectroscopy, have arbitrary units. The method described here is a

measurement of the relative uptake of fluorophores. Other alternative methods (e.g., HPLC, radiolabeled drugs) have to be used to quantify the absolute amount of drug inside cells.

3. The protocol can be applied to other cell lines.
4. If a buffer solution other that PBS is used (e.g. HBSS, HEPES), make sure it does not contain Ca^{2+} and Mn^{2+} ions to minimize cell clumping.
5. The fluorescence intensity measured from a single cell depends not only on the intracellular drug concentration, but also on cell size. As a result, comparison of accumulation of the same drug in different cell lines should be done with caution.
6. The flow cytometric method does not provide information about whether the drug locates inside cell or just binds to the cell membrane. It is always recommended to examine fluorophore-stained cells under a microscope. A confocal microscope is the best choice if possible for determining intracellular localization.

References

1. Darzynkiewicz, Z., Robinson, J. P., and Crissman, H. A. (1994) in *Flow Cytometry*, 2nd. ed., Academic, New York, pp. xxv.
2. Krishan, A. and Ganapathi, R. (1980) Laser flow studies on the intracellular fluorescence of anthracyclines. *Cancer Res.* **40,** 3895–3900.
3. Slapak, C. A., Lecerf, J-M., Daniel, J. C., and Levy, S. B. (1992) Energy-dependent accumulation of daunorubicin into subcellular compartments of human leukemia cells and cytoplasts. *J. Biol. Chem.* **267,** 10,638–10,644.
4. Herweijer, H., Sonneveld, P., Baas, F., and Nooter, K. (1990) Expression of mdr1 and mdr3 multidrug-resistance genes in human acute and chronic leukemias and association with stimulation of drug accumulation by cyclosporine. *J. Natl. Cancer Inst.* **82,** 1133–1140.
5. Minderman, H., Linssen, P., Wessels, J., and Haanen, C. (1993) Cell cycle related uptake, retention and toxicity of idarubicin, daunorubicin and doxorubicin. *Anticancer Res.* **13,** 1161–1166.
6. Bichay, T. L., Adams, E. G., Inch, W. R., Adams, W. J., Brewer, J. E., and Bhuyan, B. K. (1990) HPLC and flow cytometric analyses of uptake of adriamycin and menogaril by monolayers and multicell spheroids. *Selective Cancer Therapeutics* **6,** 153–166.
7. Feller, N., Kuiper, C. M., Lankelma, J., Ruhdal, J. K., Scheper, R. J., Pinedo, H. M., et al. (1995) Functional detection of *MDR*1/P170 and *MRP*/P190-mediated multidrug resistance in tumor cells by flow cytometry. *Br. J. Cancer* **72,** 543–549.
8. Porter, C. W., Ganis, B., Rustum, Y., Wrzosek, C., Kramer, D. L., and Bergerson, R. J. (1994) Collateral sensitivity of human melanoma multidrug-resistant variants to the polyamine analogue, N^1,N^{11}-diethylnorspermine. *Cancer Res.* **54,** 5917–5924.
9. Frey, T., Yue, S., and Haugland, R. P. (1995) Dyes providing increased sensitivity in flow-cytometric dye-efflux assays for multidrug resistance. *Cytometry* **20,** 218–227.

10. Doral, M. S., Ho, A. C., Jackson-Stone, M., Fu, Y. F., Goolsby, C. L., and Winter, J. N. (1995) Flow cytometric assessment of the cellular pharmacokinetics of fluorescent drugs, *Cytometry* **20,** 307–314.
11. Smith, P. J., Sykes, H. R., Fox, M. E., and Furlong, I. J. (1992) Subcellular distribution of the anticancer drug mitoxantrone in human and drug-resistant murine cells analyzed by flow cytometry and confocal microscopy and its relationship to the induction DNA damage. *Cancer Res.* **52,** 4000–4008.
12. Fox, M. E. and Smith, P. J. (1995) Subcellular localization of the antitumour drug mitoxantrone and induction of DNA damage in resistance and sensitive human carcinoma cells. *Cancer Chemother. Pharmacol.* **35,** 403–410.
13. Assaeaf, Y. G., Feder, J. N., Sharma, R. C., Wright, J. E., Rosowsky, A., Shane, B., et al. (1992) Characterization of the coexisting multiple mechanisms of methotrexate resistance in mouse 3T6 R50 fibroliasts. *J. Biol. Chem.* **267,** 5776–5784.
14. Fathi, R., Huang, Q., Coppola, G., Delaney, W., Teasdale, R., Krieg, A. M., et al. (1994) Oligonucleotides with novel, cationic backbone substituents: aminoethylphosphonates. *Nucleic Acids Res.* **22,** 5416–5424.
15. Gentry, M. J. Snitily, M. U., and Preheim, L. C. (1995) Phagocytosis of *Streptococcus pneumoniae* measured *in vitro* and *in vivo* in a rat model of carbon tetrachloride-induced liver cirrhosis. *J. Infect. Dis.* **171,** 350–355.
16. Wenisch, C., Parschalk, P., Hasenhundl, M., Griesmacher, A., and Graninger, W. (1995) Polymorphonuclear leucocyte dysregulation in patients with gram-negative septicaemia assessed by flow cytometry. *Eur. J. Clin. Invest.* **25,** 418–424.
17. Bandres, J. C., Trail, J., Musher, D. M., and Rossen, R. D. (1993) Increased phagocytosis and generation of reactive oxygen products by neutrophils and monocytes of men with stage 1 human immunodeficiency virus infection. *J. Infect. Dis.* **168,** 75–83.
18. Wall, J. E., Buijs-Wilts, M., Arnold, J. T., Wang, W., White, M. M., Jennings, L. K., et al. (1995) A flow cytometric assay using mepacrine for study of uptake and release of platelet dense granule contents. *Br. J. Haematol.* **89,** 380–385.
19. Gift, E. A. and Weaver, J. C. (1995) Observation of extremely heterogeneous electroporative molecular uptake by *Saccharomyces cerevisiae* which changes with electric field pulse amplitude. *Biochim. Biophys. Acta* **1234,** 52–62.
20. Stringer B., Imrich, A., and Kobzik, L., (1995) Flow cytometric assay of lung macrophage uptake of environmental particles. *Cytometry* **20,** 23–32.
21. Ranganathan, S., Hattori, H., and Kashyap, M. L. (1995) A rapid flow cytometric assay for low-density lipoprotein receptors in human peripheral blood mononuclear cells. *J. Lab. Clin. Med.* **125,** 479–486.
22. Slichenmyer, W. J., Rowinsky, E. K., Donehower, R. C., and Kaufmann, S. H. (1993) The current status of camptothecin analogues as antitumor agents. *J. Natl. Cancer Inst.* **85,** 271–291.
23. Potmesil, M. (1994) Camptothecins: from bench research to hospital wards. *Cancer Res.* **45,** 1431–1439.

24. Mi, Z. and Burke, T. G. (1994) Differential interactions of camptothecin lactone and carboxylate forms with human blood components. *Biochemistry* **33,** 10,325–10,335.
25. Mi, Z., Malak, H., and Burke, T.G. (1995) Reduced albumin binding promotes the stability and activity of topotecan in human blood. *Biochemistry* **34,** 13,722–13,728.
26. Burke, T. G., Mishra, A. K., Wani, M. C., and Wall, M. E. (1993) Lipid bilayer partioning and stability of camptothecin drugs. *Biochemistry* **32,** 5352–5364.

22

Cell-Cycle Analysis of Drug-Treated Cells

Frank Traganos, Gloria Juan, and Zbigniew Darzynkiewicz

1. Introduction

A variety of flow cytometric methods have been developed over the past 25 yr to study how treatment with chemotherapeutic agents affects cell-cycle progression. One of the most commonly used measurements relies on a single-time analysis of the DNA distribution of a cell population *(1)*. This analysis may also be multivariate, for instance, when another cell feature is measured in addition to DNA. The additional feature(s) often provides information about a particular metabolic or molecular feature of the cell that often correlates with the rate of cell progression through the cycle or cell quiescence. Therefore, although such measurements *per se* cannot reveal whether or not the cell actually progresses through the cycle, the kinetic information is inferred from the DNA content (cell-cycle position) and from the metabolic or molecular profile of that cell.

A second group of multiparametric techniques involve the analysis of DNA replication in conjunction with DNA content measurements. Generally, the incorporation of a DNA replication marker, such as bromodeoxyuridine (BrdUrd), is related to the position of the cell in the cycle. Although incorporation of BrdUrd can be detected by cytochemical methods based on the use of DNA dyes whose fluorescence is quenched by BrdUrd *(2)*, most often BrdUrd is detected with antibodies directed against the analog after DNA *in situ* has been partially denatured by acid or heat *(3)*.

With either approach, single isolated measurements provide little kinetic information. If, however, one makes time-lapse measurements of the cohort of BrdUrd-labeled cells, for instance, it is possible to estimate their rate of progression through different points of the cell cycle.

From: *Methods in Molecular Biology, Vol. 95: DNA Topoisomerase Protocols, Part II: Enzymology and Drugs*
Edited by N. Osheroff and M.A. Bjornsti © Humana Press Inc., Totowa, NJ

The methods described in this chapter all utilize flow cytometry to quantitate cellular constituents stained with fluorescent ligands. Flow cytometry has the advantage of allowing for the very accurate quantitation of small amounts of intracellular components in individual cells at rates of hundreds to thousands of cells per second. Since data are generally collected in a correlated (list mode) fashion, multiple features of individual cells can be characterized. Only a few selected methods are presented in this chapter. More detailed descriptions of these and other methods, their applicability to different cell systems, advantages, and limitations are presented elsewhere (*4–6*).

1.1. Single Parameter—DNA Content Measurements

Virtually all DNA content measurements utilize fluorescent probes with a high degree of specificity for DNA. Often these dyes intercalate into the double helix of DNA *in situ*, are excited by light of the appropriate wavelength emanating either from a laser or mercury arc lamp, and fluoresce in the visible spectra with an intensity proportional to the amount of dye bound. This type of approach provides the investigator with a DNA frequency distribution (**Fig. 1**) in which the intensity of fluorescence of the individual cells is proportional to their DNA content. The frequency distribution or histogram can than be deconvoluted into its component parts consisting of postmitotic G_1 cells with a DNA index of 1.0, post-DNA synthetic, premitotic G_2 and mitotic M-cells, which have twice the DNA content and, therefore, twice the fluorescence intensity of G_1 cells (DNA index = 2.0), and cells with a DNA index between 1.0 and 2.0, which represent the DNA-synthesizing S-phase cells (**Fig. 1**). Treatment of cells in culture with agents that perturb the progression of cells through the cycle will result in changes in the DNA distribution manifested as changes in the proportion of cells in individual cell-cycle phases (**Fig. 1**).

1.2. Dual Parameter—DNA vs Cyclin Measurements

The expression of various proteins often varies during the cell cycle as well as in cycling and quiescent cells. The cellular content of such proteins may provide information on the proliferative status of the cell. One such set of proteins are the cyclins, the regulatory subunits within the holoenzyme of cyclin–cyclin-dependent kinases, which are key components of the cell-cycle machinery. These complexes, when activated by appropriate modifications, phosphorylate a distinct set of proteins at different phases or checkpoints of the cell cycle, thereby driving the cell through the cycle (reviewed in **refs.** *7–9*). Among the family of nine cyclin types (A–I) are specific cyclins that are expressed discontinuously throughout the cycle. The timing of synthesis and degradation of particular cyclins (e.g., D-type cyclins in G_0/G_1; cyclin E in G_1

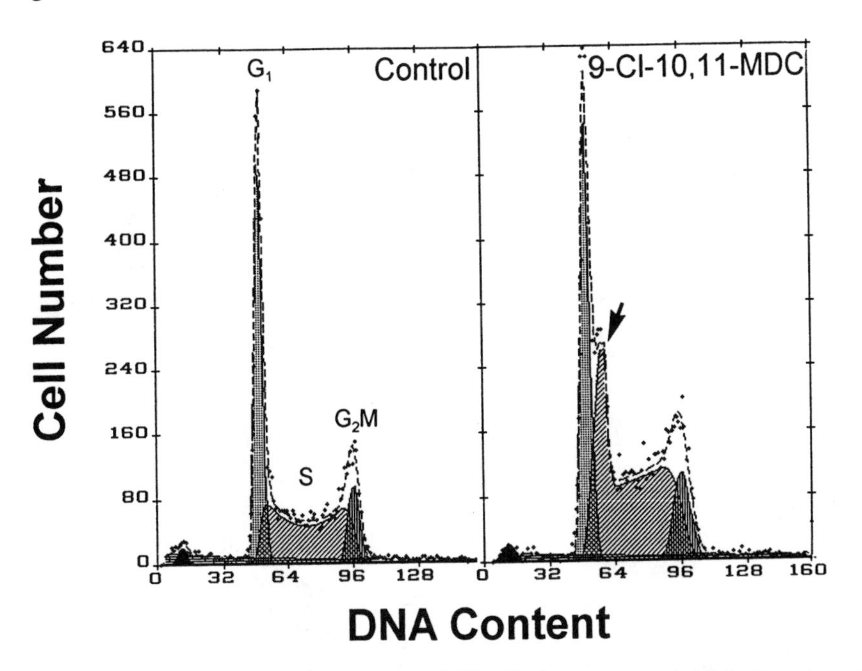

Fig. 1. DNA frequency histograms of HL-60 human myeloid leukemic cells untreated (control) and treated with the DNA topoisomerase I inhibitor 9-chloro-10, 11-dimethylcamptothecin (9-Cl-10,11-MDC). Exponentially growing cultures of HL-60 cells were either untreated or exposed to a 2.5-n*M* concentration of the camptothecin analog 9-Cl-10,11-MDC kindly provided by M. Potmesil of the New York University School of Medicine (New York, NY). The cells were harvested after 4 h of culture, fixed in ethanol, stained as described in the text, and measured by flow cytometry. The relative fluorescence of the DNA stain was proportional to DNA content and produced a characteristic distribution in which the first peak represented postmitotic cells with a G_1 DNA content, a second, minor peak at twice the fluorescence intensity of the first, which represented premitotic G_2 and mitotic (M) phase cells, as well as cells with intermediate fluorescence representing DNA synthesizing S-phase cells. Each frequency histograms contains 1×10^5 cells and has been scaled to the height of the major peak. Note that in the drug-treated distribution, the camptothecin analog caused an accumulation of cells in early S-phase (arrow), which represents an 11% shift in the distribution; the S-phase cells increased from 48% in control to 59% in the drug-treated culture with a concomitant decrease in G_1 phase cells from 41 to 30%.

and early S; cyclin A in S and G_2, and cyclin B1 in late S, G_2 and mitosis), in addition to DNA replication and mitosis, provides characteristic landmarks of cell-cycle progression. Utilizing the multiparameter approach of analyzing DNA content vs cyclin expression not only provides information on the specific

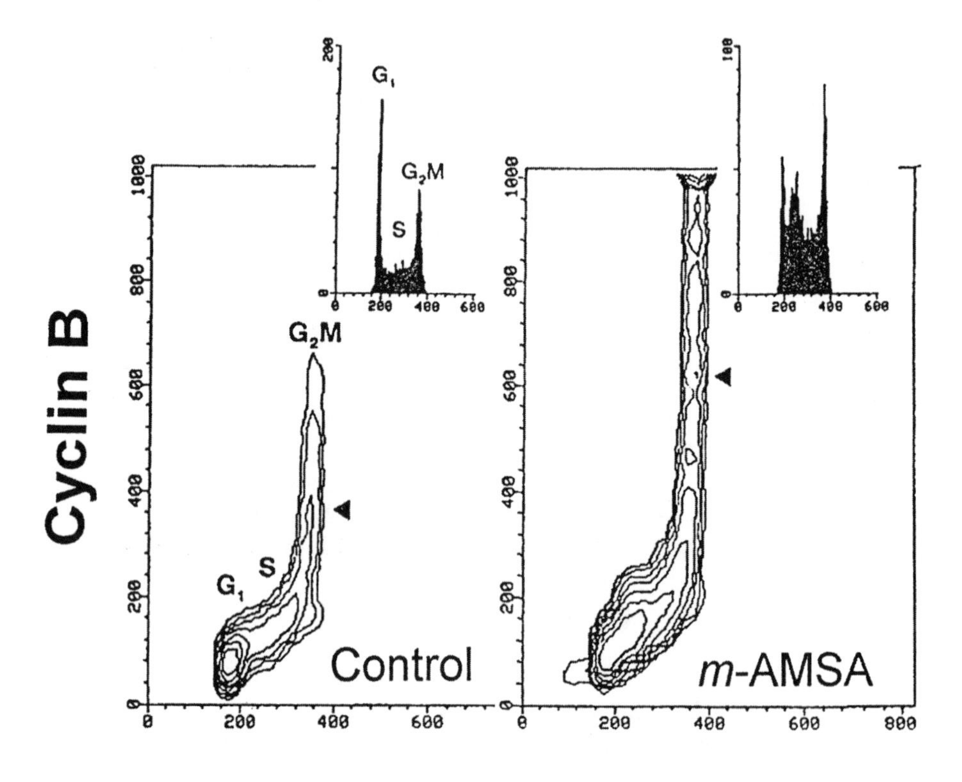

Fig. 2. Bivariate contour maps of cyclin B expression vs DNA content for untreated MOLT-4 human lymphoid leukemic cells (control) and the same cells exposed for 3 h to 0.125 µ*M* *m*-AMSA followed by 6 h growth in drug-free medium. Cyclin B1 was expressed only late in S, during G_2 phase, and in mitosis up to metaphase. As can be seen in the distribution of the control MOLT-4 cells, cyclin B expression was high in G_2M with a mean of about 375 arbitrary units (triangle). Cells exposed to the DNA topoisomerase II inhibitor *m*-AMSA were blocked in early S-phase (*see* DNA histogram inset at upper right) and in G_2M. The mean cyclin B expression in the G_2 cells of the drug-treated culture increased to over 600 arbitrary units. In addition, there was a small increase in cyclin B expression of the cells blocked in early S-phase, which was the result of unbalanced growth; although the cells failed to progress through the DNA synthesis cycle, they continued to synthesize protein as if on schedule.

cell-cycle phase that has been perturbed by a drug, but by determining whether the expression of a particular cyclin is inhibited or increased, helps to pinpoint within cell-cycle phases the point of action of a drug (**Fig. 2**) ***(10)***.

1.3. Dual Parameter—DNA vs DNA Synthesis Measurements

Incubation of cells in medium containing BrdUrd results in incorporation of this analog in place of thymidine in S-phase cells during DNA replication. The incorporated BrdUrd can be detected either as a result of its ability to quench (diminish) the fluorescence of several DNA fluorochromes, such as Hoechst 33258 or ethidium bromide *(11)*, or immunocytochemically using polyclonal or monoclonal antibodies developed against this precursor *(3)*. Continuous or pulse labeling with BrdUrd simultaneous with determination of cellular DNA content by flow cytometry provides bivariate data allowing one to estimate a variety of cell-cycle parameters (**Fig. 3**). Continuous exposure to BrdUrd can provide an estimate of the fraction of noncycling cells (cells that fail to label), whereas pulse-chase experiments can provide quantitative measures of cell-cycle phase duration, S-phase percentage, and so forth. With the introduction of halogen-selective antibodies (e.g., selective for BrdUrd or IdUrd), double-labeling experiments can be performed and allow for determination of total cell-cycle time and the duration of the different cell-cycle compartments. The action of drugs that interfere either directly or indirectly with DNA replication will affect the transit of cells through S-phase and be especially obvious on plots of DNA vs BrdUrd incorporation.

The scope of this chapter makes it impossible to describe all possibilities for analysis of the various cell-cycle parameters based on BrdUrd incorporation, either following pulse-chase or continuous cell labeling; the readers are advised to consult Gray et al. *(12)*, Crissman and Steinkamp *(13)*, and Darzynkiewicz et al. *(14)* for a more detailed description of these methods and the quantitative analysis of the data.

2. Materials

2.1. Flow Cytometry

1. A variety of models of flow cytometers can be used to measure cell fluorescence following staining of intact individual cells according to the procedures listed below. The manufacturers of the most common flow cytometers are:
   a. FACScan and FACS Vantage: Becton Dickinson Immunocytometry Systems, San Jose, CA; FAX (408) 954-2009.
   b. EPICS Elite and EPICS XL: Coulter Corporation, Miami, FL; FAX (305) 883-6881.
   c. PARTEC GmbH, Munster, Germany, FAX: (0251) 82979.
2. The software to deconvolute the DNA content frequency histograms to estimate the proportions of cells in the respective phases of the cycle is available from:
   a. Phoenix Flow Systems, San Diego, CA; FAX: (602)259-2568.
   b. Verity Software House, Topsham, ME; FAX: (207) 729-5443.

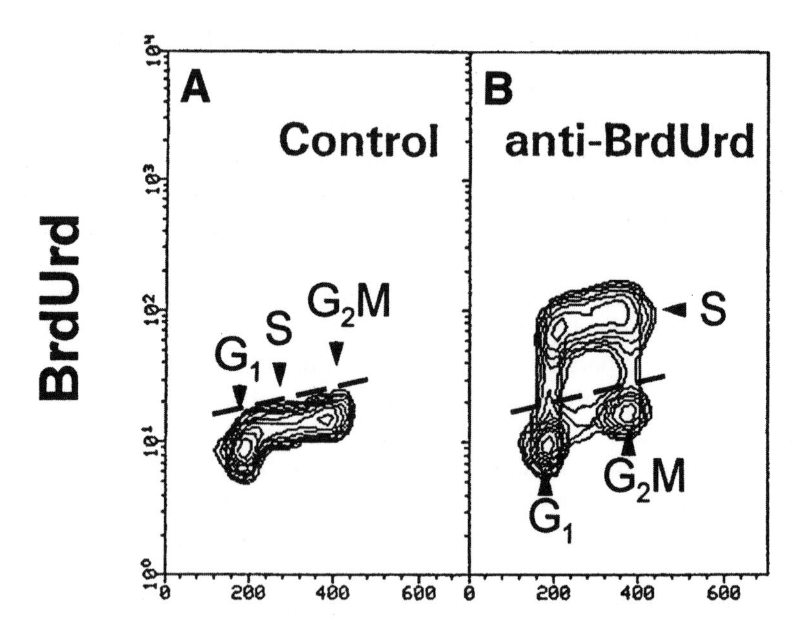

Fig. 3. Immunocytochemical detection of S-phase cells in HL-60 cell cultures using the anti-BrdUrd antibody. Exponentially growing HL-60 cells were incubated in the absence **(A)** and presence **(B)** of 50 µ*M* BrdUrd for 1 h. The incorporated BrdUrd was detected immunocytochemically using the BrdUrd antibody following acid denaturation of DNA *in situ* with 2.0 *M* HCl. Note that the intense labeling of S-phase cells (the green fluorescence emitted by fluorescein bound to the antiBrdUrd antibody) is presented using an exponential scale.

These same sources also provide software for the analysis and display of multiparameter data.

2.2. Single-Parameter Analysis of Cellular DNA Content

1. Phosphate-buffered saline solution (PBS).
2. Fixative: ice-cold 80% ethanol.
3. Propidium iodide solution (PI; Molecular Probes, Eugene, OR; cat. no. D1306): Dissolve 200 µg of PI and 0.1% RNase in 10 mL of PBS. **Note:** Prepare fresh staining solution before each use.
4. DNase-free RNase (RNase A; Sigma Chemical Co., St. Louis, MO; cat. no. R 5000). If RNase is not DNase-free, boil for 5 min.

2.3. DNA Content vs Cyclin Measurements

1. Fixative: 80% ethanol (cyclins D1, E, A, and B1) or 100% methanol (cyclin D3).
2. Cyclin antibodies (PharMingen, San Diego, CA; Santa Cruz Biotechnology Inc., Santa Cruz, CA): Antibodies are diluted to obtain 0.25 μg of antibody/5 × 10^5 cells in a 100-μL vol (*see* **Note 1**).
3. Antibody diluent: PBS containing 1% (w/v) bovine serum albumin (BSA).
4. Secondary antibody: Fluorescein isothiocyanate-(FITC) conjugated goat antimouse IgG antibody (Sigma) diluted 1:40 in PBS containing 1% BSA.
5. Isotype-specific antibody (mouse IgG; Sigma): control for the cyclin-specific antibodies.
6. PI staining solution: 10 μg/mL PI and 0.1% RNase A in PBS.
7. Triton X-100 solution: 0.25% (v/v) in PBS.

2.4. DNA Content vs DNA Synthesis Measurements

1. BrdUrd (Sigma) stock solution: 2 m*M* in Tris-HCl buffer, pH 7.5.
2. FITC-conjugated anti-BrdUrd MAb (Becton Dickinson Immunocytometry Systems, clone B44) stock solution: 0.3 μg of the antibody, 0.3% (v/v) Triton X-100, 1% (w/v) BSA in 100 μL PBS.
3. Acid buffer I: 0.1 *M* HCl containing 0.1% (v/v) Triton X-100 (Sigma). Acid buffer II: 2.0 *M* HCl.
4. FITC-labeled isotype-specific antibody (mouse IgG; Sigma): control for anti-BrdUrd made up at the appropriate dilution in the same Triton X-100/BSA buffer as above.
5. PI staining solution: 5 μg/mL PI and 0.1% RNase A in PBS.
6. DNA denaturation buffer: 0.1 m*M* Na-EDTA in 1 m*M* Na-cacodylate buffer; final pH 6.0.
7. Diluting buffer: PBS containing 0.1% (v/v) Triton X-100 and 0.5% (w/v) BSA.
8. DNase-free RNase.

3. Methods

3.1. Single-Parameter DNA Content Distributions

1. Control cells and samples exposed to various topoisomerase inhibitors should be washed by centrifugation and resuspended in PBS (*see* **Note 2**).
2. Fixation is carried out by pipeting the cells (1–5 × 10^6), suspended in 1.0 mL of PBS, into 9.0 mL of ice-cold (–20°C) 80% ethanol. Cells can be kept for several weeks at –20°C (*see* **Note 3**).
3. To rehydrate, centrifuge cells, decant ethanol, suspend cells in 10 mL of PBS, and centrifuge again.
4. Suspend cell pellet in sufficient volume of PI staining solution to provide a cell concentration of 0.5–1.0 × 10^6/mL.

5. Incubate in the dark for 20–30 min at room temperature to allow for digestion of RNA.
6. Analyze cells by flow cytometry (*see* **Note 4**): Use 488-nm laser line or blue light (BG12 filter) for excitation, and measure red fluorescence (>600 nm) and forward light scatter (if available) of 5–10 × 10^5 individual cells.

3.2. DNA Content vs Cyclin Expression

1. Rinse harvested cells (controls and drug-treated) in PBS, and fix at –20°C for a minimum of 2 h.
2. Rehydration from alcohol fixation is accomplished by centrifuging the cells and resuspending them in PBS.
3. Incubate in Triton X-100 solution for 5 min on ice.
4. Add 5 mL of PBS, centrifuge, and incubate overnight at 4°C with the appropriate mouse MAb (*see* **Notes 1** and **5**) to human cyclin proteins.
5. Rinse cells with PBS, and incubate with FITC-conjugated goat antimouse IgG antibody for 30 min.
6. Wash cells again, resuspend in PI staining solution, and incubate at room temperature for 20–30 min prior to measurement.
7. A control is prepared as above, although isotype-specific mouse IgG antibody is used in place of the mouse monoclonal cyclin-specific antibody (*see* **Note 6**).
8. Cellular fluorescence is measured flow cytometrically: blue (488 nm) light is used for excitation, and cyclin-specific green fluorescence from FITC is measured in a band from 515 to 530 nm, whereas DNA-specific PI fluorescence is measured at wavelengths above 630 nm.

3.3. DNA Contents vs DNA Synthesis

3.3.1. Heat Denaturation of DNA

1. Incubate cells with 10–30 µg/mL of BrdUrd under light proof conditions. Length of exposure to BrdUrd can vary and depends on the aim of the experiment. Pulse exposures are generally performed for 30 min to 1 h.
2. Fix cells in suspension in 70% ethanol (*see* **Note 7**).
3. Centrifuge cells (1–2 × 10^6) at 300*g* for 5 min, resuspend cell pellet in 1 mL of diluting buffer containing 100 U of RNase A, and incubate at 37°C for 30 min.
4. Centrifuge cells (400*g*, 5 min), and suspend cell pellet in 1 mL of ice-cold acid buffer I (*see* **Note 8**). After 1 min, centrifuge cells again.
5. Drain cell pellet thoroughly, and resuspend in 5 mL of DNA denaturation buffer.
6. Centrifuge cells again, resuspend cell pellet in 1 mL of DNA denaturation buffer.
7. Heat cells at 90 or 95°C for 5 min, and then place on ice for 5 min.
8. Add 5 mL of diluting buffer, and centrifuge (5 min at 400*g*).
9. Drain well, and suspend cells in 100 µL of anti-BrdUrd antibody, dissolved in dilution buffer, for 30 min at room temperature (follow the instructions provided by the supplier regarding the dilution, time, and temperature of incubation with anti-BrdUrd).

10. Add 5 mL of dilution buffer, and centrifuge (*see* **Note 9**).
11. Suspend in 100 µL of goat antimouse IgG labeled with fluorescein (dissolved in dilution buffer), and incubate for 30 min at room temperature.
12. Add 5 mL of dilution buffer, centrifuge, drain, and resuspend in 1 mL of this buffer containing PI staining solution.
13. Measure green (FITC) fluorescence and DNA-associated red (PI) fluorescence as described above in **Subheading 3.2.**

3.3.2. Acid Denaturation of DNA

1. Follow **steps 1–3** as described in **Subheading 3.3.1.** for thermal denaturation of DNA.
2. Centrifuge cells (300*g*, 5 min), and resuspend cell pellet in 1 mL of acid buffer II.
3. Incubate for 20 min at room temperature, and then add 5 mL of PBS, centrifuge, and drain well.
4. Resuspend cells in 5 mL of 0.2 *M* phosphate buffer at pH 7.4 to neutralize traces of the remaining HCl.
5. Follow **steps 9–13** as described in **Subheading 3.3.1.** for thermal denaturation of DNA.

4. Notes

1. The following clones have proven useful for flow cytometric analysis: G124-326, cyclin D1; G132-43, cyclin D2; G107-565, cyclin D3; HE 12, cyclin E; BF683, cyclin A; GNS-1, cyclin B1. It should also be noted that the D-type cyclins are tissue-specific: hematoopoietic cells express cyclin D3, but most other tissues express cyclins D1 and/or D2.
2. Each approach to cell kinetic analysis has different types of limitations and possible pitfalls. As mentioned, the univariate DNA content measurements, unless combined with analysis of cell-growth curves, do not reveal cell kinetic information, such as cell cycle or S-phase duration. Kinetic information is generally inferred from the fraction of cells in S-, G_2-, and M-phases, under the assumption that all cells progress through the cell cycle (there are no G_0 cells) and that the length of G_1 is variable, whereas the duration of S, G_2, and M remains relatively constant. This is often a good assumption when working with tissue-culture cells where the growth fraction is often 100%, but it is not always true for cultures under modified growth conditions (low serum, growth inhibitors, differentiating agents, and so on) and almost never true for tumors in vivo, where there can be a substantial number of differentiated cells, hypoxic or anoxic cells, or cells out of the cycle (G_0 cells).
3. Use of high drug concentrations, which may lead to cell death, also complicates the interpretation of single-parameter DNA histograms. Thus, one must take care to keep track of the live cell number. The DNA distribution, with the exception of apoptotic cells (*see* Chapter 33 by Darzynkiewicz et al.), provides information only on intact cells. Cells that have disintegrated because of the toxic effects of drug treatment will fall below the lower threshold of the flow cytometer and not

register on histograms. It is important to know whether the cells observed in the drug-treated DNA histogram represent 100 or 1% of the cells in the control. This can only be accomplished by keeping careful track of cell number and viability. Apoptotic cells represent a special case in which appropriate processing of samples can provide information both about the percentage of cells undergoing apoptosis and the DNA distribution of the remaining, presumably unaffected cells.

4. It should be noted that there are no perfect DNA stains. All dyes, including PI, when used correctly, bind with some degree of stoichiometry to DNA. That is, the number of dye molecules bound and their resultant fluorescence are often proportional to the amount of DNA present. However, some dye molecules are also sensitive to the structure of chromatin *in situ* so that the amount of dye bound can be affected by the extent of chromatin condensation. Thus, it has been demonstrated that if one were to dissociate all the histone proteins from DNA *in situ*, increasing the number of sites in the DNA that could bind dye molecules, a UV-excited dye, such as DAPI, would show an approx 20% increase in fluorescence/cell, whereas PI staining would double, although binding of 7-amino actinomycin D, a red fluorescent DNA intercalator, would increase by 1300% ***(14)***.
5. Detection of cyclins in intact cells by flow cytometry is predicated on the fact that antibodies often designed to detect proteins on Westerns will actually bind to the protein *in situ*. We and others ***(15,16)*** have demonstrated that at least some antibodies against cyclin proteins can be used for flow cytometry. It should be noted, however, that cyclins are involved in forming complexes with several molecules including a cdk, an inhibitor, and often PCNA. It is not known whether binding of some or all of these molecules changes the affinity of any of the several cyclin antibodies for its cyclin or whether "activation" or "inactivation" of cyclin–cdk complexes by phosphorylation and/or dephosphorylation events effects cyclin antibody binding.
6. Not all cell types exhibit the expected pattern of cyclin expression. Normal cells have fairly expected and reproducible patterns of expression for each of the cyclins whose synthesis is confined to a limited portion of the cell cycle. Thus, normal fibroblasts only express cyclin D1 in G_1, but fibroblasts and proliferating normal lymphocytes express cyclin E in late G_1, and early S-phase, cyclin A in S-phase and G_2-phase, and cyclin B1 in late S, G_2, and mitosis up to metaphase ***(7–9)***. However, many tumor cell lines express these cyclins inappropriately in that they may have high cyclin B1 levels in G_1 or high cyclin E levels in G_2, and so forth ***(17)***. Therefore, it is necessary to test each cell line against another line with "normal" cyclin expression to determine how to interpret the results of drug treatment.
7. Use of fixatives, such as formaldehyde, that crosslink DNA interferes both with the denaturation of DNA and the binding of intercalating dyes. As such, they should be avoided.
8. All the antibodies available to date require that DNA be denatured, since they only recognize the incorporated analog in single-stranded DNA. However, extensive DNA denaturation is neither required nor desired. Totally denatured

DNA will not bind the intercalating PI dye molecules, resulting in loss of DNA content information. In addition, conditions that lead to total DNA denaturation also cause unacceptable cell loss. Therefore, there must be a balance between the need to denature DNA for binding of the anti-BrdUrd antibody and the need to retain double-stranded DNA to bind the PI molecule to detect total DNA content.

9. The harsh conditions generally required to denature DNA often result in cell loss, which occurs as a result of both clumping and adherence of cells to the sides of test tubes used during the procedure. Cell clumps may sometimes be disaggregated by careful vortexing or syringing. Clumping occurring during centrifugation can be minimized by limiting the time and speed of centrifugation (<600*g*). Adherence of cells to the sides of test tubes can be minimized either by siliconizing tubes or minimizing the surface area to which cells can be exposed by keeping the sample volumes and sizes of test tubes small.

Acknowledgment

This work was supported by NCI grant R01 28704, the Chemotherapy Foundation, and the Robert Welke Cancer Research Foundation. Gloria Juan on leave from the University of Valencia, Spain is supported by a fellowship from the "This Close" for Cancer Research Foundation.

References

1. Crissman, H. A. and Tobey, R. A. (1974) Cell cycle analysis in 20 minutes. *Science* **184,** 1297–1298.
2. Rabinovitch, P. S., Kubbies, M., Chen, Y. C., Schindler, D., and Hoehn, H. (1988) BrdU-Hoechst flow cytometry: A unique tool for quantitative cell cycle analysis. *Exp. Cell Res.* **174,** 309–318.
3. Dolbeare, F., Gratzner, H., Pallavicini, M., and Gray, J. W. (1983) Flow cytometric measurements of total DNA content and incorporated bromodeoxyuridine. *Proc. Natl. Acad. Sci. USA* **80,** 5573–5577.
4. Gray, J. W. and Darzynkiewicz, Z. (eds.) (1987) *Techniques in Cell Cycle Analysis.* Humana Press, Clifton, NJ.
5. Darzynkiewicz, Z. and Crissman, H. A. (eds.) (1990) *Flow Cytometry.* Academic, San Diego.
6. Darzynkiewicz, Z., Robinson, J. P., and Crissman, H. A. (eds.) (1994) *Methods in Cell Biology.* Academic, San Diego.
7. Celis, J. E., Bravo, R., Larsen, P. M., and Fey, S. J. (1984) Cyclin: A nuclear protein whose level correlates directly with proliferative state of normal as well as transformed cells. *Leukemia Res.* **8,** 143–157.
8. Hartwell, L. H. and Weinert, T. A. (1989) Checkpoints: controls that ensure the order in cell cycle events. *Science* **246,** 629–634.
9. Pines, J. (1994) Arresting developments in cell-cycle control. *Trends Biochem. Sci.* **19,** 143–145.

10. Darzynkiewicz, Z., Traganos, F., and Gong, J. (1994) Expression of cell cycle specific proteins cyclins as a marker of the cell cycle independent of DNA content. *Methods Cell Biol.* **41,** 422–435.
11. Latt, S. A. (1977) Fluorometric detection of deoxyribonucleic acid synthesis: possibilities for interfacing bromodeoxyuridine due techniques with flow fluorometry. *J. Histochem. Cytochem.* **25,** 915–926.
12. Gray, J. W., Dolbeare, F., and Pallavicini, M. G. (1990) Quantitative cell cycle analysis, in *Flow Cytometry and Sorting* (Melamed, M. R., Lindmo, T., and Mendelsohn, M. L., eds.), Wiley-Liss, New York., pp. 445–467.
13. Crissman, H. A. and Steinkamp (1990) Cytochemical techniques for multivariate analysis of DNA and other cellular constituents, in *Flow Cytometry and Sorting* (Melamed, M. R., Lindmo, T., and Mendelsohn, M. L. eds.), Wiley-Liss, New York, pp. 227–247.
14. Darzynkiewicz, Z., Traganos, F., Kapuscinski, J., Staiano-Coico, L., and Melamed, M. R. (1984) Accessibility of DNA in situ to various fluorochromes: Relationship to chromatin changes during erythroid differentiation of Friend leukemia cells. *Cytometry* **5,** 355–363.
15. Darzynkiewicz, Z., Gong, J., Juan, G., Ardelt, B., and Traganos, F. (1996) Cytometry of cyclin proteins. *Cytometry* **25,** 1–13.
16. Kung, A. L., Sherwood, S. W., and Schimke, R. T. (1993) Differences in the regulation of protein synthesis, cyclin B accumulation, and cellular growth in response to the inhibition of DNA synthesis in Chinese hamster ovary and HeLa S3 cells. *J. Biol. Chem.* **268,** 23,072–23,080.
17. Gong, J., Ardelt, B., Traganos, F., and Darzynkiewicz, Z. (1994) Unscheduled expression of cyclins B1 and E in several leukemic and solid tumor cell lines. *Cancer Res.* **54,** 4285–4288.

23

Assaying Drug-Induced Apoptosis

Zbigniew Darzynkiewicz, Gloria Juan, and Frank Traganos

1. Introduction

Apoptosis, frequently referred to as "programmed cell death" or "cell suicide," is an active and physiological mode of cell death, where the cell itself prepares and executes the program of its own demise. A complex, multistep mechanism regulates the cell's response to various stimuli by apoptosis (reviewed in **refs. *1,2***). The regulatory system involves the presence of at least two distinct checkpoints, one controlled by *bcl-2/bax* family of proteins ***(3)***, and another by cysteine and possibly serine proteases ***(4)***. There is a close association between apoptosis and regulation of cell proliferation as well as DNA repair.

Apoptosis is the mode of cell death observed following exposure of cells to pharmacological concentrations of DNA topoisomerase inhibitors ***(5)***. The evidence also is mounting that in vivo antitumor modalities employing DNA topoisomerase inhibitors result in apoptotic death of the target cells ***(6)***. Modulation of the apoptotic regulatory machinery, therefore, begins to play an increasingly important role in designing antitumor strategies, whereas the detection of apoptosis has been widely applied in both basic research aimed at investigating drugs mechanisms and in the clinical setting in evaluating drug treatment efficacy ***(6)***.

A cell undergoing apoptosis activates a cascade of molecular events that lead to its disintegration. Since many of these changes are very characteristic to apoptosis, they have become markers used to identify this mode of cell death biochemically, by microscopy or cytometry. A multitude of methods for identification of apoptotic cells have been described ***(7,8)***. This chapter lists four different approaches. Their applicability, advantages, and limitations are briefly reviewed below.

From: *Methods in Molecular Biology, Vol. 95: DNA Topoisomerase Protocols, Part II: Enzymology and Drugs*
Edited by N. Osheroff and M.A. Bjornsti © Humana Press Inc., Totowa, NJ

1.1. Cell Morphology

Apoptosis was originally defined as a specific mode of cell death based on very characteristic changes in cell morphology *(9)*. One of the early events is cell dehydration, which leads to condensation of the cytoplasm followed by a change in cell shape and size: the originally round cells may become elongated and generally are smaller. Perhaps the most characteristic change is condensation of nuclear chromatin. The condensation starts at the nuclear periphery with the condensed chromatin often taking on a concave shape. The chromatin becomes uniform and smooth in appearance, with no evidence of any texture normally seen in the nucleus. DNA in the condensed (pycnotic) chromatin exhibits hyperchromasia, staining strongly with fluorescent or light-absorbing dyes. The nuclear envelope disintegrates, and lamin proteins undergo degradation, followed by nuclear fragmentation (karyorrhexis). Nuclear fragments, which stain uniformly with DNA dyes and thereby resemble DNA droplets of different sizes, are scattered throughout the cytoplasm (**Fig. 1**; *see* **Note 1**). The nuclear fragments, together with constituents of the cytoplasm (including intact organelles), are then packaged and enveloped by fragments of the plasma membrane. These structures, called "apoptotic bodies," are then shed from the dying cell. When apoptosis occurs in vivo, apoptotic bodies are phagocytized by neighboring cells, including those of epithelial or fibroblast origin, without triggering an inflammatory reaction in the tissue.

Another characteristic feature of apoptosis is the preservation, at least during the initial phase of cell death, of the structural integrity and most of the function of the plasma membrane. Also, cellular organelles, including mitochondria and lysosomes remain preserved, although the the mitochondrial transmembrane potential drops early during apoptosis. Other features of apoptosis include mobilization of intracellular ionized calcium, activation of transglutaminase, which crosslinks cytoplasmic proteins, loss of the microtubules, and loss of asymmetry of the phospholipids on the plasma membrane leading to exposure of phosphatidylserine on the outer surface. The latter preconditions remnants of the apoptotic cell to become a target for phagocytizing cells. The duration of apoptosis varies depending on the cell type and whether it occurs in vivo, within the tissue, or in vitro. Under conditions of tissue homeostasis, when the rate of cell death is balanced by the rate of cell proliferation, the mitotic index may exceed the index of apoptosis, which indicates that in these instances, duration of apoptosis is shorter than duration of mitosis.

Although apoptosis is characterized by an active participation of the affected cell in its own demise, even to the point of triggering (in some cell systems) the *de novo* synthesis of the effectors of cell death, the alternative mode of cell death, necrosis, is a passive, catabolic, and degenerative process *(10)*. Necrosis

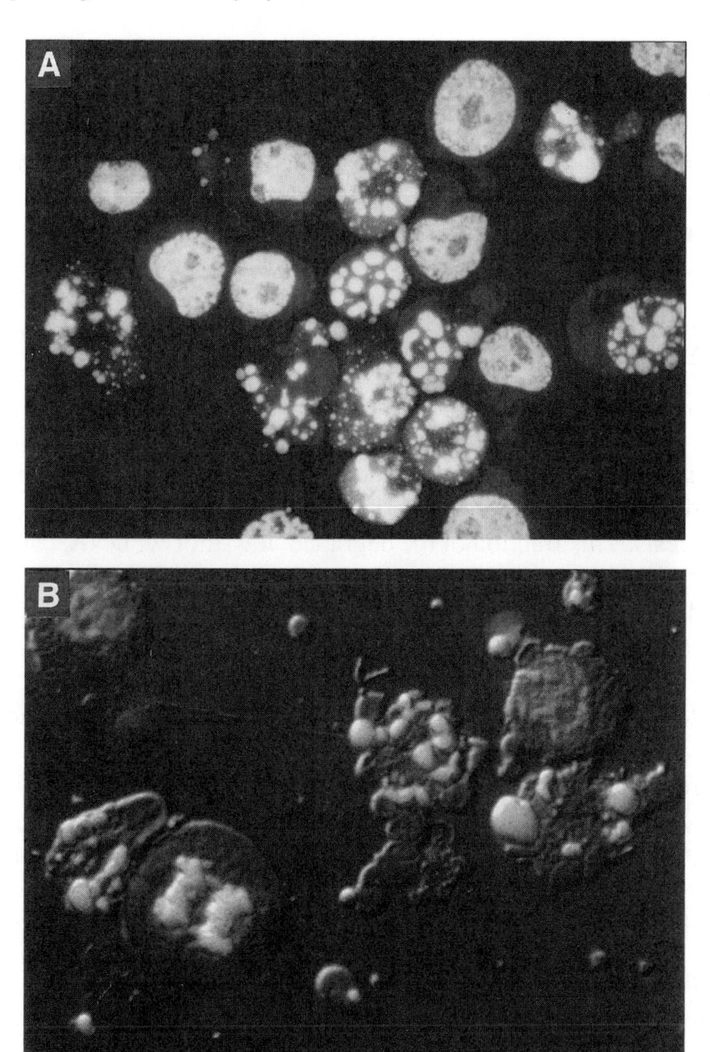

Fig. 1. Detection of apoptotic cells by staining with DAPI and sulforhodamine (Method 1). HL-60 cells were induced to undergo apoptosis by treatment with 0.15 μ*M* camptothecin for 4 h. The cytospun cells were fixed for 15 min in 1% formaldehyde and then in 70% ethanol, stained with 1 μg/mL of DAPI and 20 μg/mL sulforhodamine 101 in PBS, and viewed either under UV light illumination **(A)** or under combined UV light and the transmitted light interference contrast illumination **(B)**. Note typical appearance of apoptotic cells having fragmented nuclei and the presence of individual apoptotic bodies.

generally represents a cell's response to gross injury and can be induced by an overdose of cytotoxic agents. The early event of necrosis is mitochondrial swelling followed by rupture of the plasma membrane and release of cytoplasmic constituents, which include proteolytic enzymes. Nuclear chromatin shows patchy areas of condensation, and the nucleus undergoes slow dissolution (karyolysis). Necrosis triggers an inflammatory reaction in the tissue and often results in scar formation.

1.2. DNA Degradation: Electrophoresis of DNA Extracted from Apoptotic Cells

Activation of endonuclease(s), which preferentially cleaves DNA between nucleosomes, is another characteristic event of apoptosis ***(11,12)***. The products of this DNA degradation are nucleosomal and oligonucleosomal DNA sections, which generate a characteristic "ladder" pattern during agarose-gel electrophoresis. This fraction of the degraded, low-mol-wt DNA, can be easily extracted from apoptotic cells after their permeabilization ***(12–15)***. It should be pointed out, however, that in many cell types, DNA degradation does not proceed to nucleosomal-sized fragments, but rather results in 300- to 50-kb size DNA fragments ***(16)***.

DNA degradation is not as extensive during necrosis as in the case of apoptosis, and the products of degradation are heterogenous in size, failing to form discrete bands on electrophoretic gels.

The electrophoretic method presented in this chapter is based on the *in situ* precipitation of DNA with ethanol (by cell's fixation in 70% ethanol), followed by selective extraction of low-mol-wt DNA from the prefixed with high molarity phosphate–citric acid buffer ***(15)***.

1.3. Identification of Apoptotic Cells by Their Fractional DNA Content

Extensive DNA cleavage, which occurs during apoptosis, provided a basis for several flow cytometric assays to identify apoptotic cells. Two approaches are frequently used to detect DNA cleavage. One is based on extraction of low-mol-wt DNA prior to cell staining. The other relies on fluorochrome labeling of DNA strand breaks *in situ*.

In the first approach, the cellular DNA content is measured following cell permeabilization with detergents or prefixation with precipitating fixatives, such as alcohols, or acetone. Cell permeabilization or alcohol fixation does not fully preserve the degraded DNA within apoptotic cells: this fraction of DNA leaks out during subsequent cell rinsing and staining. As a consequence, apoptotic cells contain reduced DNA content and therefore can be recognized,

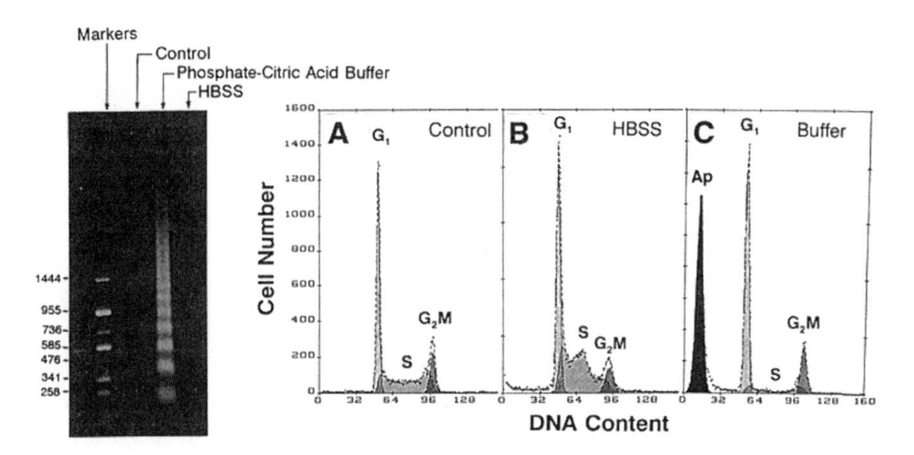

Fig. 2. Combined analysis of apoptosis by DNA gel electrophoresis and flow cytometry of the same cells (Methods 2 and 3). DNA gel electrophoresis and DNA content frequency histograms of corresponding HL-60 cells untreated (Control, **A**) and treated with 0.15 μ*M* DNA topoisomerase I inhibitor camptothecin, which selectively induces apoptosis of S-phase cells **(B,C)**. The cells represented by B histogram were extracted with Hanks' balanced salt solution (HBSS), whereas those represented by C histogram were extracted by 0.2 *M* phosphate-citrate buffer at pH 7.8. DNA extracted with HBSS or with the buffer was analyzed by gel electrophoresis, as shown in respective lanes, whereas the same cells after the extraction were stained with DAPI and subjected to flow cytometry. Note the presence of degraded DNA extracted from the camptothecin-treated cells by the buffer and its absence in the HBSS extract and in the extract from untreated (control) cells. The DNA marker size is expressed in basepairs.

following staining of cellular DNA, as cells with DNA stainability ("sub-G_1"peak) often lower than that of G_1 cells *(13–15)*.

The degree of DNA degradation varies depending on the stage of apoptosis, cell type, and often, the nature of the apoptosis-inducing agent. The extractability of DNA during the staining procedure (and thus separation of apoptotic from live cells by this assay) also varies. As mentioned, addition of high-molarity phosphate–citric acid buffer to the rinsing solution enhances extraction of the degraded DNA *(15)*. This approach can be used to control the extent of DNA extraction from apoptotic cells to the desired level to obtain the optimal separation of apoptotic cells by flow cytometry. **Figure 2** illustrates an application of this approach to extract DNA from apoptotic cells for their subsequent identification by flow cytometry (*see* **Note 2**) as well as for electrophoretic analysis of DNA extracted from those cells *(15)*.

1.4. In Situ DNA Strand-Break Labeling

As mentioned, endonucleolytic DNA cleavage during apoptosis results in the presence of extensive DNA breakage. The 3'-OH termini in DNA breaks are detected by attaching to them biotin- or digoxygenin-conjugated nucleotides, in a reaction catalyzed by exogenous TdT ("end labeling," "tailing," "TUNEL") or DNA polymerase (nick translation) ***(17–20)***. Fluorochrome-conjugated avidin or digoxygenin antibody has often been used in the second step of the reaction to label DNA strand breaks. A new method was recently introduced in which BrdUTP, incorporated by TdT, is used as the marker of DNA strand breaks, which is simpler, more sensitive and costs less compared with the digoxygenin or biotin labeling (*see* **Note 3**). Commercial kits designed to label DNA strand breaks for identification of apoptotic cells are offered by ONCOR Inc. (Gaithersburg, MD; double-step assay using digoxygenin) and Phoenix Flow Systems (San Diego, CA; single-step and BrdUTP-labeling assays).

2. Materials

2.1. Morphological Identification of Apoptotic Cells

1. Cytospin centrifuge (e.g., Cytospin 3; Shandon, Pittsburgh, PA).
2. UV light microscope.
3. Microscope slides and cover slips.
4. 1% Formaldehyde solution in phosphate-buffered saline (PBS).
5. 70% Ethanol solution.
6. Stain solution contains: 1 µg/mL of 4,6-diamidino-2-phenylindole (DAPI) and 20 µg/mL of sulforhodamine (both available from Molecular Probes, Eugene, OR) dissolved in PBS.

2.2. DNA Gel Electrophoresis

1. Phosphate–citric acid buffer: Mix 192 mL of 0.2 *M* Na_2HPO_4 with 8 mL of 0.1 *M* citric acid; the final pH is 7.8.
2. Nonidet NP-40 (Sigma, St. Louis, MO): Dissolve 0.25 mL of Nonidet NP-40 in 100 mL of distilled water.
3. RNase A (Sigma): Dissolve 1 mg of DNase-free RNase A in 1 mL of distilled water.
4. Proteinase K (Boehringer Mannheim, Indianapolis, IN): Dissolve 1 mg of proteinase K in 1 mL of distilled water.
5. Loading buffer: Dissolve 0.25 g of bromophenol blue and 0.25 g of xylene cyanol FF in 70 mL of distilled water. Add 30 mL of glycerol (dyes are available, e.g., from Bio-Rad Laboratories, Richmond, CA).
6. DNA mol-wt standards: Use the standards containing DNA between 100 and 1000 bp in size (e.g., from Integrated Separation Systems, Natick, MA).

7. Electrophoresis buffer (TBE, 10X concentrated): Dissolve 54 g Tris base and 27.5 g boric acid in 980 mL of distilled water; add 20 mL of 0.5 *M* EDTA (pH 8.0).
8. Agarose gel (0.8%): Dissolve 1.6 g of agarose in 200 mL of TBE.
9. Ethidium bromide (EB, Polysciences Inc., Warrington, PA): Stock solution, dissolve 1 mg of EB in 1 mL of distilled water; for staining gels (working solutions), add 100 µL of the stock solution to 200 mL of TBE.

2.3. Flow Cytometry: Detection of Cells with Fractional DNA Content

1. DNA extraction buffer: Mix 192 mL of 0.2 *M* Na_2HPO_4 with 8 mL of 0.1 *M* citric acid; pH of this buffer is 7.8.
2. DNA staining solution (*Prepare fresh staining solution before each use*): Dissolve 200 µg of PI in 10 mL of PBS and add 2 mg of DNase-free RNase A (boil RNase for 5 min if it is not DNase-free).
3. Flow cytometer with blue light source (e.g., FACScan, Becton Dickinson Immunocytometry Systems, San Jose, CA).

2.4. Flow Cytometry: In Situ DNA Strand-Break Labeling

1. Prepare fixatives:
   a. First fixative: 1% methanol-free formaldehyde (available from Polysciences Inc., Warrington, PA) in PBS, pH 7.4.
   b. Second fixative: 70% ethanol.
2. The TdT reaction buffer (5X concentrated; Boehringer Mannheim, Indianapolis, IN) contains: 1 *M* potassium (or sodium) cacodylate, 125 m*M* Tris-HCl (pH 6.6), and 1.25 mg/mL bovine serum albumin (BSA).
3. Cobalt chloride ($CoCl_2$; Boehringer), 10 m*M*.
4. TdT (Boehringer) in storage buffer, 25 U in 1 µL.
5. BrdUTP stock solution: BrdUTP (Sigma) 2 m*M* (100 nmol in 50 µL) in 50 m*M* Tris-HCl, pH 7.5.
6. FITC-conjugated anti-BrdUrd MoAb solution (per 100 µL of PBS): 0.3 µg of anti-BrdUrd FITC-conjugated MoAb (available from Becton Dickinson), 0.3% Triton X-100, and 1% BSA.
7. Rinsing buffer: 5 mg/mL BSA and 0.1% (v/v) Triton X-100 in PBS.
8. PI staining buffer: Dissolve 5 µg/mL PI and 200 µg/mL DNase-free RNase A in PBS.

3. Methods

3.1. Microscopy

1. Add 300 µL of cell suspension containing approx 20,000 cells into a cytospin chamber.
2. Centrifuge at 1000 rpm for 6 min.

3. Prefix the cells in 1% formaldehyde in PBS for 15 min.
4. Transfer the slides to 70% ethanol, and fix for at least 1 h; the cells can be stored after fixation for weeks.
5. Rinse the slides in PBS, and stain cells for 10 min in the solution of DAPI and sulforhodamine.
6. View under UV light microscope. Use 40X or higher objective magnification (*see* legend to **Fig. 1**).

(*See* **Notes 3–6**.)

3.2. DNA Agarose-Gel Electrophoresis

1. Fix 10^6–10^7 cells in suspension in 5 mL of 80% ethanol on ice.
2. Cells can be subjected to DNA extraction and analysis after 4 h of fixation in ethanol, or stored in fixative at –20°C for several weeks.
3. Centrifuge cells at 300*g* for 5 min. *Thoroughly remove ethanol.* Resuspend cell pellets in 40 µL of phosphate–citric acid buffer, and transfer to 0.5-mL vol Eppendorf tubes. Keep at room temperature for at least 30 min, occasionally shaking.
4. Centrifuge at 300*g* for 5 min. Transfer the supernatant to new Eppendorf tubes and concentrate by vacuum, e.g., in SpeedVac concentrator (Savant Instruments, Inc., Farmingdale, NY) for 15 min.
5. Add 5 µL of 1% Nonidet NP-40 and 5 µL of RNase A solution. Close the tube to prevent evaporation, and incubate at 37°C for 30 min.
6. Add 5 µL of proteinase K solution, and incubate for an additional 30 min at 37°C.
7. Add 5 µL of the loading buffer, and transfer entire contents of the tube to 0.8% horizontal agarose gel.
8. Load a sample of DNA standards in the mol-wt standard lane of the gel.
9. Run electrophoresis at 2 V/cm for 16 h.
10. To visualize the bands, stain the gel with 5 µg/mL of ethidium bromide, and illuminate with UV light.

3.3. Cellular DNA Content Measurement

1. Fix cells in suspension in 70% ethanol by adding 1 mL of cells suspended in PBS (10^6–5×10^6 cells) into 9 mL of 70% ethanol in a tube on ice. *Cells can be stored in fixative at –20°C for several weeks.*
2. Centrifuge cells, decant ethanol, suspend cells in 10 mL of PBS, and centrifuge.
3. Suspend cells in 0.5 mL of PBS, into which you may add 0.2–1.0 mL of the DNA extraction buffer.
4. Incubate at room temperature for 5 min, and centrifuge at 300*g* for 5 min.
5. Suspend cell pellet in 1 mL of DNA-staining solution.
6. Incubate cells for 30 min at room temperature.
7. Analyze cells by flow cytometry: Use 488-nm laser line or blue light (BG12 filter) for excitation, and measure red fluorescence (>600 nm) and forward light scatter.

Alternative methods: Cellular DNA may be stained with other fluorochromes instead of PI, and other cell constituents may be counterstained in addition to DNA. The following is the procedure used to stain DNA with DAPI:

1. After **step 4**, suspend the cell pellet in 1 mL of a staining solution that contains: 0.1% (v/v) Triton X-100, 2 m*M* $MgCl_2$, 0.1 *M* NaCl, 10 m*M* PIPES (Sigma Chemical Co.) buffer (final pH 6.8), and DAPI (final concentration 1 µg/mL; Molecular Probes, Inc.).
2. Analyze cells by flow cytometry: Excite with UV light (e.g., 351-nm argon ion line, or UG1 filter for mercury lamp illumination) and measure the blue fluorescence of DAPI in a band from 460 to 500 nm. (*See* **Notes 2–6**.)

3.4. DNA Strand-Break Labeling

1. Fix cells in suspension in 1% formaldehyde for 15 min on ice.
2. Centrifuge, resuspend cell pellet in 5 mL of PBS, centrifuge at 300*g* for 5 min, and resuspend cells (approx 10^6 cells) in 0.5 mL of PBS.
3. Add the above 0.5-mL aliquot of cell suspension into 5 mL of ice-cold 70% ethanol. *The cells can be stored in ethanol at –20°C for several weeks.*
4. Centrifuge at 300*g*, remove ethanol, resuspend cells in 5 mL of PBS, and centrifuge.
5. Resuspend the pellet (not more than 10^6 cells) in 50 µL of a solution, which contains: 10 µL of the reaction buffer, 2.0 µL of BrdUTP stock solution, 0.5 µL (12.5 U) of TdT in storage buffer, 5 µL of $CoCl_2$ solution, and 33.5 µL distilled H_2O.
6. Incubate cells in this solution for 40 min at 37°C (*alternatively, incubation can be carried at 22–24°C overnight*).
7. Add 1.5 mL of the rinsing buffer, and centrifuge at 300*g* for 5 min.
8. Add 100 µL of FITC-conjugated anti-BrdUrd MoAb solution.
9. Incubate at room temperature for 1 h.
10. Resuspend the cell pellet in 1 mL of PI staining solution containing RNase.
11. Incubate for 30 min at room temperature in the dark.
12. Analyze cells by flow cytometry: Illuminate with blue light (488-nm laser line or BG12 excitation filter), and measure green fluorescence of FITC-anti BrdUrd MoAb at 530 ± 20 nm and the red fluorescence of PI at >600 nm (*see* **Notes 3–6**).

Commercial kits: Phoenix Flow Systems (San Diego, CA) provides a kit (ApoDirect™) to identify apoptotic cells based on a single-step procedure utilizing TdT and FITC-conjugated dUTP. A description of the method, which is nearly identical to the above, is included with the kit. Another kit (ApopTag™), based on two-step DNA strand-break labeling with digoxygenin-16- dUTP by TdT, is provided by ONCOR (Gaithersburg, MD).

4. Notes

1. Optimal preparations for light microscopy require cytospinning of live cells followed by their fixation and staining on slides. The cells are then flat, and their

morphology is easy to assess (**Fig. 1**). On the other hand, when the cells are initially fixed and stained in suspension, then transferred to slides and analyzed under the microscope, their morphology is obscured by the unfavorable geometry: the cells are spherical and thick, and require confocal microscopy to reveal details, such as early signs of apoptotic chromatin condensation.

We have noticed that differential staining of cellular DNA and protein with DAPI and sulforhodamine 101 of the cells on slides, which is very rapid and simple, gives a very good morphological resolution of apoptosis and necrosis (**Fig. 1**). Other DNA fluorochromes, such as PI or 7-AMD, or the DNA/RNA fluorochrome AO, can be used as well.

2. Owing to their fractional DNA content (after extraction of the degraded, low-mol-wt DNA) apoptotic cells are characterized by a decreased PI or DAPI fluorescence (and a diminished forward light scatter) compared to the cells in the main (G_1) peak (**Fig. 2**). It should be emphasized that the degree of extraction of DNA from apoptotic cells, and, consequently, the content of DNA remaining in the cell for flow cytometric analysis may vary depending on the degree of DNA degradation (duration of apoptosis), the number of cell washings, pH and molarity of the washing, and staining buffers. Therefore, in **Subheading 3.3., step 3** less or no extraction buffer (e.g., 0–0.2 mL) should be added if DNA degradation in apoptotic cells is extensive (late apoptosis) and more (up to 1.0 mL of extraction buffer) if DNA is not markedly degraded (early apoptosis) and there are problems with separating apoptotic cells from G_1 cells owing to their overlap on DNA content frequency histograms. Necrotic cells do not show a peak containing cells with diminished DNA staining, such that their DNA histograms are remarkably similar to those of untreated cells.

   DNA content analysis is a simple and inexpensive procedure to identify apoptotic cells, and in addition to reveal DNA ploidy and/or the cell-cycle distribution of the nonapoptotic cell population. Another advantage of this method is its applicability to any DNA fluorochrome or instrument. Degraded DNA extracted from the ethanol-prefixed apoptotic cells with phosphate–citrate buffer can be directly analyzed by gel electrophoresis for "laddering" ***(14)***.

   In this method objects with a fractional DNA content are assumed to be apoptotic cells. It should be stressed, however, that the "sub-G_1" peak can represent, in addition to apoptotic cells, mechanically damaged cells, cells with lower DNA content (e.g., in a sample containing cell populations with different DNA indices), or cells with different chromatin structure (e.g., cells undergoing erythroid differentiation) in which the accessibility of DNA to the fluorochrome (DNA stainability) is diminished.

   It also should be stressed that to be subjected to DNA content analysis, the cells have to be prefixed with precipitating fixatives, such as alcohols or acetone, rather than treated with detergent or hypotonic solutions. Detergent or hyopotonic treatments cause cell lysis. Because the nucleus of the apoptotic cell is fragmented, several distinct fragments may be present in a single cell. The percentage of objects with a fractional DNA content (represented by the "sub-G_1" peak) released from a lysed cell, thus, does not correspond to the apoptotic index. Furthermore, lysis of mitotic cells or cells with micronuclei releases individual chromosomes, chromosome aggregates, and micronuclei, which also erroneously can be identified as individual apoptotic cells.

Being simple and of low cost, the method is most applicable for screening large numbers of samples to quantify apoptotic cells. However, because of low specificity, this method when used alone, cannot provide evidence of apoptosis. It should be used, therefore, in conjunction with analysis of cell morphology by light or electron microscopy or with detection of DNA strand breaks.

3. The detection of DNA strand breaks requires cell prefixation with a crosslinking agent, such as formaldehyde, which, unlike ethanol prevents the extraction of degraded DNA. Thus, despite the sequential cell washings during the procedure, the DNA content of early apoptotic cells (and with it the number of DNA strand breaks) is not markedly diminished compared to unfixed cells.

   Identification of apoptotic cells is simple owing to their intense labeling with FITC-anti BrdUrd MoAb, which frequently requires use of an exponential scale (logarithmic photomultipliers) for data acquisition and display (**Fig. 3**). Because cellular DNA content of both apoptotic and nonapoptotic cell populations is measured, the method offers the unique possibility to analyze the cell-cycle position, and/or DNA ploidy, of apoptotic cells. The method also appears to be useful for clinical material, in leukemias, lymphomas, and solid tumors ***(6)***, and can be combined with surface immunophenotyping. In the latter case, the cells are first immunotyped, then fixed with formaldehyde (which stabilizes the antibody on the cell surface), and subsequently subjected to the DNA strand-break assay using different colored fluorochromes than those used for immunophenotyping.

   The procedure of DNA strand-break labeling is rather complex and involves many reagents. Negative results, therefore, may not necessarily mean the absence of DNA strand breaks, but be a result of some methodological problems, such as the loss of TdT activity, degradation of triphosphonucleotides, and so forth. It is always necessary, therefore, to include a positive and negative control. An excellent biological control consist of HL-60 cells treated (during their exponential growth) for 3–4 h with 0.2 μ*M* of the DNA topoisomerase I inhibitor camptothecin (CAM). Because CAM induces apoptosis selectively during S-phase, the populations of G_1 and G_2/M cells may serve as negative populations (background), whereas the S-phase cells in the same sample serve as the positive control.

4. Apoptosis is of short and variable duration. The time interval during which apoptotic cells can be recognized before their total disintegration varies depending on the method used, cell type, and nature of the inducer of apoptosis. Some inducers may slow down or accelerate the apoptotic process by affecting the rate of formation and shedding of apoptotic bodies, endonucleolysis, proteolysis, and so on. An observed twofold increase in apoptotic index may either indicate that twice as many cells were dying by apoptosis, compared to control, or that the same number of cells were dying, but that the duration of apoptosis was pro-

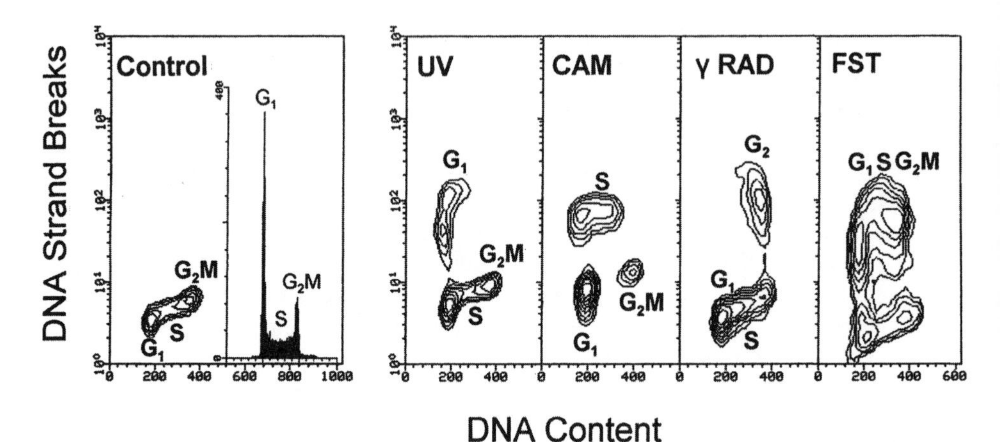

Fig. 3. Detection of the apoptosis-associated DNA strand breaks by flow cytometry (Method 4). Exponentially growing HL-60 cells were either untreated (control) or exposed for 20 s to UV radiation followed by 3 h of culture (UV), exposed to 0.15 μ*M* CAM for 3 h (CAM), exposed to 20 gy γ-irradiation followed by 3 h culture (γ RAD) or treated with 100 μ*M* of the DNA topoisomerase II inhibitor fostriecin (FST) for 3 h. Cells were then fixed as described, and digoxygenin-labeled dUTP incorporated at the sites of apoptosis-induced DNA strand breaks by TdT ***(8)***. The incorporated dUTP was labeled with FITC antidigoxygenin antibody (ApopTag™ kit; ONCOR, Inc.). Total DNA content was obtained following staining with PI after degradation of RNA with RNase. Exposure to UV caused apoptosis predominately of G_1 cells, whereas CAM and γ-radiation caused apoptosis of S- and G_2-phase cells, respectively. FST caused apoptosis of cells in all cell-cycle phases.

longed twofold. No method exists to obtain cumulative estimates of the rate of cell entrance to apoptosis as there is, for example, for mitosis, which can be arrested by microtubule poisons in a stathmokinetic experiment. Thus, the percentage of apoptotic cells in a cell population estimated by a given method is not a measure of the rate of cells dying by apoptosis.

The rate of cell death can be estimated when the absolute number (not the percent) of live cells is measured in the control and the treated culture, together with the rate of cell proliferation. The latter may be obtained from the rate of cell entrance to mitosis or the cell doubling rate. The observed deficit in the actual number of live cells from the expected number of live cells estimated based on the rate of cell birth provides the cumulative measure of cell loss (death).

5. Cells may die by a process resembling apoptosis, which lacks one or more typical apoptotic features ***(20–23)***. In some instances, DNA degradation is incomplete, resulting in 50- to 300-kb fragments rather than internucleosomal fragmentation. Any method of identification of apoptosis based on the detection of a single feature that happens to be missing (e.g., DNA laddering on gels) will fail to identify

atypical apoptosis in such a situation. Application of more than one method, each based on a different principle (i.e., detecting a different cell feature) stands a better chance of detecting atypical apoptosis than any single method. Analysis of cell morphology is generally needed to detect atypical apoptosis.

6. Apoptotic cells detach from the surface of the culture flasks and float in the medium. The standard procedure of discarding the medium, trypsinization, or EDTA treatment of the attached cells and their collection results in selective loss of apoptotic cells, which may vary from flask to flask depending on the handling of the culture (e.g., the degree of mixing or shaking, efficiency in discarding the old medium, and so on). Such an approach, therefore, cannot be used for quantitative analysis of apoptosis. To estimate the apoptotic index in cultures of adherent cells, the floating cells have to be pooled with the trypsinized cells and measured together.

   Density separation of cells (e.g., Ficoll-Hypaque or Percoll gradients) may result in selective loss of dying and dead cells. Knowledge of any selective loss of dead cells in cell populations purified by such approaches is essential.

Acknowledgments

This work was supported by NCI grant CA RO1 28704, the Chemotherapy Foundation, the Robert A. Welke Cancer Research Foundation, and "This Close" Foundation for Cancer Research.

References

1. Kerr, J. F. R., Winterford, C. M., and Harmon, V. (1994) Apoptosis. Its significance in cancer and cancer therapy. *Cancer* **73,** 2013–2026.
2. Vaux, D. L. (1993) Toward an understanding of the molecular mechanisms of physiological cell death. *Proc. Natl. Acad. Sci. USA* **90,** 786–789.
3. Oltvai, Z. N. and Korsmeyer, S. J. (1994) Checkpoints of dueling dimers foil death wishes. *Cell* **79,** 189–192.
4. Lazebnik, Y. A., Kaufmann, S. H., Desnoyers. S., Poirier, G. G., and Earnshaw, W. C. (1994) Cleavage of poly (ADP-ribose) polymerase by a proteinase with properties like ICE. *Nature* **371,** 346–347.
5. Weaver, V. M., Lach, B., Walker, P. R., and Sikorska, M. (1993) Role of proteolysis in apoptosis: involvement of serine proteases in internucleosomal DNA fragmentation in immature lymphocytes. *Biochem. Cell Biol.* **71,** 488–500.
6. Kaufmann, S. H. (1989) Induction of endonucleolytic DNA cleavage in human acute myelogenous leukemia cells by etoposide, camptothecin, and other cytotoxic anticancer drugs. A cautionary note. *Cancer Res.* **49,** 5870–5878.
7. Gorczyca, W., Bigman, K., Mittelman, A., Ahmed, T., Gong, J., Melamed, M. R., et al. (1993) Induction of DNA strand breaks associated with apoptosis during treatment of leukemias. *Leukemia (Baltimore)* **7,** 659–670.
8. Darzynkiewicz, Z., Juan, G., Li, X., Gorczyca, W., Murakami, T., and Traganos, F. (1997) Cytometry in cell necrobiology. Analysis of apoptosis and accidental cell death (necrosis). *Cytometry* **27,** 1–20.

9. Darzynkiewicz, Z., Li, X., and Gong, J. (1994) Assays of cell viability. Discrimination of cells dying by apoptosis. *Methods Cell Biol.* **41,** 16–39.
10. Kerr, J. F. R., Wyllie, A. H., and Curie, A. R. (1972) Apoptosis: a basic biological phenomenon with wide-ranging implications in tissue kinetics. *Br. J. Cancer* **26,** 239–257.
11. Arends, M. J., Morris, R. G., and Wyllie, A. H. (1990) Apoptosis: The role of endonuclease. *Am. J. Pathol.* **136,** 593–608.
12. Compton, M. M. (1992) A biochemical hallmark of apoptosis: Internucleosomal degradation of the genome. *Cancer Metastasis. Rev.* **11,** 105–119.
13. Umansky, S. R., Korol, B. R., and Nelipovich, P. A. (1981) In vivo DNA degradation in the thymocytes of gamma-irradiated or hydrocortisone-treated rats. *Biochim. Biophys. Acta* **655,** 281–290.
14. Nicoletti, I., Migliorati, G., Pagliacci, M. C., Grignani, F., and Riccardi, C (1991) A rapid and simple method for measuring thymocyte apoptosis by propidium iodide staining and flow cytometry. *J. Immunol. Methods* **139,** 271–280.
15. Gong, J., Traganos, F., and Darzynkiewicz, Z. (1994) A selective procedure for DNA extraction from apoptotic cells applicable for gel electrophoresis and flow cytometry. *Anal. Biochem.* **218,** 314–319.
16. Oberhammer, F., Wilson, J. M., Dive, C., Morris, I. D., Hickman, J. A., Wakeling, A. E., et al. (1993) Apoptotic death in epithelial cells: Cleavage of DNA to 300 and/or 50 kb fragments prior to or in the absence of internucleosomal fragmentation. *The EMBO J.* **12,** 3679–3684.
17. Gorczyca. W., Bruno, S., Darzynkiewicz, R. J., Gong, J., and Darzynkiewicz, Z. (1992) DNA strand breaks occurring during apoptosis: Their early in situ detection by the terminal deoxynucleotidyl transferase and nick translation assays and prevention by serine protease inhibitors. *Int. J. Oncol.* **1,** 639–648.
18. Gold, R., Schmied, M., Giegerich, G., Breitschopf, H., Hartung, H. P., Toyka, K. V., et al. (1994) Differentiation between cellular apoptosis and necrosis by the combined use of in situ tailing and nick translation techniques. *Lab. Invest.* **71,** 219–225.
19. Li, X. and Darzynkiewicz, Z. (1995) Labelling DNA strand breaks with BrdUTP. Detection of apoptosis and cell proliferation. *Cell Prolif.* **28,** 571–579.
20. Li, X., Melamed, M. R., and Darzynkiewicz, Z. (1996) Detection of apoptosis and DNA replication by differential labeling of DNA strand breaks with fluorochromes of different color. *Exp. Cell Res.* **222,** 28–37.
21. Cohen, G. M., Su, X.-M., Snowden, R. T., Dinsdale, D., and Skilleter, D. N. (1992) Key morphological features of apoptosis may occur in the absence of internucleosomal DNA fragmentation. *Biochem. J.* **286,** 331–334.
22. Collins, R. J., Harmon, B. V., Gobe, G. C., and Kerr, J. F. R. (1992) Internucleosomal DNA cleavage should not be the sole criterion for identifying apoptosis. *Int. J. Radiat. Biol.* **61,** 451–453.
23. Catchpoole, D. R. and Stewart, B. W. (1993) Etoposide-induced cytotoxicity in two human T-cell leukemic lines: Delayed loss of membrane permeability rather than DNA fragmentation as an indicator of programmed cell death. *Cancer Res.* **53,** 4287–4296.

24

Single-Strand Conformational Polymorphism Analysis of DNA Topoisomerases I and II

Mary K. Danks

1. Introduction

1.1. Rationale

Single-strand conformational polymorphism (SSCP) analysis is a method for screening for mutations. This method is based on observations of Orita and coworkers that single strands of DNA assume unique conformations depending on their primary sequence. The uniqueness of each three-dimensional conformation, in turn, dictates a unique migration rate following electrophoresis through a nondenaturing polyacrylamide gel of the single-stranded fragments *(1,2)*. Shown schematically in **Fig. 1**, differences in migration rates reflect differences in primary sequences. Since complementary strands of DNA have different primary sequences, under optimal conditions, each fragment of double-stranded DNA analyzed by SSCP will appear as two distinct bands that migrate at different rates through a nondenaturing gel (**Fig. 2**). Usually, double-stranded radiolabeled PCR or RT-PCR products are generated, and then electrophoresed on polyacrylamide gels using a sequencing apparatus and detected by autoradiography. Nondenaturing conditions are essential to allow each DNA fragment to retain its three-dimensional conformation.

SSCP has been used extensively to screen for mutations in genes coding for the tumor suppressor genes p53 and Rb, the *ras* family of oncogenes, and genes involved in hereditary diseases, such as neurofibromatosis, familial adenomatous polyposis, and Wilms' tumor (reviewed in **ref. *4***). SSCP has also been successfully applied to analyze genes or messages coding for DNA topoisomerase I *(5)* or topoisomerase II *(3,6–9)*. In the case of topoisomerases,

From: *Methods in Molecular Biology, Vol. 95: DNA Topoisomerase Protocols, Part II: Enzymology and Drugs*
Edited by N. Osheroff and M.A. Bjornsti © Humana Press Inc., Totowa, NJ

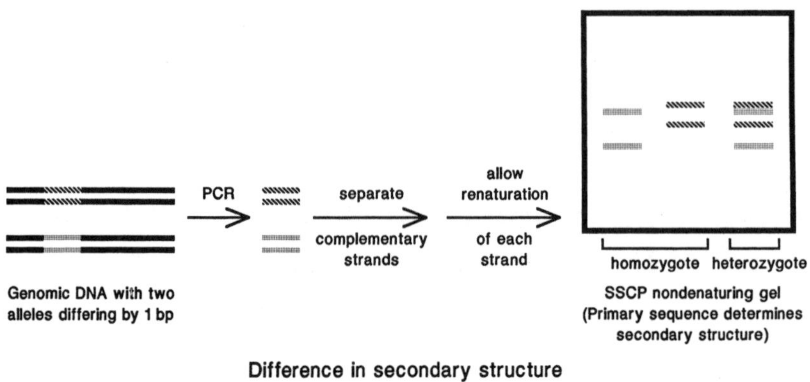

Fig. 1. Schematic representation of detection of mutations by SSCP analysis. Difference in secondary structure of mutant allele causes mobility shift.

SSCP has been applied to detect mutations associated with resistance to inhibitors of these enzymes. With topoisomerase II, as with p53, there are apparent "hot spots" associated with mutations conferring mutant phenotypes, and these regions are readily analyzed by SSCP. With topoisomerase I, no mutagenic or resistance-associated hot spots have been identified, but SSCP analysis of the entire message is feasible.

1.2. SSCP as the Method of Choice

Several different types of methods are available for screening for mutations. These methods include: denaturing gradient gel electrophoresis (DGGE) of PCR products attached with a GC clamp; electrophoresis of heteroduplex DNA; chemical mismatched cleavage; allele-specific oligonucleotide hybridization (ASO); and restriction fragment-length polymorphism (RFLP) analysis (reviewed in **ref. *3***). Each method is suitable for assessing or comparing multiple regions of multiple samples, and each has advantages and disadvantages. However, none of these methods is considered a substitute for sequencing. SSCP is a simple, sensitive, rapid screening assay, which indicates the regions of individual samples that should be sequenced. The disadvantage of SSCP has been reported to be a lack of sensitivity under some conditions, resulting in failure to detect mutations. However, recent studies have identified the variables that determine the efficiency of the assay and accurate detection of mutations can be expected to be >90–100% ***(10)***.

The SSCP protocol chosen will depend on the research question being asked: If the goal of a given series of experiments is to determine whether a specific

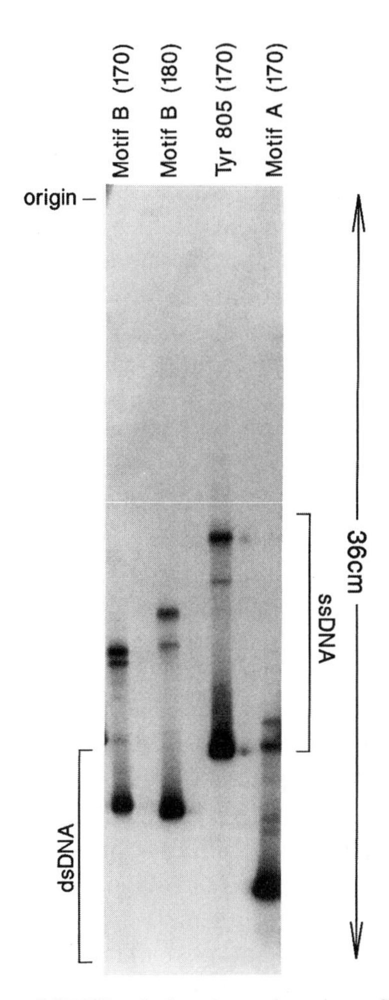

Fig. 2. Autoradiogram of SSCP gel showing migration of each of several RT-PCR products from CEM cells: Lane 1, "Motif B" dinucleotide binding sequence of topoisomerase IIα; Lane 2, Motif B sequence of topoisomerase IIβ; Lane 3, region adjacent to tyrosine 805 of topoisomerase IIα; and, Lane 4, the Motif A sequence of topoisomerase IIα ***(3)***. Reprinted with permission of *Cancer Res.*

mutation is present in a panel of samples, then assay conditions are tailored to detect that particular mutation, resulting in 100% detection efficiency. If the goal is to determine whether any of many possible mutations is present within a given nucleic acid sequence, then the conditions of the assay must be optimized to detect all mutations. The protocol detailed in **Subheading 3.** of this chapter is an example of a protocol used to screen for many possible mutations.

As noted above, the most commonly used method for detection of DNA bands in SSCP gels is by incorporation of radionuclides into PCR products and subsequent detection by autoradiography. Nonisotopic detection of bands in SSCP gels has also been done. Silver staining ***(11,12)*** or ethidium bromide (EtBr) staining ***(13)*** can be used to detect DNA bands in smaller gels (up to ~14 cm). This adaptation of the more often used method is excellent if sufficient starting material is available to generate relatively large amounts of PCR product and if differences in migration are apparent in smaller gels. However, if longer migration paths (sequencing-sized gels) are necessary to optimize detection of subtle differences in migration rates of the DNA fragments, silver staining of sequencing-sized gels may be unwieldy. The protocol outlined in **Subheading 3.** is applicable to PCR products generated from the DNA or RNA extracted in limited quantities from clinical material ***(3,14)***. It is reproducible and sufficiently sensitive to detect mutations by SSCP in samples in which the nucleic acid concentration of the original sample is too low to be quantitated by fluorometry.

1.3. Assay Efficiency

Several factors affect the efficiency of the assay. These factors include: the distance that the DNA fragments migrate, the concentration of the DNA in the sample, the temperature at which electrophoresis is done, the percent of acrylamide and glycerol in the gels, and the rate at which the electrophoresis is carried out ***(3,4,10)***. It has also been reported that DNA fragments should be limited to 200 bp or less; however, our observations (unpublished) and reports from other laboratories ***(10)*** show that SSCP analysis can be done successfully on DNA fragments up to ~375 bp in length.

In general, the two variables that most affect the sensitivity and reproducibility of the assay are the distance that the DNA fragments migrate and the concentration of the DNA in the sample. If differences in migration rates between a normal and a mutant sequence are small, it is logical that these differences will become more apparent with longer migration distances. The sensitivity of the assay has been directly related to the distance of migration of the DNA fragments. Fan and coworkers recommend that the minimum distance that the leading bands of single-stranded DNA (ssDNA) be run is 16–18 cm ***(10)***. The second variable that affects migration of DNA in SSCP gels is the concentration of the DNA in the sample. The total amount of DNA does not influence the migration rate of the DNA fragments (Danks, unpublished observation), but the concentration of the DNA of some PCR products can have a profound influence on migration (**Fig. 3**).

Other factors reported to influence the sensitivity of the assay include aberrant band migration resulting from a nonuniform temperature of the gel after

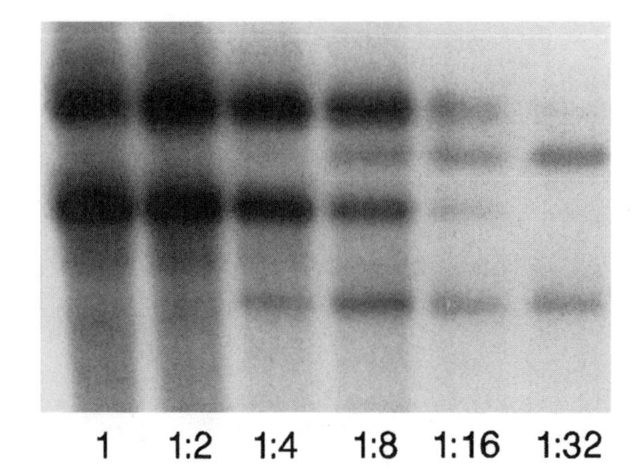

Fig. 3. Effect of DNA concentration on migration of PCR products containing the Motif B/DNBS region of topoisomerase IIα from CEM cells. A single PCR product was diluted as indicated, and electrophoresed on an SSCP gel as outlined in **Subheading 3.** *(3)*. Dilution of PCR product motif B/DNBS. Reprinted with permission of *Cancer Res.*

long run times or when high wattage or voltage is used. Simple solutions to minimize this problem are to direct a fan toward the electrophoresis apparatus during the run, use lower power settings, or carry out the elctrophoresis procedure at 4°C to maintain constant gel temperature. Alternatively, temperature-controlled electrophoresis units are now commercially available. However, we have found that mutations readily detected in gels run at room temperature are not always apparent in gels run at 4°C. We have also found the use of a fan unnecessary. The percentages of acrylamide and glycerol in the gel also affect SSCP results. Optimal concentrations are usually determined empirically, with the most important consideration being inclusion of at least 5% glycerol.

It is interesting that the type or location of the mutation within the DNA fragment does not affect the sensitivity of the method; single-base deletions or substitutions are readily detected, irrespective of their location in the sequence being analyzed.

1.4. Interpretation of Results

As shown in **Fig. 2**, in each lane of an SSCP gel containing one double-stranded (dsDNA) sequence, three bands are evident: residual dsDNA, and each strand of the dsDNA PCR product. In the simplest case, each dsDNA sample to be analyzed contains the same sequence and each of the complementary strands of that sequence has a single stable conformation under the electrophoresis conditions chosen. Two ssDNA bands will be seen in each lane.

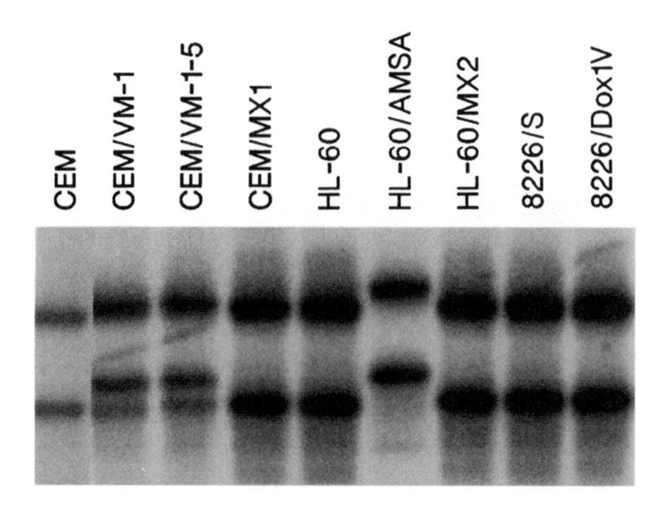

Fig. 4. Autoradiogram of an SSCP gel showing migration of PCR products of the Motif B/DNBS consensus dinucleotide binding sequences generated from RNA of the indicated cell lines. Motif B/DNBS DNA topoisomerase IIα.

Differences in migration of either or both bands of the control compared to the unknown sample indicate the presence of a mutant sequence (**Fig. 4**).

If more than two ssDNA bands are seen in a given lane on an SSCP gel, several explanations are possible: 1) Under the conditions of the assay (DNA concentration, temperature), one or each of the ssDNA fragments can assume more than one stable three-dimensional conformation. If this is the case, two, three, or four ssDNA bands may be evident for a single dsDNA PCR product (**Fig. 3**). 2) If the original DNA or RNA sample codes for or contains a message from two distinct sequences, as would be seen if a mutant and wild-type protein were coexpressed in a cell, a minimum of four bands would be seen in each lane. Similarly, if two populations of cells were represented in the original DNA or RNA sample, such as contamination of normal tissue with tumor tissue that expressed a mutant protein, more than two bands would be evident on an SSCP gel.

2. Materials

1. Thermal cycler.
2. Horizontal gel apparatus suitable for DNA agarose gels.
3. Sequencing apparatus with 0.4-mm spacers.
4. Comb with 1-cm wells for sequencing apparatus.
5. Large sheets of Whatman 3MM paper (or equivalent).
6. Stock solution for making 6% acrylamide, 0.12% bis-acrylamide gels:

90 mL 30% acrylamide.
27 mL 2% bis-acrylamide.
45 mL glycerol.
90 mL of 5X electrophoresis buffer.
198 mL deionized water.

Store in a dark bottle at room temperature. No precipitates should be evident. For each gel: Use 90 mL of stock solution + 787 µL 10% ammonium persulfate (make fresh daily) + 30 µL TEMED. The gel will polymerize quickly; pour and insert comb within ~3 min after adding ammonium persulfate and TEMED.

7. Electrophoresis buffer 5X: Tris-borate 89 m*M*, pH 8.2 with 4 m*M* EDTA. To make 4 L:
   216 g Tris base.
   110 g boric acid.
   160 mL 0.5 *M* EDTA, pH 8.0.

   Store at room temperature; no precipitates should be evident.
8. Sample buffer, 5X:
   750 µL formamide (*see* **Note 2**).
   250 µL glycerol.
   40 µL 0.5*M* EDTA, pH 8.0.
   2.5 µg bromophenol blue.
   2.5 µg xylene cyanol.

3. Method

3.1. Outline of Method

1. Generate radioactive PCR or RT-PCR product 75–350 bp in length.
2. Verify that PCR conditions produce a single dsDNA product of appropriate size on an agarose gel.
3. Put aliquot of PCR product into formamide buffer.
4. Boil samples to separate complementary strands of DNA.
5. Put boiled samples on ice immediately.
6. Electrophorese samples on SSCP polyacrylamide gel until bands have migrated 15–20 cm from origin.
7. Immobilize gel on a piece of Whatmann filter paper.
8. Wrap in plastic wrap.
9. Develop autoradiograph for 1–24 h.
10. Compare position of bands from different samples. Differences in migration represent differences in primary sequences.

3.2. Detailed Protocol

1. Generate a radioactive dsDNA PCR or RT-PCR product by any standard method. Radiolabeling may be done by incorporating ^{32}P into the primers used for amplification or by labeling the PCR product metabolically by including 1 µCi ^{32}P dNTP in each PCR reaction mixture. A single dNTP, e.g., ^{32}P[dCTP] or

^{32}P[dATP], or a mixture of dNTPs may be used. Using 1 µg DNA or RNA as the initial substrate, followed by 15 min of reverse-transcriptase reaction, if appropriate, and 30 cycles of amplification by DNA, polymerase should yield sufficient product to run a minimum of 10 SSCP gels *(3)* (*see* **Note 8**).

2. Confirm by agarose-gel electrophoresis that a single PCR product of the appropriate size was produced. If a single PCR product has been generated, it is not necessary to purify the product further, i.e., residual primers do not interfere with interpretation of SSCP results.
3. Pour SSCP gel. Gel polymerizes relatively rapidly; work quickly after ammonium persulfate and TEMED have been added to the gel solution. Allow to polymerize (1–2 h). Place gel in electrophoresis apparatus, with buffer in upper and lower chambers. Samples will be underlaid into wells.
4. Prepare duplicate samples of each PCR product to be analyzed. In each of two microfuge tubes, place 4 µL PCR product and 1 µL sample buffer. One tube will be used to determine where dsDNA migrates in the SSCP gel. The other tube is for SSCP analysis. When the migration of the dsDNA for each set of primers has been determined not to interfere with interpretation of results with the ssDNA fragments, this control may be omitted.
5. Heat one of each pair of samples to 99°C for 6 min to separate the complementary strands of dsDNA.
6. Place the heated samples on ice *immediately* after heating, and allow them to remain at 4°C for a minimum of 8 min to allow formation of stable conformations of the single strands of DNA.
7. Load samples. Combs with 1-cm wells are suggested, since the thin flat bands of DNA that result from combs of this dimension yield easily interpretable results. The bands will assume the shape of the well and are relatively unforgiving (unpublishable) if the well is distorted when the sample is loaded. A needle or flat plastic strip (the blank end of a plastic pH indicator strip works well) can be used to remove any air bubbles caught in the wells or to remove any remnants of polymerized acrylamide.
8. Elecrophorese at 30 W for 8–10 h (or until the ssDNA bands have migrated 15–20 cm). Electrophoresis at constant wattage minimizes problems of uneven heating of the gel. Constant voltage or amperage can also be used, depending on the power supply available. Most reproducible and convenient are power supplies that can be programmed for constant power for a predetermined time.
9. When electrophoresis is complete, separate the gel plates such that the entire gel remains associated with one of the plates. This is accomplished fairly easily by separating the plates very slowly. Undue haste in separating gel plates invariably has disastrous results.
10. Cut Whatman 3MM paper to the size of the gel, and "attach" gel to the paper by gently pressing or patting the paper onto the gel. Gloves should be worn from this step until the completion of the procedure (*see* **Note 5**).
11. Peel paper and gel from glass plate. Wrap in plastic wrap (*see* **Note 6**).
12. Expose to X-ray film, and develop film after appropriate exposure at –70°C.

13. If differences in migration of ssDNA bands from a control sample and a sample of unknown sequence are detected, run a second SSCP gel to compare migration of bands in serial dilutions of each of the samples. As shown in **Fig. 3**, dilutions of 1:2 to 1:32 should be sufficient.
14. If differences in migration rates are apparent at all DNA concentrations, repeat the PCR reaction with the original nucleic acid sample to rule out misincorporation of nucleotide by DNA polymerase as the source of the "mutation".
15. If differences in migration rates are still apparent after the second PCR reaction and serial dilution, a mutation is likely, and DNA sequencing is necessary to determine the precise change in nucleotide sequence.

4. Notes

1. On the autoradiograph of an SSCP gel, the intensity of the two bands derived from complementary strands of dsDNA may or may not be of equal intensity. Since the primary sequence of each strand is different, the intensity of the labeling of each strand will depend on which dNTP was chosen to label the PCR products and how many residues of the original sequence dictated incorporation of a labeled nucleotide into the PCR products. A mixture of all four radiolabeled dNTPs usually yields bands of approximately equal intensity, if this is desirable.
2. Do not use dimethylformamide in the sample buffer instead of formamide.
3. If bands appear fuzzy, clean gel plates sequentially with soap and water, methanol, and acetone; purchase new ammonium persulfate and TEMED; and make a fresh stock solution of Tris buffer.
4. After electrophoresis, separate plates slowly.
5. Removal of the SSCP gel from the glass plate with Whatman paper should be very easy. If the gel continues to adhere to the plate instead of the paper, the composition of the gel is probably not as specified in the protocol. Old pieces of X-ray film do not work as well as Whatman paper.
6. It is not necessary to remove every crease or wrinkle from the plastic wrap when the gel is prepared for autoradiography. However, if the plastic wrap or the gel has been touched (without gloves), fingerprints may be visible on the autoradiograph.
7. If appropriate samples (cells, DNA, RNA) are not available to serve as positive controls, a primer with a point mutation can be used to generate PCR products with a mutant sequence.
8. PCR products can be stored at 4°C and used with good results for up to a week. After a week, the signal:noise ratio increases, and results are not always of publishable quality. Samples analyzed within 24 h after the DNA polymerase reaction give the cleanest results.

References

1. Orita, M., Iwahana, H., Kanazawa, H., Hayashi, K., and Sekiya, T. (1989) Detection of polymorphisms of human DNA by gel electrophoresis as single-strand conformation polymorphisms. *Proc. Natl. Acad. Sci. USA* **86,** 2766–2770.
2. Orita, M., Suzuki, Y., Sekiya, T., and Hayashi, K. (1989) Rapid and sensitive

detection of point mutations and DNA polymorphisms using the polymerase chain reaction. *Genomics* **5,** 874–879.

3. Danks, M. K., Warmoth, M. R., Friche, E., Granzen, B., Bugg, B. Y., Harker, W. G., et al. (1993) Single strand conformational polymorphism (SSCP) analysis of the 170 kDa isozyme of DNA topoisomerase II in human tumor cells. *Cancer Res.* **53,** 1373–1379.
4. Hayashi, K., (1992) PCR-SSCP: A Method for detection of mutations. *GATA* **9,** 73–79.
5. Kubota, N., Kanzawa, F., Nishio, K., Takeda, Y., Ohmori, T., Fujiwara, Y., et al. (1992) Detection of topoisomerase I gene point mutation in CPT-11 resistant lung cancer cell line. *Biochem. Biophys. Res. Commun.* **188,** 571–577.
6. Hashimoto, S., Danks, M. K., Beck, W. T., Chatterjee, S., and Berger, N. A. (1995) A Novel point mutation in the 3' flanking region of the DNA-binding domain of topoisomerase IIα associated with acquired resistance to topoisomerase II active agents. *Oncol. Res.* **7,** 21–29.
7. Kohno, K., Danks, M. K., Matsuda, T., Nitiss, J. L., Kuwano, M., and Beck, W. T. (1995) A Novel mutation of DNA topoisomerase IIα gene in an etoposide-resistant human cancer cell line. *Cell Pharmacol.* **2,** 87–90.
8. Ritke, M. K., Allan, W. P., Fattman, C., Gunduz, N. N., and Yalowich, J. C. Reduced phosphorylation of topoisomerase II in etoposide-resistant human leukemia K562 cells. *Mol. Pharmacol.* **46,** 58–66.
9. Rizvi, N. A., Ng, S. W., Sullivan, D., Eder, J. P., Schnipper, L. E., and Chan, V. T. (1993) Identification of a point mutation in topoisomerase II cDNA in a mitoxantrone resistant Chinese hamster ovary cell line, MXN4. *Proc. Am. Assoc. Cancer Res.* **34,** A1984.
10. Fan, E., Levin, D. B., Glickman, B. W., and Logan, D. M. (1993) Limitations in the use of SSCP analysis. *Mutat. Res.* **288,** 85–92.
11. Ainsworth, P. J., Surh, L. C., and Colter-Mackie, M. B. (1991) Diagnostic single strand conformation polymorphism: a simplified non-radioisotopic method as applied to a Tay-Sachs B1 variant. *Nucleic Acids Res.* **19,** 405–406.
12. Tooke, N. (1993) Rapid detection of point mutations and polymorphisms by PCR-SSCP analysis on PhastSystem. *J. NIH Res.* **5,** 78.
13. Hongyo, T., Buzard, G. S., Calvert, R. J., and Weghorst, C. M. (1993) "Cold SSCP": A Simple, rapid and non-radioactive method for optimized single-strand conformation polymorphism analyses. *Nucleic Acids Res.* **21,** 3637–3642.
14. Danks, M. K., Beck, W. T., and Suttle, D. P. (1993) Topoisomerse IIα mutation in leukemic cells from a patient with lineage switch AML. *Proc. Amer. Assoc. Cancer Res.* **34,** A1982.

25

PCR-Based Cloning of DNA Topoisomerase Genes

Wai Mun Huang

1. Introduction

Numerous type I and type II DNA topoisomerase genes have now been identified and sequenced from both the eukaryotic and prokaryotic sources. Amino acid sequence alignments of these genes firmly establish that all type II topoisomerase genes belong to one family regardless of source, whereas among the type I topoisomerase genes, there are two subtypes: one typified by the bacterial *topA* gene and the other by the eukaryotic *TOP1* gene ***(1,2)***. The high degree of sequence conservation among these families of genes provides a straightforward approach to identifying new genes by PCR methodologies. These genes may be used as specific DNA probes for genomic organization studies, for topoisomerase-targeting drug resistance studies, or they may be overexpressed as recombinant proteins for structure–function and comparative analyses.

The use of reverse translation to convert a block of seven or more contiguous amino acids into degenerate oligonucleotide mixtures, which can in turn be used as probes for library screening, is a common method to identify genes when partial amino acid sequence information is available. This approach is applicable to identifying topoisomerase genes, since oligonucleotide probes can be generated from any one of the invariant or conserved regions, and this method was used to clone human and trypanosome *TOP2* genes ***(3,4)***. However, a more specific and foolproof method is to select two conserved regions in a topoisomerase gene, derive opposing degenerate oligonucleotides from these two amino acid blocks, and use them in a PCR reaction to generate a longer specific DNA fragment from the organism of interest. Screening of a library using such a specific and longer DNA fragment as a probe allows more stringent screening to identify specific genes with fewer false positives. This

From: *Methods in Molecular Biology, Vol. 95: DNA Topoisomerase Protocols, Part II: Enzymology and Drugs*
Edited by N. Osheroff and M.A. Bjornsti © Humana Press Inc., Totowa, NJ

method has been used to clone gyrase genes from a number of bacteria ***(5,6)***. When using this approach, it is important to verify that the PCR product generated by the degenerate primers is actually the fragment of a topoisomerase gene. It is common to find incorrect PCR fragments generated from nonspecific sites when degenerate primers are used, especially when the annealing temperature of the PCR reaction is not optimal. Thus, it is advisable first to clone the PCR fragment into an *Escherichia coli* high-copy-number plasmid, and then to sequence the inserted PCR fragment using primers from the flanking plasmid sequences. In doing so, the presence of the two initial PCR-primer pairs can be verified, and more importantly, other invariant amino acids located between the two selected conserved blocks can provide confirmation that the amplified fragment is the gene of interest.

When a well-represented library is not available for screening, in order to obtain the remaining part of the gene, gene-walking techniques involving inverse PCR *(7)* can be used to extend gene sequences flanking either the 5'- or the 3'-side of the characterized fragment. In the inverse PCR procedure, genomic DNA is first digested with a restriction enzyme, followed by ligation at a low DNA concentration that favors the formation of monomeric circles. The cyclized DNA is then used as template for the PCR. For example, if an outward pointing oligonucleotide primer pair located on the 3'-side of a selected restriction enzyme site is chosen, the inverse PCR product can be used as template to sequence the remaining portion of the restriction fragment harboring the primer pair, thereby extending the gene sequence 3' to the next restriction site. From the sequence of the newly extended region, another restriction enzyme and another pair of outward pointing primers can be chosen, and inverse PCR can be repeated until a complete open reading frame is obtained. In principle, any restriction enzyme uniquely located 5' or 3' from the pair of outward-pointing primers chosen for the genomic walking experiment may be used as the digestive enzyme to cyclized the template DNA for inverse PCR. Well-behaved restriction enzymes having six or more symmetric base recognition sites, which cut the genomic DNA to approx 1- to 3-kb fragments with cohesive ends, are preferred, since these products are more likely to form monomeric circles on ligation. Since the products of the inverse PCR reactions are rearranged gene fragments, it is best to prepare a new round of forward (or standard) PCR to isolate the full-length gene after the sequences of the inverse PCR products are determined.

In this chapter, the cloning of a bacterial *gyrA* gene from an extreme thermophile is given as an example. The method is generally applicable to other highly conserved genes, provided suitable PCR primers can be selected for the initial amplification reaction. Bacterial DNA topoisomerase genes of either type I and type II enzymes are especially well suited for this application because of their ubiquitous presence in all eubacteria, the absence of introns, and the presence of a number of blocks of conserved amino acids that have already been identified from known sequence alignments *(1)*. Because of the likely presence of introns, cDNA may be used as the source of starting template in the cloning of eukaryotic topoisomerase genes. Bacterial gyrases are encoded by two genes, *gyrB* and *gyrA*. Both genes can be identified using different pairs of opposing degenerate primers derived from consensus regions in each gene *(8)*. In many bacteria, the *gyrB* gene is located immediately upstream of the *gyrA* gene. Attempts can be made to test this possibility in new cases with a PCR reaction using a forward primer near the C-terminus of *gyrB* and a reverse primer near the N-terminus of *gyrA* gene.

Eubacteria encode two sets of type II topoisomerases, namely gyrase and topoisomerase IV. Among the genes encoding their subunits, *gyrA* and *parC* are homologs as are *gyrB* and *parE (1)*. Depending on the region of conserved amino acids chosen to design the degenerate primers, both types of gyrase-like gene sequences may be simultaneously amplified. By comparing the derived amino acid sequences from each type of these cloned sequences, it is often possible to obtain both homologous genes using one set of degenerate primers *(5)*. By analogy, it may be possible to obtain DNA clones harboring fragments from the two *TOP2* genes in mammalian cells, since two highly homologous *TOP2* genes have been identified in human and in mouse cells *(2)*.

The amino acid sequences of bacterial *gyrA* genes share about 50% identity *(1)*. Longer runs of highly conserved amino acids are more abundant near the N-terminal third of these molecules. In the remaining region of the gene, invariant amino acids are found in smaller clusters or as single amino acids. Few useful regions of conserved amino acid blocks suitable to generate PCR primers are present within the C-terminal half of the molecule. In principle, an oligonucleotide with a 12- to 15-basepairing capacity (a block of 4–5 codons) may reasonably serve as a PCR primer. However, in light of the wide range of G + C contents among eubacterial DNA, longer oligonucleotide primers are recommended in order to compensate for the higher denaturing temperature of the higher G + C bacterial DNA. Two degenerate primers derived from codons 39–45 (DGLKPVH) and 173–179 (GIAVGMA) (according to the *E. coli gyrA* residue number) have been found to serve well as PCR primers for the amplification of a *gyrA* gene fragment from diverse bacteria *(9)*.

2. Materials

1. Oligonucleotide primers were prepared using an Applied Biosystems DNA synthesizer and were used without purification. The locations of the primers used in this example are shown in **Fig. 1**. The sequences of three pairs of primers are given below with the noncoding restriction sites used for cloning purposes

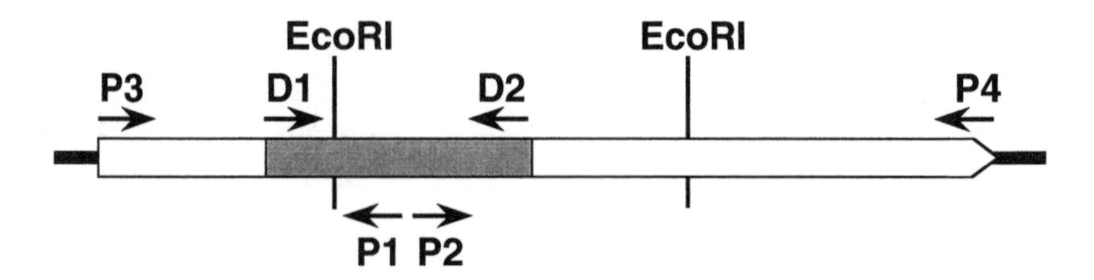

Fig. 1. A schematic diagram of the locations of the primers used in amplifying the *T. maritima gyrA* gene. The shaded area represents the initial amplification product. P1 and P2 were used in the first round of inverse PCR (*see* **Subheading 2.** for detail).

underlined. In the pair of degenerate primers, N represents all four bases. The other two pairs of primers are specific for *Thermotoga maritima* used in this example ***(10)***.

Degenerate primers spanning the regions DGLKPVH (*E. coli* gyra codons 39–45) and GIAVGMA (codons 173–179):

D1: 5' ACCGGTACCGANGGGNTNAANCCNGTNCA (*Kpn*I)

D2: 5' CGGAAGCTTGCCATACCNACNGCNATNCC (*Hin*dIII)

Two internal primers for inverse PCR:

P1: 5' GGCCGAGTTCGTACATTCCATAG

P2: 5' AGCCTGAAGTCCTCCCATCCAAAGTG

Two primers at the beginning and end of the *Thermotoga gyrA* gene:

P3 5' GTTAAACATATGCCAGAGATCCTGATAAACAAACC (*Nde*I)

P4 5' CGGGATCCTCAATCTTTCACCACCGCCACCTTCG (*Bam*HI)

2. Genomic DNA from *T. maritima.*
3. 100-m*M* solution of dATP, dTTP, dCTP, and dGTP mix (from Pharmacia, Piscataway, NJ).
4. Light mineral oil.
5. Plasmid vector pBluescript (Stratagene, La Jolla, CA) is a high-copy-number plasmid, and pET21a (Novagen, Madison, WI) is a lower-copy-number plasmid used to clone the full-length gene.
6. *Taq* DNA polymerase and buffer (e.g., from Gibco BRL, Gaithersburg, MD) and KlentaqLA DNA polymerase and buffer (Ab Peptides, St. Louis, MO).
7. DNA thermal cycler (Perkin Elmer) or similar.
8. Restriction enzymes and buffers provided by supplier.
9. T4 DNA ligase and buffer provided by supplier.

3. Methods

3.1. Amplification by PCR of Genomic DNA and Cloning of the Product

1. Set up PCR reaction in 1X PCR buffer containing 0.2 m*M* each dNTP, 20–50 ng of genomic DNA, 1 µ*M* each primer, and 5 U of *Taq* DNA polymerase in 50 µL. Overlay the mixture with 50 µL of mineral oil (*see* **Note 1**).

2. Perform the PCR reaction in a thermal cycler using the following program: 1 cycle at 95°C for 3 min linked to 30 cycles at 94°C for 1 min, 55°C for 2 min, 72°C for 3 min, and finally 1 cycle at 72°C for 10 min.
3. Analyze 3 µL of the PCR product by electrophoresis on a 2.5% Nusieve (FMC bioProducts, Rockland, ME) agarose gel to check the reaction product, which should be 450 bp in length (*see* **Note 2**).
4. Add an equal volume of phenol/chloroform to the PCR product to extract the aqueous phase. Repeat once more, and ethanol-precipitate the PCR product. Resuspend the DNA pellet in 20 µL of sterile water.
5. Separately digest 2–3 µg of the purified PCR DNA and 0.5 µg of pBluescript plasmid DNA with *Kpn*I and *Hin*dIII for 2 h at 37°C in the buffer supplied by the manufacturer. Check for completion of enzyme digestions by analyzing small aliquots of the reactions in Nusieve or agarose-gel electrophoresis (*see* **Note 3**).
6. Phenol/chloroform-extract digested DNA, and ethanol-precipitate DNA. Resuspend DNA in 15 µL of water.
7. Set up ligation reaction containing 1X ligase buffer provided by manufacturer, PCR product and cleaved vector with a molar ratio of 4 PCR fragment molecules to 1 cut vector and 1 U of T4 ligase in a volume of 20 µL. Incubate at 15°C for 4 h.
8. Transform competent *E. coli* XL1-Blue cells. Clones carrying the *gyrA* fragment can be identified by agarose-gel analysis of plasmids having larger size than the vector (*see* **Note 4**).

3.2. Sequence Confirmation of Insert

1. Sequence candidate clones carrying the *gyrA* gene fragment using T7 and T3 primers, which are located in the pBluescript plasmid flanking the insertion (*see* **Note 5**).
2. Translate the sequence into encoding amino acids, and align the amino acid sequence with known sequences of the *gyrA* gene between residues 39 and 179 of the *E. coli* gene.
3. Confirm the presence of the consensus amino acids from which the degenerate primers are derived, namely the sequence DGLKPVH at the 5'-end and the sequence GIAVGMA at the 3'-end of the insertion.
4. Confirm the presence of sequence AAXRYTE at about the middle of the insertion (*see* **Note 6**). Numerous other conserved amino acids should also be evident by using an amino acid sequence-alignment program.

3.3. Amplification by Inverse PCR and Extension of Sequence

1. Digest 1–2 µg of genomic DNA with 50 U of *Eco*RI for 4 h at 37°C in 1X buffer supplied by the manufacturer (*see* **Note 7**).
2. Heat digested genomic DNA to 65°C for 20 min to inactivate *Eco*RI.
3. Set up ligation reaction containing cut DNA at 5–10 µg/mL in 1X ligation buffer supplied by the manufacturer and 1 U of T4 ligase at 14°C for 2 h to cyclize the digested genomic DNA.

4. Set up PCR reaction as described in **Subheading 3.1.** using 40 ng of the cyclized genomic DNA as template, 0.1 μ*M* each of the internal primers P1 (antisense) and P2 (sense), which are located 3' to the unique *Eco*RI site, 0.2 m*M* d3NTP, and 3U of *Taq* DNA polymerase in 1X PCR buffer. Use the same cycling program as described in **Subheading 3.1.** to perform the reaction.
5. Analyze 3 μL of the inverse PCR product on 1% agarose-gel electrophoresis to check the purity and yield of the reaction. The product is 1.3 kb in size for the *Thermotoga* gene.
6. Sequence the inverse PCR product using P1 and P2 as sequencing primers.
7. Confirm that sequences generated from the P1 and P2 primers begin with known sequences determined in **Subheading 3.2.** and extending beyond.
8. Select appropriate sequencing primers for subsequent sequence extension until the sequence of the entire region is determined. Beware of the presence of *Eco*RI sites on which the actual gene sequence continuity becomes interrupted owing to the use of the cyclized DNA template. If there are more than one *Eco*RI site in the inverse PCR fragment, any sequence from beyond the first *Eco*RI site from either side of the P1 or P2 primer should be ignored, because it may represent the joining of random linear fragments during the ligation step (*see* **Note 8**).
9. Identify another restriction enzyme for repeated application of inverse PCR reaction and subsequent DNA sequencing of the product. Repeat the inverse PCR for both the 5'- and 3'-regions for as many rounds as necessary until one large open reading frame encoding approx 800 residues is identified.
10. Perform amino acid sequence alignment with other known bacterial *gyrA* genes to confirm that consensus amino acids are present.

3.4. Cloning of the Full-Length Gene

1. Set up a PCR reaction in 1X PCR buffer with 0.2 m*M* dNTP, 20–50 ng of genomic DNA, 0.1 μ*M* each of primers P3 and P4, and 3 U of KlenTaqLA DNA polymerase in 50 μL. Overlay the mixture with 50 μL of mineral oil (*see* **Note 9**).
2. Perform PCR reaction as described in **Subheading 3.1.**
3. Check purity and quantity of PCR yield by analyzing 3 μL of product on a 1% agarose-gel electrophoresis.
4. Follow **steps 4–8** as described in **Subheading 3.1.** to digest the PCR product and the vector pET21a with *Nde*I and *Bam*HI. Ligate the cut fragment into the cut plasmid vector, and transform the ligated product into competent *E. coli* cells to obtain clones that carry the full-length *gyrA* gene. Confirmation of the clone may be done by sequencing across the insertion sites and by restriction analysis, since the complete sequence of the gene has already been determined (*see* **Note 10**).

4. Notes

1. A higher concentration of degenerate primers is required in this PCR reaction than that commonly used for nondegenerate primers. We recommend that *Taq* DNA polymerase be used in this reaction because of its reliability. A slightly

higher concentration of the enzyme is used to compensate for the initial heating of the reaction mixture in the linked program used for the amplification. The degenerate primers are suitable for the amplification of a wide variety of eubacterial DNAs. Therefore, extreme care must be exercised to ensure that contaminating template is not inadvertently introduced during the setup of this reaction. The use of aerosol-resistant pipet tips is strongly recommended.

2. Depending on the G + C content of the bacterial DNA, it is frequently necessary to adjust the program used in the thermal cycler. For bacterial DNA with G + C content higher than 55%, it is recommended that a series of increasing annealing temperatures be tested for optimization of reaction yield and specificity of product.
3. The PCR product may contain an internal site within the coding region for one of the cloning enzymes. This will be evident by the reduction in size of the PCR fragment after enzyme digestion. (Restriction enzyme digestion of the PCR fragment to expose the noncoding cloning sites should not alter the size of the fragment significantly at the resolution of the agarose gel.) One may select a different cloning enzyme by redesigning the degenerate primer using a different enzyme at the cloning site at the 5'-end of the primers. Alternatively, one may try to insert the digested PCR fragment into the vector with both the asymmetric cloning sites via *Kpn*I and *Hin*dIII as designed as well as with the one enzyme that cuts the PCR product internally. In the latter case, a shorter fragment may be inserted into the vector in both orientations. Therefore, the composite sequence of the region may have to be sorted out from all the candidates.
4. One simple method to identify the presence of the expected insert in the pBluescript derivative clones by size is to amplify the plasmid DNA using T7 and T3 primers in a PCR reaction using a setup similar to that described in **step 1** under **Subheading 3.1.** The source of the template DNA in this case comes from a colony of transformants, which is resuspended in 20 μL of sterile water and boiled for 5 min. The PCR reaction product is analyzed on an agarose gel. With this method, 20–30 plasmid-containing colonies can be analyzed easily and accurately at one time.
5. We recommend that at least three candidates be sequenced from plasmid primers, since *Taq* DNA polymerase is prone to generate errors. Isolated differences uniquely found in the sequence of one of the clones may be ignored. Although one may be able to obtain sequence information by directly sequencing the PCR product generated in **Subheading 3.1.**, thereby eliminating the need to sequence multiple samples, we have found that degenerate primers serve poorly as sequencing primers. Therefore, the PCR product is first cloned and sequenced from unique primers present in the plasmid.
6. The Y in this conserved block of amino acids is the active tyrosine that forms the covalent protein–DNA intermediate during topoisomerization reaction.
7. In this example of cloning the *Thermotoga gyrA* gene, a unique *Eco*RI site is present in the PCR product amplified in **Subheading 3.1.** It is a reasonable choice to generate cyclized DNA for genomic walking toward both the 5'- and 3'-directions,

depending on the appropriate choice of the outward-pointing pair of primers for the inverse PCR reaction. This enzyme is also heat-labile. It can be inactivated by heat, as described here. If a heat-resistant enzyme is chosen, phenol extraction should be used as described in **step 6** of **Subheading 3.1.** to inactivate the restriction enzyme before ligation.

8. If the inverse PCR reaction product consists of one major band and some minor bands, a third nested primer should be chosen for the sequencing reaction. In the case where the inverse PCR product does not give one major discrete product, one should perform a Southern hybridization experiment using a C-terminal region of the first PCR fragment as a probe to determine the size of the restriction fragment spanning the region in order to provide credible information to judge the feasibility of the choice of the restriction enzyme in the inverse PCR step for genomic walking.
9. At this point, the sequence of the *gyrA* gene of interest has been determined as a composite from sequences generated from DNA fragments derived from forward and inverse PCR reactions. A clone carrying the full-length gene can be generated using the primers spanning the beginning and the end of the gene in another forward PCR reaction. We recommend that a thermal, stable DNA polymerase with lower misincorporation rate be used. A number of commercially available thermal, stable enzymes with 3'–5' exonuclease activity may all be suitable. The use of a KlentaqLA enzyme ***(11)***, which is a mixture of a derivative of *Taq* DNA polymerase and a *Pfu* DNA polymerase (Stratagene, La Jolla, CA) is described here. It also has the advantage of being able to amplify long DNA fragments.
10. The full-length *gyrA* fragment can be cloned into a low-copy-number plasmid under the control of an inducible promoter. In this case, a T7-driven promoter is used. This precaution is recommended, because many full-length topoisomerase genes are found to be toxic to the *E. coli* host ***(6)***. Furthermore, the clone can be used to express the *gyrA* gene product when it is transformed into a T7 RNA polymerase-producing host ***(12)***.

References

1. Huang, W. M. (1994) Type II DNA topoisomerase genes, in *DNA Topoisomerases, Biochemistry and Molecular Biology* (Liu, L. F., ed.), Academic, San Diego, pp. 201–222.
2. Wang, J. C. (1996) DNA topoisomerases. *Annu. Rev. Biochem.* **65,** 635–692.
3. Tsai-Pfluglelder, M., Liu, L., Liu, A., Tewey, K., Whang-Peng, J., Knutsen, T., et al. (1988) Cloning and sequencing of cDNA encoding human DNA topoisomerase II and localization of the gene to chromosome region 17q21-22. *Proc. Natl. Acad. Sci. USA* **85,** 7177–7181.
4. Strauss, P. R. and Wang, J. C. (1990) The *TOP2* gene of *Trypanosoma brucei*: a single-copy gene that shares extensive homology with other *TOP2* genes encoding eukaryotic DNA topoisomerase II. *Mol. Biochem. Parasitol.* **36,** 141–150.

5. Belland, R., Morrison, S., Ison, C., and Huang, W. M. (1994) *Neisseria gonorrhoeae* acquires mutations in analogous regions of *gyrA* and *parC* in fluoroquinolone-resistant isolates. *Mol. Microbiol.* **14,** 371–380.
6. Wang, Y., Huang, W. M., and Taylor, D. E. (1993) Cloning and nucleotide sequence of the *Campylobacter jejuni gyrA* gene and characterization of quinolone resistance mutations. *Antimicrob. Agents Chemother.* **37,** 457–463.
7. Ochman, H., Medhora, M., Garza, D., and Hartl, D. (1990) Amplification of flanking sequences by inverse PCR, in *PCR Protocols, a Guide to Methods and Applications* (Innis, M., Gelfand, D., Sninsky, J., and White, T., eds.), Academic, San Diego, pp. 219–227.
8. Huang, W. M. (1993) Multiple DNA gyrase-like genes in eubacteria, in Molecular Biology of DNA Topoisomerases and Its Application to Chemotherapy. (Andoh, T., Ikeda, H., and Oguro, M., eds.), CRC, Baca Raton, FL, pp. 39–48.
9. Huang, W. M. (1996) Bacterial diversity based on type II DNA topoisomerase genes. *Annu. Rev. Genet.* **30,** 79–109.
10. Huber, R., Langworthy, T., Konig, H., Thomm, M., Woese, C., Sleytr, U., et al. (1986) *Thermotoga maritima* sp. nov. represents a new genus of unique extremely thermophilic eubacteria growing up to 90°C. *Arch. Microbiol.* **144,** 324–333.
11. Barnes, W. M. (1994) PCR amplification of up to 35-kb DNA with high fidelity and high yield from λ bacteriophage templates. *Proc. Natl. Acad. Sci. USA* **91,** 2216–2220.
12. Studier, F. W., Rosenberg, A. H., Dunn, J. J., and Dubendorff, J. W. (1990) Use of T7 RNA polymerase to direct the expression of cloned genes. *Methods Enzymol.* **185,** 60–89.

26

Topoisomerase II-Catalyzed Relaxation and Catenation of Plasmid DNA

John M. Fortune and Neil Osheroff

1. Introduction

Type II topoisomerases act by making a double-stranded break in a DNA segment, passing an intact DNA segment through the break, and resealing the break *(1–4)*. The DNA strand passage reaction of topoisomerase II can be either intra- or intermolecular. The enzyme relaxes supercoiled DNA in an intramolecular reaction, while it catenates and decatenates closed circular DNA in an intramolecular reaction *(1,5)*. This chapter describes the relaxation and catenation reactions of topoisomerase II.

Relaxation of supercoiled DNA can be carried out by topoisomerase I or topoisomerase II *(6–10)*. The type I enzyme relaxes DNA by making a single-stranded break in the double helix and passing (or allowing passage of) the intact strand through the break *(10,11)*. Each strand passage event alters the DNA-linking number by one *(12)*. DNA relaxation catalyzed by topoisomerase II differs from that catalyzed by the type I enzyme in two important respects. First, the topoisomerase II-catalyzed passage of an intact DNA segment through a double-stranded break alters the DNA linking number by two *(5,13,14)*. Second, topoisomerase II requires the high energy cofactor ATP for catalysis *(7,8,14,15)*.

The catenation reaction of topoisomerase II creates a topological linkage between two free molecules of covalently closed circular DNA *(7,14,16)*. To monitor catenation under experimental conditions, a DNA condensing agent must be included in the reaction. Typically, the positively-charged protein histone H1 is used as a condensing agent. Since catenation requires creation of a double-stranded DNA break, it can only be performed by a type II topoiso-

From: *Methods in Molecular Biology, Vol. 95: DNA Topoisomerase Protocols, Part II: Enzymology and Drugs*
Edited by N. Osheroff and M.A. Bjornsti © Humana Press Inc., Totowa, NJ

merase. Thus, catenation is a powerful diagnostic tool for detection of topoisomerase II activity.

The assays described in this chapter are based on the methods of Osheroff *et al.* **(8)** and Shelton *et al.* **(16)**. Although they are described with optimal conditions for *Drosophila* topoisomerase II, they can be adapted for use with other type II topoisomerases. The adaptations for performing these assays with human topoisomerase IIα are given (*see* **Note 1**).

2. Materials

2.1. Relaxation of Plasmid DNA

1. Topoisomerase II.
2. 5X assay buffer stock: 50 m*M* Tris–HCl, pH 7.9, 250 m*M* NaCl, 250 m*M* KCl, 0.5 m*M* NaEDTA, pH 8.0, 25 m*M* $MgCl_2$, 12.5% glycerol (Stored at 4°C).
3. ATP (20 m*M* stock in H_2O, stored at –20°C).
4. Negatively supercoiled plasmid DNA, such as pBR322.
5. Topoisomerase II diluent: 10 m*M* NaP, pH 7.1, 50 m*M* NaCl, 0.1 m*M* NaEDTA, pH 8.0, 0.5 mg/mL BSA, 10% glycerol (Stored at 4°C).
6. Stop solution: 0.77% SDS, 77 m*M* NaEDTA, pH 8.0.
7. Loading buffer: 30% sucrose, 0.5% bromophenol blue, and 0.5% xylene cyanole FF in 10 m*M* Tris–HCl, pH 7.9.
8. 10X TBE buffer: 1 *M* Tris-borate, pH 8.3, 20 m*M* EDTA.
9. Agarose.
10. Ethidium bromide (10 mg/mL stock in H_2O, stored in the dark at 4°C).
11. Agarose gel electrophoresis apparatus.
12. Microwave oven.
13. Water bath, set at 60°C.
14. Ultra violet (UV) light box.
15. Camera or digital imaging system (Optional).
16. Densitometer (Optional).

2.2. Catenation of Plasmid DNA

All of the above materials are used, in addition to the following:

1. Histone H1 (50 μg/mL stock in H_2O, stored at –20°C).
2. Plasmid DNA, such as pBR322 (either supercoiled or relaxed may be used).

3. Methods

3.1. Relaxation of Plasmid DNA

1. Prepare each sample on ice. Combine 4 μL of 5X assay buffer, 1 μL of ATP (1 m*M* final ATP concentration), negatively supercoiled plasmid DNA to a final concentration of 5 to 10 n*M*, and H_2O such that the final reaction volume will be 20 μL upon addition of topoisomerase II.

2. Add topoisomerase II to a final concentration of ~1.5 n*M* to initiate the reaction. The topoisomerase II stock may need to be diluted prior to this addition. Incubate the sample for 15 min at 30°C (*see* **Note 2**). Terminate the reaction by adding 3 µL of stop solution (final concentrations of SDS and NaEDTA are 0.1% and 10 m*M*, respectively).
3. Add 2 µL of agarose gel loading buffer, then heat samples for 2 min at 70°C immediately prior to agarose gel electrophoresis.
4. Prepare an agarose gel for resolution of the samples. Make a 1% agarose solution (w/v) in 1X TBE buffer. Melt the agarose by heating in microwave oven, then allow the gel solution to cool to approximately 60°C in a water bath. Pour the gel in an electrophoresis apparatus and allow the gel to solidify (this should take 15 to 30 min).
5. Load all samples on the agarose gel. Resolve the products by electrophoresis in 1X TBE buffer at ~5 V/cm for ~2.5 h (until the bromophenol blue dye approaches the end of the gel and the xylene cyanole dye is approximately halfway to the end).
6. Stain the gel in a solution of 1 µg/mL ethidium bromide in 1X TBE buffer. Gently agitate the gel on a shaker or rocker for 30 min to allow the ethidium bromide to bind the DNA.
7. Visualize the DNA products by placing the gel on a UV light box. The products can be quantified by one of two methods. First, a digital imaging system may be used to capture and analyze an electronic image of the gel. Alternatively, a photograph of the gel may be taken and the photographic negative analyzed on a densitometer.
8. The percentage of the initial supercoiled plasmid DNA that has been relaxed can be determined from the supercoiled DNA band on the agarose gel (**Fig. 1**). The total amount of supercoiled DNA is quantified in a sample that contains no topoisomerase II and compared to the supercoiled DNA that remains in topoisomerase II-containing samples. The percentage of supercoiled DNA that is lost corresponds to the percentage of the DNA substrate that has been relaxed.

3.2. Catenation of Plasmid DNA

1. Prepare each sample on ice. Combine 4 µL of 5X assay buffer, 1 µL of ATP (1 m*M* final ATP concentration), relaxed or supercoiled plasmid DNA (*see* **Note 3**) to a final concentration of 5 to 10 n*M*, histone H1, and H_2O such that the final reaction volume will be 20 µL upon addition of topoisomerase II. The amount of histone H1 to be added must be determined by performing a histone H1 titration according to the catenation procedure described here (*see* **Note 4**).
2. Add topoisomerase II to a final concentration of ~3 n*M* to initiate the reaction. The topoisomerase II stock may need to be diluted prior to this addition. Incubate the sample for 15 min at 30°C (*see* **Note 2**). Terminate the reaction by adding 3 µL of stop solution (final concentrations of SDS and NaEDTA are 0.1% and 10 m*M*, respectively).

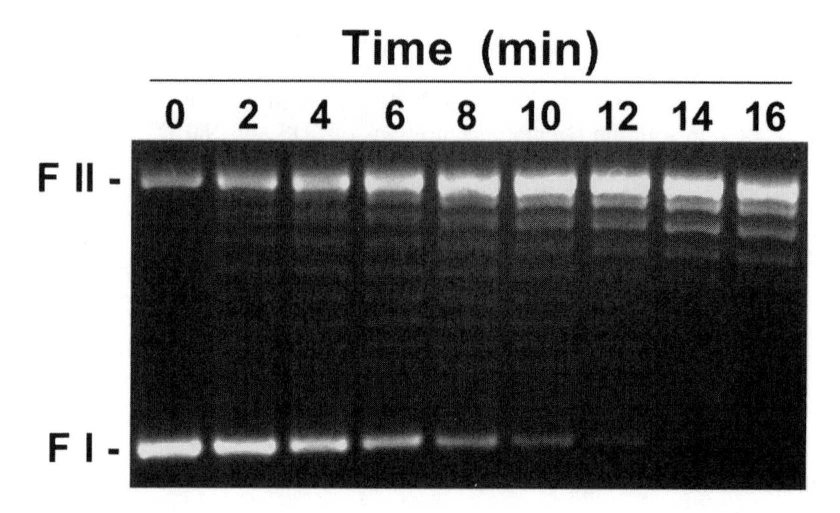

Fig. 1. Relaxation of supercoiled DNA by topoisomerase II. An ethidium bromide-stained agarose gel of a time course of topoisomerase II-catalyzed DNA relaxation is shown. The substrate was negatively supercoiled closed circular pBR322 plasmid DNA. The positions of negatively supercoiled DNA (Form I, FI) and nicked DNA (Form II, FII) are indicated. Since no ethidium bromide was present during electrophoresis (the gel was stained with ethidium following electrophoresis), the mobility of fully relaxed covalently closed circular DNA is identical to that of nicked plasmid (FII).

3. Add 2 μL of agarose gel loading buffer, then heat samples for 2 min at 70°C immediately prior to agarose gel electrophoresis.
4. Prepare an agarose gel for resolution of the samples. Make a 1% agarose solution (w/v) in 1X TBE buffer that contains 0.5 μg/mL ethidium bromide. Melt the agarose by heating in microwave oven, then allow the gel solution to cool to approximately 60°C in a water bath. Pour the gel in an electrophoresis apparatus and allow the gel to solidify (this should take 15 to 30 min).
5. Load all samples on the agarose gel. Resolve the products by electrophoresis in 1X TBE buffer containing 0.5 μg/mL ethidium bromide at ~5 V/cm for ~2.5 h (until the bromophenol blue dye approaches the end of the gel and the xylene cyanole dye is approximately halfway to the end).
6. Visualize the DNA products by placing the gel on a UV light box. Catenated plasmid DNA remains at the origin and does not enter the agarose gel (**Fig. 2**).

4. Notes

1. The following adaptations should be made in the protocols for relaxation and catenation when human topoisomerase IIα is utilized ***(17,18)***. The 5X assay buffer for human topoisomerase IIα is as follows: 50 m*M* Tris–HCl, pH 7.9, 500 m*M* KCl, 0.5 m*M* NaEDTA, pH 8.0, 25 m*M* $MgCl_2$, 12.5% glycerol (Stored

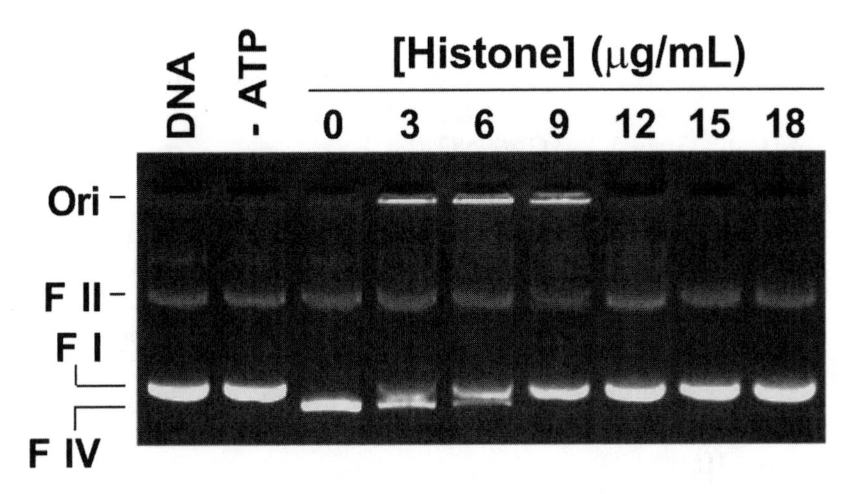

Fig. 2. Catenation of circular plasmid DNA by topoisomerase II. The substrate in this experiment was negatively supercoiled pBR322 plasmid. A histone titration is shown. Control samples include DNA alone (DNA) and a reaction that does not contain ATP (– ATP). Catenated DNA does not enter the gel, but remains at the origin (Ori). The positions of negatively supercoiled plasmid DNA (Form I, FI), nicked circular plasmid DNA (Form II, FII), and relaxed DNA (Form IV, FIV) are indicated. In contrast to the data shown in **Fig. 1**, the present agarose gel was subjected to electrophoresis in the presence of ethidium bromide. Under these conditions, the mobility of relaxed DNA is slightly greater than that of supercoiled substrate.

at 4°C). Dilutions of human topoisomerase IIα are made using the enzyme storage buffer: 50 m*M* Tris–HCl, pH 7.7, 0.5 m*M* dithiothreitol, 0.1 m*M* NaEDTA, pH 8.0, 750 m*M* KCl, 40% glycerol (Stored at 4°C). Finally, reactions with the human enzyme are carried out at 37°C.

2. The incubation time can be varied to examine the time dependence of DNA relaxation or catenation by topoisomerase II. The amount of time required for complete relaxation or catenation depends on the specific concentration of topoisomerase II utilized and the activity of a particular enzyme preparation.
3. Both relaxed and supercoiled plasmid DNA are acceptable substrates for catenation assays. Although topoisomerase II binds more efficiently to supercoiled DNA *(8,19)*, relaxed DNA is actually a better substrate for the catenation reaction *(20,21)*. Relaxed DNA can be obtained by treating supercoiled DNA with topoisomerase I, then heat inactivating the type I enzyme.
4. Catenation reactions typically require histone H1 concentrations between 3 μg/mL and 18 μg/mL. The optimal concentration is best determined by performing a histone H1 titration (0, 3, 6, 9, 12, 15, 18 μg/mL are good concentrations to use in the titration) with a given histone H1 stock (50 μg/mL stock in H_2O) prior to any other catenation assays.

Acknowledgment

This protocol was developed in part under the auspices of Grant GM33944 from the National Institutes of Health. J.M.F. is a trainee under National Institutes of Health Grant 5 T32 CA09582.

References

1. Watt, P. M. and Hickson, I. D. (1994) Structure and function of type II DNA topoisomerases. *Biochem. J.* **303,** 681–695.
2. Burden, D. A. and Osheroff, N. (1998) Mechanism of action of eukaryotic topoisomerase II and drugs targeted to the enzyme. *Biochim. Biophys. Acta* **1400,** 139–154.
3. Wang, J. C. (1998) Moving one DNA double helix through another by a type II DNA topoisomerase: the story of a simple molecular machine. *Quart. Rev. Biophys.* **31,** 107–144.
4. Fortune, J. M. and Osheroff, N. (2000) Topoisomerase II as a target for anticancer drugs: when enzymes stop being nice. *Prog. Nucleic Acid Res. Mol. Biol.* **64,** 221–253.
5. Hsieh, T. (1990) Mechanistic aspects of type-II DNA topoisomerases, in *DNA Topology and Its Biological Effects* (Cozzarelli, N. R. and Wang, J. C., eds.), Cold Spring Harbor Laboratory Press, Cold Spring Harbor, pp. 242–263.
6. Pulleyblank, D. E., Shure, M., Tang, D., Vinograd, J., and Vosberg, H. (1975) Action of nicking-closing enzyme on supercoiled and nonsupercoiled closed circular DNA: Formation of a Boltzman distribution of topological isomers. *Proc. Natl. Acad. Sci. USA* **72,** 4280–4284.
7. Hsieh, T.-S. and Brutlag, D. L. (1980) ATP-dependent DNA topoisomerase from *D. melanogaster* reversibly catenates duplex DNA rings. *Cell* **21,** 115–121.
8. Osheroff, N., Shelton, E. R., and Brutlag, D. L. (1983) DNA topoisomerase II from *Drosophila melanogaster*. Relaxation of supercoiled DNA. *J. Biol. Chem.* **258,** 9536–9543.
9. Nitiss, J. L. (1998) Investigating the biological functions of DNA topoisomerases in eukaryotic cells. *Biochim. Biophys. Acta* **1400,** 63–81.
10. Pommier, Y., Pourquier, P., Fan, Y., and Strumberg, D. (1998) Mechanism of action of eukaryotic DNA topoisomerase I and drugs targeted to the enzyme. *Biochim. Biophys. Acta* **1400,** 83–106.
11. Champoux, J. J. (1998) Domains of human topoisomerase I and associated functions. *Prog. Nucleic Acid Res. Mol. Biol.* **60,** 111–132.
12. Champoux, J. J. (1990) Mechanistic Aspects of Type-I Topoisomerases, in *DNA Topology and Its Biological Effects* (Cozzarelli, N. and Wang, J. C., eds.), Cold Spring Harbor Laboratory Press, Cold Spring Harbor Laboratory, pp. 217–242.
13. Brown, P. O. and Cozzarelli, N. R. (1979) A sign inversion mechanism for enzymatic supercoiling of DNA. *Science* **206,** 1081–1083.
14. Liu, L. F., Liu, C. C., and Alberts, B. M. (1980) Type II DNA topoisomerases: enzymes that can unknot a topologically knotted DNA molecule via a reversible double-strand break. *Cell* **19,** 697–707.

15. Miller, K. G., Liu, L. F., and Englund, P. T. (1981) A homogeneous type II DNA topoisomerase from HeLa cell nuclei. *J. Biol. Chem.* **256,** 9334–9339.
16. Shelton, E. R., Osheroff, N., and Brutlag, D. L. (1983) DNA topoisomerase II from *Drosophila melanogaster*. Purification and physical characterization. *J. Biol. Chem.* **258,** 9530–9535.
17. Kingma, P. S., Greider, C. A., and Osheroff, N. (1997) Spontaneous DNA lesions poison human topoisomerase IIα and stimulate cleavage proximal to leukemic 11q23 chromosomal breakpoints. *Biochemistry* **36,** 5934–5939.
18. Fortune, J. M. and Osheroff, N. (1998) Merbarone inhibits the catalytic activity of human topoisomerase IIα by blocking DNA cleavage. *J. Biol. Chem.* **273,** 17,643–17,650.
19. Osheroff, N. (1986) Eukaryotic topoisomerase II. Characterization of enzyme turnover. *J. Biol. Chem.* **261,** 9944–9950.
20. Holden, J. A. and Low, R. L. (1985) Characterization of a potent catenation activity of HeLa cell nuclei. *J. Biol. Chem.* **260,** 14,491–14,497.
21. Rybenkov, V. V., Vologodskii, A. V., and Cozzarelli, N. R. (1997) The effect of ionic conditions on the conformations of supercoiled DNA. II. Equilibrium catenation. *J. Mol. Biol.* **267,** 312–323.

27

Topoisomerase II-Mediated Cleavage of Plasmid DNA

D. Andrew Burden, Stacie J. Froelich-Ammon, and Neil Osheroff

1. Introduction

Integral to all catalytic functions of topoisomerase II is the ability of the enzyme to generate transient double-stranded breaks in the backbone of DNA *(1–3)*. When topoisomerase II cleaves DNA, it maintains the topological integrity of the genetic material by forming covalent bonds between its active site tyrosyl residues (one on each subunit of the homodimeric enzyme) and the newly generated 5'-terminal phosphates *(4–6)*. The enzyme acts by making two coordinated nicks on opposite strands of the double helix *(7–9)*. These points of cleavage are staggered by 4 bases, such that scission results in 5'-overhanging cohesive ends *(4)*.

Although critical to the activity of the enzyme, topoisomerase II-cleaved DNA complexes are usually fleeting intermediates in the catalytic cycle of the enzyme and hence are present at very low steady-state concentrations *(1,3)*. However, in the presence of a variety of clinically relevant anticancer drugs, levels of these cleavage complexes increase dramatically *(3,10–13)*. Since it was first discovered that the DNA cleavage/religation equilibrium of topoisomerase II is the primary target for these drugs *(14)*, interest in the DNA cleavage activity of the enzyme has soared.

Several methods can be used to monitor the in vitro DNA cleavage activity of type II topoisomerases. They all take advantage of the fact that topoisomerase II-cleaved DNA complexes can be trapped by rapidly denaturing the complexed enzyme with SDS. The plasmid-based system was first utilized by Sander and Hsieh *(4)* and Liu et al. *(5)*. The assay which is described in this chapter is based on the method of Osheroff and Zechiedrich *(15)*. Although it has been optimized for Drosophila melanogaster topoisomerase II,

From: *Methods in Molecular Biology, Vol. 95: DNA Topoisomerase Protocols, Part II: Enzymology and Drugs*
Edited by N. Osheroff and M.A. Bjornsti © Humana Press Inc., Totowa, NJ

it can easily be adapted for use with virtually any type II topoisomerase. This assay quantifies enzyme-mediated double-stranded DNA cleavage by monitoring the conversion of covalently-closed circular DNA to linear molecules. Furthermore, it can simultaneously quantify topoisomerase II-mediated single-stranded DNA cleavage by monitoring the production of nicked plasmids. The assay is simple, straightforward, and does not require the use of radiolabeled DNA. However, it requires relatively high concentrations of topoisomerase II and is not as sensitive as methods that utilize radiolabeled nucleic acid substrates.

2. Materials

1. Topoisomerase II.
2. 5X assay buffer stock: 50 m*M* Tris–HCl, pH 7.9, 250 m*M* NaCl, 250 m*M* KCl, 0.5 mM EDTA, 25 m*M* $MgCl_2$, 12.5% glycerol (Stored at 4°C).
3. 20 mM stock of ATP or a nonhydrolyzable ATP analog such as adenyl-5'-yl imidodiphosphate (Optional).
4. Plasmid DNA such as pBR322.
5. 10% (w/v) SDS.
6. 250 m*M* Na_2EDTA.
7. 4 mg/mL stock solution of proteinase K in 50 m*M* Tris–HCl, pH 7.9, 1 m*M* $CaCl_2$ (Stored at 4°C) diluted with the same buffer to 0.8 mg/mL just prior to use.
10. Loading buffer: 10 m*M* Tris–HCl, pH 7.9, 60% w/v sucrose, 0.5% (w/v) bromophenol blue, 0.5% (w/v) xylene cyanole FF.
11. 50X TAE buffer: 2 *M* Tris, 5.7% (v/v) glacial acetic acid, 100 m*M* Na_2EDTA (Stored at 4°C).
12. Agarose.
13. 10 mg/mL stock solution of ethidium bromide in H_2O (Stored in the dark at 4°C).
14. Agarose gel electrophoresis apparatus.
15. Microwave oven.
16. 50°C H_2O bath.
17. Ultra Violet (UV) light box.
18. Camera (Optional).
19. Densitometer (Optional).

3. Methods

1. Prepare samples on ice for 20 µL reactions as follows: mix 4 µL of 5X assay buffer, 1 µL of ATP or APP(NH)P if desired (*see* **Note 6**), plasmid DNA to a final concentration of 5-10 n*M* (this is equivalent to ~20–40 µM base pairs when pBR322 plasmid DNA is used), and H_2O to a final volume that will result in a 20 µL reaction after the addition of topoisomerase II.
2. Initiate the reaction by the addition of topoisomerase II to a final concentration of 100 n*M* and incubate the sample at 30°C for 6 min. Terminate the reaction by the addition of 2 µL of 10% SDS.

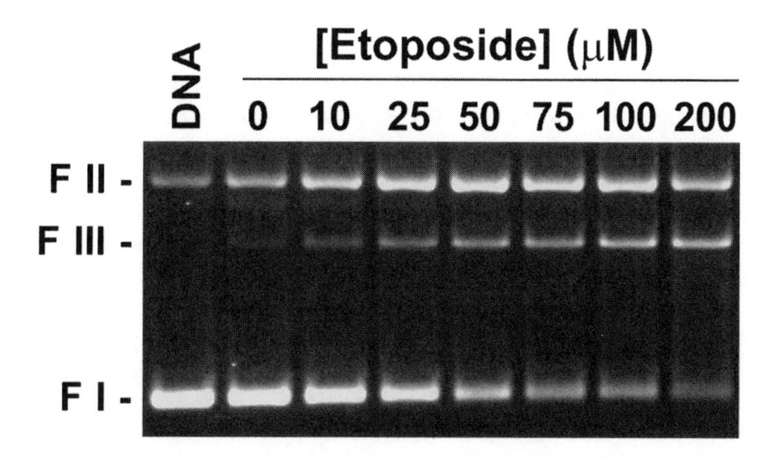

Fig. 1. Topoisomerase II-mediated DNA cleavage. The effects of the anticancer drug etoposide on enzyme-mediated DNA cleavage are shown. Double-stranded DNA cleavage by topoisomerase II converts negatively supercoiled plasmid (Form I, FI) to linear DNA (Form III, FIII). The position of nicked circular DNA (Form II, FII) is shown for reference.

3. Add 1.5 μL of 250 m*M* Na_2EDTA followed by 2 μL of 0.8 mg/mL proteinase K and incubate the samples at 45 °C for 30 min to digest the topoisomerase II.
4. Add 2 μL of loading buffer and heat samples for 2 min at 70°C immediately prior to electrophoresis in an agarose gel.
5. Prepare the agarose gel by making a 1% agarose solution (w/v) in 1X TAE buffer. Melt the agarose by heating the solution in a microwave oven and cool it to ~50°C in a water bath. Pour the gel and allow it to solidify (this should take 15–30 min).
6. Load the samples in the agarose gel and resolve the reaction products by electrophoresis in 1X TAE buffer at ~5 V/cm for ~3 h (until the bromophenol blue dye has run approximately to the end of the gel and the xylene cyanole FF has run approximately half as far).
7. Stain the gel in 1 μg/mL ethidium bromide (in H_2O or 1X TAE buffer) for ~30 min with continuous agitation on a shaker or rocker. The gel may be destained in H_2O for 60 min, but generally it is not necessary.
8. Visualize the DNA reaction products by transillumination with medium wave (~300 nm) ultraviolet light. DNA cleavage products may be quantified either by photography and densitometry of the fluorescent nucleic acid bands or by digital imaging.
9. Topoisomerase II-mediated double-stranded or single-stranded DNA cleavage results in the conversion of covalently closed supercoiled plasmid substrates (form I DNA, FI) to linear (form III DNA, FIII) or nicked (form II DNA, FII) molecules, respectively (**Fig. 1**).

4. Notes

1. Although reaction conditions described have been optimized for *Drosophila* topoisomerase II, concentrations of enzyme and DNA as well as the specific buffer components and final reaction volume can be changed as needed to optimize individual systems.
2. Reactions can be run at temperatures other that those specified. For yeast and *Drosophila* topoisomerase II, temperatures up to 30°C may be used ***(4,15,16)***. For topoisomerase II from mammalian species, temperatures up to 37°C may be used ***(5)***.
3. Plasmid DNA can be prepared from bacteria by a number of procedures. Protocols described in *Molecular Cloning* ***(17)*** and *Current Protocols in Molecular Biology* ***(18)*** routinely work well.
4. The DNA cleavage/religation equilibrium of topoisomerase II greatly favors the religation event. Therefore, at any given instant, only a small proportion (<1%) of the topoisomerase II-DNA complex exists with DNA in the cleaved state. Consequently, in order to visualize DNA cleavage products, the ratio of topoisomerase II to plasmid molecule is generally in excess of 10:1.
5. The affinity of topoisomerase II for its DNA substrate varies with sequence content ***(19)***. Therefore, the cleavage activity of the enzyme may depend on the plasmid utilized. Enzyme affinity is also dependent on the topological state of the plasmid DNA, with supercoiled substrates being preferred over relaxed, nicked, or linear molecules ***(20)***.
6. Topoisomerase II establishes two distinct DNA cleavage/religation equilibria: one prior to its double-stranded DNA passage event (which corresponds to the ATP-free form of the enzyme) and one following its DNA passage event (which corresponds to the ATP-bound form) ***(2,21)***. Hence, assays carried out in the absence of a nucleotide triphosphate monitor pre-strand passage DNA cleavage, those in the presence of a non-hydrolyzable ATP analog monitor post-strand passage DNA cleavage, and those in the presence of ATP monitor a mixture of the two equilibria. It should be noted that levels of DNA cleavage observed following strand passage generally are 3–5 times higher than those observed prior to this step of the topoisomerase II catalytic cycle ***(21)***.
7. As a result of the high concentrations of topoisomerase II employed in cleavage assays, DNA catentation may be observed in the presence of ATP or nonhydrolyzable ATP analogs ***(22)***. If uncleaved (i.e., nonlinear) plasmid molecules become part of catenated networks, they remain at the point of origin in agarose gels.
8. The requirement of topoisomerase II for a divalent cation may be fulfilled in cleavage assays by Ca^{2+} ***(15)***. Although Ca^{2+} will not support the interactions of the enzyme with its ATP cofactor (and consequently will not support DNA strand passage ***[23]***), it promotes higher levels of prestrand passage DNA cleavage than does Mg^{2+} and does not significantly alter the site specificity of topoisomerase II.
9. It is critical to terminate DNA cleavage reactions with SDS prior to the addition of EDTA, as chelation of the Mg^{2+} in the assay buffer inhibits topoisomerase

II-mediated DNA scission and leads to the rapid loss of cleavage complexes *(4,15)*. The ability of SDS to trap cleavage complexes diminishes considerably when the final concentration of the detergent drops below 0.5%.

10. The mobility of linear DNA molecules in agarose gels changes relative to those of nicked and negatively supercoiled circular plasmids depending on the running buffer that is used. In TAE buffer, linear molecules migrate intermediate to nicked and supercoiled DNA (supercoiled molecules have the highest electrophoretic mobility). However, in TBE (Tris/borate/EDTA) buffers, the mobility of linear DNA is often greater than that of supercoiled plasmid.
11. DNA bands tend to resolve more sharply if the comb is not removed from the gel until shortly before the samples are loaded.
12. If gels are stained with ethidium bromide after electrophoresis, covalently closed relaxed forms of DNA (generated in the presence of ATP or nonhydrolyzable ATP analogs) are visible. If ethidium bromide is included in the gel and running buffer prior to electrophoresis, relaxed molecules have a mobility similar to that of supercoiled DNA. The relative electrophoretic mobilities of nicked, linear, and supercoiled DNA molecules (in TAE buffer) are not affected by the presence of ethidium bromide.
13. Since topoisomerase II-mediated DNA cleavage is the target reaction for a number of chemotherapeutic agents, the plasmid-based assay described above represents a straightforward technique for analyzing the effects of drugs on the enzyme. An example of such a drug-induced DNA cleavage titration is shown in **Fig. 1**.
14. Dimethyl sulfoxide, which is often used as a solvent for drugs, alters the DNA cleavage/religation equilibrium of topoisomerase II and often increases levels of cleavage. Therefore, care must be taken to minimize the final concentration of dimethyl sulfoxide in reaction mixtures.
15. In the absence of cleavage-enhancing drugs, topoisomerase II establishes its DNA cleavage/religation equilibrium rapidly (within 5–10 s) *(24)*. In contrast, equilibrium is reached considerably more slowly in the presence of drugs (several minutes may be required to reach maximal levels of cleavage) *(25)*.

Acknowledgments

This protocol was developed in part under the auspices of Grant GM33944 from the National Institutes of Health. D.A.B. was a trainee under National Institutes of Health Grant 5 T32 CA09582. The authors are grateful to Jo Ann Wilson Byl for providing the DNA cleavage gel shown in **Fig. 1**.

References

1. Watt, P. M. and Hickson, I. D. (1994) Structure and function of type II DNA topoisomerases. *Biochem. J.* **303,** 681–695.
2. Wang, J. C. (1996) DNA topoisomerases. *Annu. Rev. Biochem.* **65,** 635–692.
3. Burden, D. A. and Osheroff, N. (1998) Mechanism of action of eukaryotic topoisomerase II and drugs targeted to the enzyme. *Biochim. Biophys. Acta* **1400,** 139–154.

4. Sander, M. and Hsieh, T. (1983) Double strand DNA cleavage by type II DNA topoisomerase from *Drosophila melanogaster. J. Biol. Chem.* **258,** 8421–8428.
5. Liu, L. F., Rowe, T. C., Yang, L., Tewey, K. M., and Chen, G. L. (1983) Cleavage of DNA by mammalian DNA topoisomerase II. *J. Biol. Chem.* **258,** 9536.
6. Rowe, T. C., Chen, G. L., Hsiang, Y. H., and Liu, L. F. (1986) DNA damage by antitumor acridines mediated by mammalian DNA topoisomerase II. *Cancer Res.* **46,** 2021–2026.
7. Muller, M. T., Spitzner, J. R., DiDonato, J. A., Mehta, V. B., Tsutsui, K., and Tsutsui, K. (1988) Single-strand DNA cleavages by eukaryotic topoisomerase II. *Biochemistry* **27,** 8369–8379.
8. Zechiedrich, E. L., Christiansen, K., Andersen, A. H., Westergaard, O., and Osheroff, N. (1989) Double-stranded DNA cleavage/religation reaction of eukaryotic topoisomerase II: evidence for a nicked DNA intermediate. *Biochemistry* **28,** 6229–6236.
9. Lee, M. P. and Hsieh, T. (1992) Incomplete reversion of double-stranded DNA cleavage mediated by *Drosophila* topoisomerase II: formation of single stranded DNA cleavage complex in the presence of an anti-tumor drug VM26. *Nuc. Acids Res.* **20,** 5027–5033.
10. Chen, A. Y. and Liu, L. F. (1994) DNA topoisomerases: essential enzymes and lethal targets. *Annu. Rev. Pharmacol. Toxicol.* **34,** 191–218.
11. Pommier, Y., Fesen, M. R., and Goldwasser, F. (1996) Topoisomerase II inhibitors: the epipodophyllotoxins, *m*-AMSA, and the ellipticine derivatives *Cancer Chemotherapy and Biotherapy: Principles and Practice.* Edited by Chabner, B. A. and Longo, D. L., 2nd Ed., pp. 435–61, Lippincott-Raven Publishers, Philadelphia.
12. Pommier, Y. (1997) DNA topoisomerase II inhibitors. *Cancer Therapeutics: Experimental and Clinical Agents* Edited by Teicher, B. A., Vol. I, pp. 153–174, Humana Press, Totowa, New Jersey.
13. Fortune, J. M. and Osheroff, N. (2000) Topoisomerase II as a target for anticancer drugs: when enzymes stop being nice. *Prog. Nuc. Acid Res. Mol. Biol.* **64,** 221–253.
14. Nelson, E. M., Tewey, K. M., and Liu, L. F. (1984) Mechanism of antitumor drug action: poisoning of mammalian DNA topoisomerase II on DNA by 4'-(9-acridinylamino)-methanesulfon-*m*-anisidide. *Proc. Natl. Acad. Sci. USA* **81,** 1361–1365.
15. Osheroff, N. and Zechiedrich, E. L. (1987) Calcium-promoted DNA cleavage by eukaryotic topoisomerase II: trapping the covalent enzyme-DNA complex in an active form. *Biochemistry* **26,** 4303–4309.
16. Elsea, S. H., Hsiung, Y., Nitiss, J. L., and Osheroff, N. (1995) A yeast type II topoisomerase selected for resistance to quinolones. Mutation of histidine 1012 to tyrosine confers resistance to nonintercalative drugs but hypersensitivity to ellipticine. *J. Biol. Chem.* **270,** 1913–1920.
17. Sambrook, J., Fritsch, E. F., and Maniatis, T. (1989) *Molecular Cloning: A Laboratory Manual*, 2nd Ed., Cold Spring Harbor Laboratory Press, Cold Spring Harbor, NY.

18. Ausubel, F. M., Brent, R., Kingston, R. E., Moore, D. D., Seidman, J. G., Smith, J. A., and Struhl, K. (eds.) (1995) *Current Protocols in Molecular Biology*. Edited by Janssen, K., John Wiley & Sons, Inc., New York.
19. Osheroff, N., Shelton, E. R., and Brutlag, D. L. (1983) DNA topoisomerase II from *Drosophila melanogaster*. Relaxation of supercoiled DNA. *J. Biol. Chem.* **258,** 9536–9543.
20. Zechiedrich, E. L. and Osheroff, N. (1990) Eukaryotic topoisomerases recognize nucleic acid topology by preferentially interacting with DNA crossovers. *EMBO J.* **9,** 4555–4562.
21. Osheroff, N. (1986) Eukaryotic topoisomerase II. Characterization of enzyme turnover. *J. Biol. Chem.* **261,** 9944–9950.
22. Corbett, A. H., Zechiedrich, E. L., Lloyd, R. S., and Osheroff, N. (1991) Inhibition of eukaryotic topoisomerase II by ultraviolet-induced cyclobutane pyrimidine dimers. *J. Biol. Chem.* **266,** 19,666–19,671.
23. Osheroff, N. (1987) Role of the divalent cation in topoisomerase II mediated reactions. *Biochemistry* **26,** 6402–6406.
24. Gale, K. C. and Osheroff, N. (1990) Uncoupling the DNA cleavage and religation activities of topoisomerase II with a single-stranded nucleic acid substrate: evidence for an active enzyme-cleaved DNA intermediate. *Biochemistry* **29,** 9538–9545.
25. Froelich-Ammon, S. J., Burden, D. A., Patchan, M. W., Elsea, S. H., Thompson, R. B., and Osheroff, N. (1995) Increased Drug Affinity as the Mechanistic Basis for Drug Hypersensitivity of a Mutant Type II Topoisomerase. *J. Biol. Chem.* **270,** 28,018–28,021.

28

Drug-induced Stabilization of Covalent DNA Topoisomerase I-DNA Intermediates

DNA Cleavage Assays

Paola Fiorani, Christine L. Hann, Piero Benedetti, and Mary-Ann Bjornsti

1. Introduction

Eukaryotic DNA topoisomerase I is a highly conserved enzyme that catalyzes the relaxation of positively and negatively supercoiled DNA *(1–4)*. The enzyme binds duplex DNA and transiently cleaves a single DNA strand. This is accompanied by the formation of a phospho-tyrosyl linkage between the active site tyrosine and the 3'-phosphate of the cleaved strand. The presence of this protein-linked nick in the DNA presumably allows the rotation of the nicked DNA end around its complementary strand to effect changes in DNA linking number. A second transesterification reaction religates the nicked DNA and restores the active site tyrosine.

Eukaryotic DNA topoisomerase I is also the cellular target of a number of antitumor agents, including camptothecin (CPT) and its analogues, topotecan (TPT), irinotecan (CPT-11) and 9 amino-camptothecin *(3,5–8)*. CPT reversibly stabilizes the covalent enzyme-DNA intermediate by inhibiting DNA religation. During S-phase, this ternary drug-enzyme-DNA complex presents an obstacle to advancing replication forks, resulting in fork breakage, double-strand DNA breaks, cell cycle arrest in G2 and cell death. Although the DNA lesions induced by CPT and the specific repair processes required for their resolution remain unclear, numerous studies indicate the cytotoxic action of these drugs derives from the stabilization of the covalent complex, rather than

From: *Methods in Molecular Biology, Vol. 95: DNA Topoisomerase Protocols, Part II: Enzymology and Drugs*
Edited by N. Osheroff and M.A. Bjornsti © Humana Press Inc., Totowa, NJ

a direct inhibition of catalytic activity. In yeast, for example, cells deleted for *TOP1* (which encodes DNA topoisomerase I) are viable and resistant to CPT ***(3,9)***. However, if either human or yeast *TOP1* is expressed in these cells from a plasmid, cell sensitivity to CPT is restored. In mammalian cells, a common mechanism of CPT resistance involves reduced levels of active enzyme ***(5)***. Thus, these drugs convert DNA topoisomerase I into a cellular poison, rather than suppress the catalytic activity of the enzyme. Similar cytotoxic mechanisms have been ascribed to other drugs that target topoisomerase I, including indolocarbazole derivatives (NB-506 and rebeccamycin) ***(10,11)***, the protoberberine, coralyne ***(12)***, and minor groove binding benzimidizole derivatives ***(13–15)***. Since DNA topoisomerase I is an essential enzyme in mouse ***(16)***, any therapeutic value of agents that suppress the catalytic activity of DNA topoisomerase I has yet to be established.

The ability of specific agents to interfere with the catalytic cycle of DNA topoisomerase I and stabilize the covalent complex may be biochemically addressed in plasmid DNA nicking assays or in linear DNA cleavage assays. The use of linear DNA substrates, uniquely labeled at a single 3' end, allows for a direct determination of the sequences of drug-induced cleavage sites (*see* refs. ***17–19***). The potency of different drugs or analogs in stabilizing the enzyme-DNA intermediate can be directly compared as can the relative sensitivity of mutant and wild-type enzymes derived from different cell line or species ***(17,18,20–24)***. As detailed later, stabilized DNA-enzyme intermediates are trapped by protein denaturants. To analyze the sequence specificity and reversibility of complexes formed, the extent of cleavage at specific sites can be assessed following resolution of the radiolabeled DNA fragments by denaturing polyacrylamide gel electrophoresis. Relative band intensity may be determined by densitometric analysis of exposed X-ray films or by PhosphorImage quantitation. Comparisons with DNA sequencing reactions will establish sequences flanking the sites of cleavage. Data obtained from such analyses may be correlated with cell based assays to establish DNA topoisomerase I as the relevant cellular target, to investigate mechanisms of drug resistance and to evaluate the relative potency of specific analogs.

2. Materials

2.1. DNA Substrate

DNA substrates containing a high affinity DNA topoisomerase I cleavage site were derived from plasmid pBluescript II containing a high affinity DNA topoisomerase I cleavage site (5' TAAAAAGACTT↓AGAAAAATTT TTAAA 3') inserted in to the *Bam*HI site of the polylinker. The arrow indicates the site of enzyme cleavage on the scissile strand. A ~450 bp *Pvu* II DNA fragment was excised from this plasmid, *Bgl*II or *Bam*HI linkers were ligated to opposite ends and the DNA was ligated into the *Bam*HI/*Bgl*II sites of plasmid pHC624 to yield plasmid pHCAK3 (*see* **Notes 1** and **2**).

1. T7 DNA polymerase (Sequenase from Amersham/USB) (*see* **Note 3**).
2. Acrylamide solution: 30% acrylamide : bis-acrylamide (19 : 1) dissolved in dH_2O and filtered.
3. 10% ammonium persulfate (freshly made).
4. 10X TBE buffer: 1.0 *M* Tris-borate, 20 m*M* EDTA, pH 8.0.
5. 0.5X TBE buffer: 50 m*M* Tris-borate, 1 m*M* EDTA, pH 8.0.
6. 7X loading buffer: 30% Ficoll (type 400), 0.1% bromophenol blue, 0.1% xylene cyanol.
7. 2.5 *M* NH_4OAc.
8. Deoxyribonucleotides: 100 m*M* solutions of dATP, dGTP, dTTP.
9. $\alpha^{32}P$-dCTP.
10. Phenol : chloroform: 1 : 1 mixture of Tris–buffered phenol and chloroform.

2.2. DNA Sequencing

Chemical (Maxam-Gilbert) sequencing of the radiolabeled DNA substrate is necessary to precisely map the sites of DNA cleavage by DNA topoisomerase I that are enhanced in the presence of drug. Although the necessary reagents are listed below, Maxam-Gilbert DNA sequencing kits are commercially available, such as that supplied by Sigma (*see* **Note 4**).

1. DNA mix: ^{32}P-labeled DNA substrate (~50,000 cpm per 50–100 ng), 3 µL sheared salmon sperm DNA (10 µg/ mL), dH_2O to a final 25 µL.
2. 10 µg/mL salmon sperm DNA dissolved in dH_2O and sonicated to reduce viscosity.
3. Saturated NaCl in dH_2O.
4. Hydrazine.
5. Piperidine: freshly diluted 1 : 10 in dH_2O.
6. Formic acid.
7. 5 *M* NaOH.
8. tRNA: 10 µg/mL dissolved in H_2O.
9. A reaction mix: 1.5 *M* NaOH, 1 m*M* EDTA.
10. Cacodylate Buffer: 50 m*M* Na~cacodylate, pH 8.0, 10 m*M* $MgCl_2$, 0.1 m*M* EDTA.
11. C + T stop buffer: 0.3 *M* NaOAc, 0.01 MgOAc, 0.1 *M* EDTA.
12. DMS stop mix: 1.0 *M* Tris-acetate, pH 7.5, 1.0*M* β-mercaptoethanol, 1.5 *M* NaOAc, 50 m*M* MgOAc.
13. Screw cap microfuge tubes.
14. Teflon tape.

2.2. DNA Cleavage Assays

Eukaryotic DNA topoisomerase I is readily obtained from commercial sources, or may be directly purified from a variety of expression systems or tissues. Regardless of the source of the enzyme, the specific activity of the protein should first be established in plasmid DNA relaxation assays (*see* **Note 5**). The protocol described here uses ~75 units of enzyme per reaction.

1. 1X G buffer: 20 m*M* Tris, pH 7.5, 10 m*M* $MgCl_2$, 0.1 m*M* EDTA, 50 m*M* KCl, 50 µg/mL gelatin.
2. 10% SDS: 10 g dissolved in 100 mL dH_2O.
3. Proteinase K: 20 mg/mL in dH_2O.
4. DMSO (stored under N_2, Aldrich) (*see* **Note 6**).
5. Camptothecin: 4 mg/mL dissolved in DMSO and stored at –20°C.
6. Sample Buffer: 38% formamide, 8 m*M* EDTA, 2 mg/mL bromophenol blue, 2 mg/mL xylene cyanol (*see* **Note 7**).
7. Acrylamide/Urea gel mix: 7.6% acrylamide, 0.4% bis-acrylamide (8 g of 19:1 acrylamide:bis acrylamide per 100 mL), 7 *M* Urea, 0.1 *M* Tris-borate, pH 8.0, 2 m*M* EDTA.
8. 1X TBE buffer: 100 m*M* Tris-borate, 2 m*M* EDTA, pH 8.0.
9. Fixer: 10% acetic acid, 10% methanol.

3. Methods

3.1. Preparation of the DNA Substrate

In order to assess drug-stabilized DNA topoisomerase I-DNA complexes, the integrity of the DNA strands in the DNA substrate is of paramount importance. Any nicks in the original plasmid DNA will contribute to the background obtained in the DNA cleavage assay and in the Maxam-Gilbert sequencing reactions. Thus, freshly prepared supercoiled plasmid DNA is recommended, as well as freshly prepared phenol:chloroform. If a significant fraction of the plasmid DNA is nicked, the DNA may be treated with T4 DNA ligase and repurified prior to the reactions detailed below.

To prepare the labeled DNA substrate, plasmid DNA is restricted at a unique *Bgl*II site, the ends treated with a DNA polymerase, cold nucleotides (dGTP, dATP and dTTP) and α^{32}P-dCTP (depicted in **Fig. 1**). As a result, a single ^{32}P-labeled deoxynucleotide is inserted at each 3' end of the molecule. To isolate a single end-labeled DNA molecule, where the ^{32}P-label is at the 3' end of the strand containing the high affinity cleavage site, the DNA is restricted at a unique *Bam*HI site. The smaller 450 bp fragment is purified as a substrate for the DNA sequencing and cleavage assays.

1. Prepare a 5% nondenaturing polyacrylamide gel: Mix 16.7 mL of 30% acrylamide:bis-acrylamide, 5 mL of 20X TBE and 78.3 mL of dH_2O. Add 430 µl of 10% ammonium persulfate and 167 µL of TEMED and pour a gel using 1.5 mm spacers and comb from a vertical electrophoresis apparatus (*see* **Notes 8** and **9**).
2. Digest 15 µg of plasmid pHCAK3 DNA with *Bgl*II (50–100 units in reaction buffer provided by the manufacturer) at 37°C for 60 min.

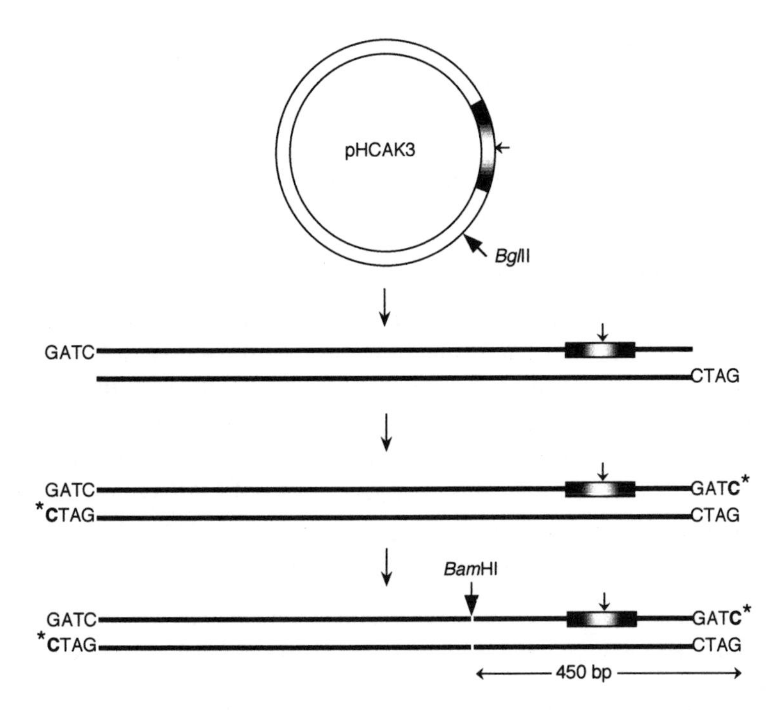

Fig. 1. Scheme depicting the isolation of a 3', uniquely end-labeled DNA substrate for topoisomerase I DNA cleavage assays. The shaded bar indicates the position of the high affinity cleavage site, while the small arrow shows the site of DNA cleavage. The large arrow heads indicate the relative positions of the unique *BglII* and *Bam*HI restriction sites. C* shows the incorporation of a ^{32}P-labeled nucleotide at each 3' end of the linear DNA molecule following treatment with T7 Sequenase.

3. Add dGTP, dATP, dTTP (to a final 1.25 m*M*), 100 µCi of ^{32}P-dCTP and 10–15 units of T7 Sequenase.
4. Incubate at 37°C for 20 min.
5. Extract with an equal volume of phenol: chloroform.
6. Precipitate the sample with 1/10 volume 3*M* NaOAc and 2.5 volumes 100% ETOH. Wash the DNA pellet with 70% EtOH and dry under vacuum.
7. Resuspend the pellet in 43 µl of dH_2), add 5 µl 10X *Bam*HI buffer (supplied by the manufacturer) and 2 µl (20–50 Units) *Bam*HI. Incubate at 37°C for 60 min.
8. Add 10 µl of 7X loading buffer and load into several wells of a 5% nondenaturing polyacrylamide gel and electrophoreses in 1X TBE at 25V for 14–16 h, or 2.5 h at 150 V.
9. Disassemble the apparatus and remove one of the glass plates. Cover the surface of the gel with plastic wrap and expose the gel to X-ray film for 10–30 s in the dark. Develop the film and, using a scalpel or razor blade, cut out the region of the gel corresponding to the 450 bp radiolabeled fragment (*see* **Note 10**).

10. Chop the bands into small pieces and place in a microfuge tube (*see* **Note 11**). Add 450 µL of 2.5*M* NH4OAc and incubate at 37°C for 5 h.
11. Transfer the supernatant to a fresh tube which contains 1 mL 100% EtOH. Mix and microfuge the contents at 18,000*g* for 20 min at 4°C.
12. Wash the pellet with 70% EtOH and dry under vacuum.
13. Resuspend the pellet in dH_2O to yield ~50,000 cpm/µL.

3.3. DNA Cleavage Reaction

This protocol details the analysis of camptothecin-stabilized enzyme-DNA complexes. However, the same approach may be used to investigate the ability of potential DNA topoisomerase I poisons to enhance the stability of the covalent complexes. In these cases, the drugs should be serially diluted in DMSO such that a range of concentrations may be compared with camptothecin-enhanced DNA cleavage. In this manner, the final DMSO concentration in each reaction will be the same. Using this approach, DNA topoisomerase I mutants may also be assessed for drug sensitivity.

1. DNA topoisomerase I (~75 units) is mixed with the single-end labeled DNA substrate (5,000 cpm per 5–10 ng) and 5 µL 10X G buffer in a final 50 µL (*see* **Note 12**).
2. The reactions are incubated at 30°C (for yeast DNA topoisomerase I) or 37°C (for mammalian DNA topoisomerase I) for 1–10 min. Longer times may be used, however, shorter incubations avoid any effects because of protein denaturation/ inactivation.
3. To stop the reaction and trap the cleavable complexes, 5 µL of 10% SDS is added and the samples are heated at 75°C for 10 min.
4. To digest the covalently attached protein, add 0.5 µL of proteinase K (20 mg/mL) and incubate the sample at 37°C for ≥ 15'.
5. Add 1 µL of 10 mg/mL tRNA, 6µL of 3*M* NaOAc and 120 µL 100% ETOH to precipitate the DNA fragments (*see* **Note 13**). Leave at –20°C for 30 min and pellet the DNA at 18,000*g* for 20 min at 4°C. Wash with 70% EtOH.
6. Remove the supernatant and dissolve the pellet in 200 µL TE buffer. Add 20 µL 3 *M* NaOAc and 500 µL 100% EtOH. Leave at –20°C for 30', spin at 18,000*g* for 20 min at 4°C (*see* **Note 14**). Wash with 70% ETOH and dry the sample.
7. Resuspend the pellet in 6 µL of the formamide containing sample buffer. The samples may be stored at –20°C.
8. Prior to loading onto a denaturing polyacrylamide gel, heat the samples at 75°C for ≥ 5 min to denature the DNA fragments.

3.5. Maxam-Gilbert Sequencing

As the DNA substrate is uniquely 3'-end labeled, a comparison of cleavage site with sequences derived by Maxam-Gilbert (chemical) sequencing of the same substrate is the most direct method for precisely mapping the sites of

Table 1
Base Modification Reactions

C + T rxn	C rxn	A rxn	G rxn
5 μl DNA 20 μl dH_2O	5 μl DNA 20 μl saturated NaCl	5 μl DNA 100 μl A reaction mix	5 μl DNA 200 μl Cacodylate buffer
↓	↓	↓	↓
Add 30 μl of hydrazine Incubate 7' at 20°C	Add 30 μl of hydrazine Incubate 20' at 20°C	Seal tube with Teflon tape Incubate 3' at 90°C	Add 1μl of DMS Incubate 2' at 20°C
↓	↓	↓	↓
Add: 200 μl C + T stop buffer 25 μg tRNA 650 μl 100% ETOH	Add: 200 μl C + T stop buffer 25 μg tRNA 650 μl 100% ETOH	Add: 150 μl cold HOAc 25 μl NaOAc 25 μg tRNA 870 μl 100% ETOH	Add: 50 μl DMS stop buffer 25 μg tRNA 750 μl 100% ETOH

cleavage. Sanger sequencing of the plasmid DNA using a primer complementary to 3'-end of scissile strand of the DNA will yield sequences complementary to the sequence of the cleavage sites. While this may be used to approximate the sites of cleavage, sequence specific changes in the mobility of the complementary sequences may complicate the assignment of bases.

1. After completing the base modification reactions described in **Table 1**, place the samples in a dry ice/EtOH bath for at least 5 min. Spin at 18,000*g* for 5 minutes at 4°C. Wash with 70% EtOH.
2. Repeat the ETOH precipitation and wash 2–3 times. Dry the sample under vacuum.
3. Resuspend each pellet in 50 μL dH_2O and transfer the samples to screw cap microfuge tubes. Add 50 μL 20% piperidine, seal the tubes with Teflon tape and heat at 90°C for 30 min (*see* **Note 15**).
4. Place tubes on ice for 2 min. Dry the samples under vacuum.
5. Resuspend the pellets in 50 μL dH_2O, transfer to a new tube and dry under vacuum.
6. Repeat 2–3 times (*see* **Note 16**).
7. Resuspend DNA in pellet in 15 μL of formamide containing sample buffer.

3.6. Resolution of DNA Cleavage Products

1. Prepare an 8% denaturing polyacrylamide gel: to 100 mL of 8% sequencing gel mix, add 400 μL 10 % ammonium persulfate and 28 μL TEMED. Any standard DNA sequencing apparatus, such as that supplied by IBI or Hoeffer, will do. Wedge gel spacers should be avoided as they preclude loading of adequate sample volumes. We typically use 1.5 mm spacers.
2. Prerun the gel in 1X TBE buffer at 100 Watts until it heats to 50–55°C.

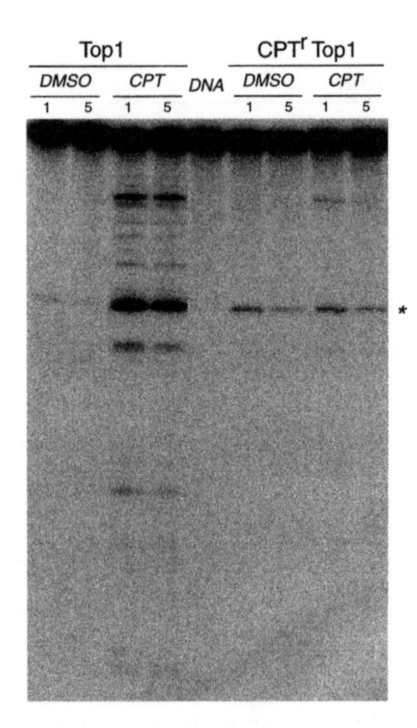

Fig. 2. DNA cleavage assay of human wild-type DNA topoisomerase I (Top1) and a camptothecin resistant mutant enzyme (CPT^r Top1). Equal concentrations of the proteins were incubated with the 3' end-labeled DNA substrate in the presence (CPT) or absence (DMSO) of the drug for 1 or 5 min. The reaction products were processed as described, resolved in a denaturing polyacrylamide gel and visualized with a PhosphorImager. The asterisk indicates the position of the high affinity cleavage site. DNA contains the DNA substrate control.

3. Heat samples to 75°C for ≥ 5 min. Load 6 μL of the cleavage reaction samples and 5 μL of each chemical sequencing sample in each well.
4. Run the gel at 100 Watts constant power for 1.5 h.
5. Disassemble the apparatus, immerse the gel in fixer for ≥ 30 min (*see* **Note 17**). Dry the gel and expose to X-ray film ≥ 16 h with intensifying screens or visualize bands with a PhosphorImager. A representative gel is shown in **Fig. 2**.

4. Notes

1. Although DNA fragments of a few hundred base pairs usually provide numerous DNA topoisomerase I cleavage sites detectable in the presence of enzyme and camptothecin, the inclusion of this high affinity site serves as a internal standard for enzyme-DNA complexes in the absence of drug.
2. Substitution of G for the A residue immediately 3' to the cleavage site may further enhance DNA topoisomerase I cleavage at this site.

3. The use of DNA polymerases lacking a 3' → 5' exonuclease activity ensures the uniform incorporation of radiolabeled deoxynucleotides at the 3' ends of the DNA molecules without additional cycles of proofreading/resynthesis by the DNA polymerase.
4. As a cautionary note, care should be exercised when doing Maxam-Gilbert sequencing. As several reagents are toxic (DMS and hydrazine, for example), gloves should be worn at all times. DMS is also volatile, so reactions should be carried out in a fume hood. In the anhydrous state, hydrazine is explosive. DMS may be inactivated with 5*M* NaOH, and hydrazine with 3*M* ferric chloride.
5. DNA topoisomerase I specific activity may be established by assaying the extent of enzyme catalyzed relaxation of supercoiled plasmid DNA under standard reaction conditions (physiological salt and temperature for 30 min). The products of reactions containing serially-diluted enzyme are resolved in agarose gels and the conversion of supercoiled plasmid DNA to relaxed DNA topoisomers assessed following staining with ethidium bromide. One unit of activity is defined as the amount of enzyme needed to completely relax 0.3 µg of supercoiled plasmid DNA. The reader is referred to ***(17)*** for more detailed description.
6. The quality of the DMSO used in the DNA cleavage reactions is important. As DMSO readily oxidizes, fresh aliquots of DMSO stored under N_2 is strongly recommended.
7. This solution is readily made with 4 parts Stop solution (USB/Amersham) and 6 parts dH_2O.
8. Any vertical electrophoresis unit, such as that supplied by BioRad or Hoeffer, will do.
9. The ammonium sulfate solution should be made fresh each time.
10. The use of fluorescent labels (available from Stratagene) on the plastic wrap will simplify the alignment of the film with the gel. The positions of the dyes may be used to approximate the relative mobility of the 450 bp fragment.
11. Care should be taken to avoid crushing the gel pieces as this will complicate the subsequent recovery of the supernatant.
12. The final KCl concentration should be 50 m*M*.
13. The addition of tRNA ensures the efficient precipitation of the DNA and will not affect the electrophoretic resolution of the reaction products.
14. Two ethanol precipitations are necessary to remove any residual SDS that may interfere with the electrophoretic separation of the DNA fragments.
15. Following the base modification reactions, treatment with piperidine results in cleavage of the phosphodiester backbone bond at these sites. In order to get sharp bands, DNA strand scission should occur at all modified sites. Screw cap tubes and teflon tape ensure the tubes are completely sealed and the concentration of piperidine is unchanged.
16. These steps are necessary to ensure that all residual piperidine is removed from the samples, as even low concentrations of piperidine will severely distort band mobility during electrophoresis.

17. Urea is extremely hygroscopic. The fixation step has the added advantage of removing the urea, which aids in the subsequent drying of the gel. This is of particular importance when using thicker spacers.

Acknowledgements

We are grateful to Jolanta Fertala and Anne Knab for helping in the development of this assay. This work was supported by NIH grant 58755 (to M-A.B.), a grant from Associazione Italiana per la Ricerca sul Cancro (to P.B.), CA21675 Cancer Center Grant and the American Lebanese Syrian Associated Charities (ALSAC).

References

1. Wang, J. C. (1996) DNA topoisomerases. *Ann. Rev. Biochem.* **65,** 635–692.
2. Bjornsti, M.-A. and Osheroff, N. (1999) Introduction to DNA topoisomerases, in *DNA topoisomerase protocols: DNA topology and enzymes*, vol. 94 (Bjornsti, M.-A. and Osheroff, N., eds.), Humana Press, Totowa, pp. 1–8.
3. Reid, R. J. D., Benedetti, P., and Bjornsti, M.-A. (1998) Yeast as a model organism for studying the actions of DNA topoisomerase-targeted drugs. *B.B.A.* **1400,** 289–300.
4. Redinbo, M. R., Stewart, L., Kuhn, P., Champoux, J. J., and Hol, W. G. J. (1998) Crystal structures of human topoisomerase I in covalent and noncovalent complexes with DNA. *Science* **279,** 1504–1513.
5. Pommier, Y., Pourquier, P., Fan, Y., and Strumberg, D. (1998) Mechanism of action of eukaryotic DNA topoisomerase I and drugs targeted to the enzyme. *B.B.A.* **1400,** 83–106.
6. Chen, A. and Liu, L. F. (1994) DNA Topoisomerases: Essential Enzymes and Lethal Targets. *Ann. Rev. Pharmacol. Toxicol.* **34,** 191–218.
7. Benedetti, P., Benchokroun, Y., Houghton, P. J., and Bjornsti, M.-A. (1998) Analysis of camptothecin resistance in yeast: relevance to cancer. *Drug Res. Updates* **1,** 176–183.
8. Thompson, J., Stewart, C. F., and Houghton, P. J. (1998) Animal models for studying the action of topoisomerase I targeted drugs. *B.B.A.* **1400,** 301–319.
9. Nitiss, J. (1998) Investigating the biological functions of DNA topoisomerases in eukaryotic cells. *B.B.A.* **1400,** 63–82.
10. Kanzawa, F., Nishio, K., Kubota, N., and Saijo, N. (1995) Antitumor activities of a new indolocarbazole substance, NB-506, and establishment of NB-506-resistant cell lines, SBC-3/NB. *Cancer Res.* **55,** 2806–2813.
11. Bailly, C., Riou, J. F., Colson, P., Houssier, C., Rodrigues-Pereira, E., and Prudhomme, M. (1997) DNA cleavage by topoisomerase I in the presence of indolocarbazole derivatives of rebeccamycin. *Biochem.* **36,** 3917–3929.
12. Gatto, B., Sanders, M. M., Yu, C., Makhey, D., Lavoie, E. J., and Liu, L. F. (1996) Identification of topoisomerase I as the cytotoxic target of the protoberberine alkaloid coralyne. *Cancer Res.* **56,** 2795–2800.

13. Chen, A. Y., Yu, C., Gatto, B., and Liu, L. F. (1993) DNA minor groove-binding ligands: a different class of mammalian DNA topoisomerase I inhibitors. *Proc. Natl. Acad. Sci. USA* **90,** 8131–8135.
14. Pilch, D. S., Xu, Z., Sun, Q., LaVoie, E. J., Liu, L. F., and Breslauer, K. J. (1997) A terbenzimidazole that preferentially binds and conformationally alters structurally distinct DNA duplex domains: a potential mechanism for topoisomerase I poisoning. *Proc. Natl. Acad. Sci. USA* **94,** 13,565–13,570.
15. Xu, Z., Li, T. K., Kim, J. S., LaVoie, E. J., Breslauer, K. J., Liu, L. F., and Pilch, D. S. (1998) DNA minor groove binding-directed poisoning of human DNA topoisomerase I by terbenzimidazoles. *Biochem.* **37,** 3558–3566.
16. Morham, S. G., Kluckman, K. D., Voulomanos, N., and Smithies, O. (1996) Targeted disruption of the mouse topoisomerase I gene by camptothecin selection. *Mol. Cell. Biol.* **16,** 6804–6809.
17. Knab, A. M., Fertala, J., and Bjornsti, M.-A. (1993) Mechanisms of camptothecin resistance in yeast DNA topoisomerase I mutants. *J. Biol. Chem.* **268,** 22,322–22,330.
18. Knab, A. M., Fertala, J., and Bjornsti, M.-A. (1995) A Camptothecin-resistant DNA Toposiomerase I Mutant Exhibits Altered Sensitivities to Other DNA Topoisomerase Poisons. *J. Biol. Chem.* **270,** 6141–6148.
19. Benedetti, P., Fiorani, P., Capuani, L., and Wang, J. C. (1993) Camptothecin resistance from a single mutation changing glycine 363 of human DNA topoisomerase I to cysteine. *Cancer Res.* **53,** 4343–4348.
20. Capranico, G. and Binaschi, M. (1998) DNA sequence selectivity of topoisomerases and topoisomerase poisons. *B.B.A.* **1400,** 185–194.
21. Tanizawa, A., Kohn, K. W., Kohlhagen, G., Leteurte, F., and Pommier, Y. (1995) Differential stabilization of eukaryotic DNA topoisomerase I cleavable complexes by camptothecin derivatives. *Biochem.* **34,** 7200–7206.
22. Jaxel, C., Capranico, G., Kerrigan, D., Kohn, K. W., and Pommier, Y. (1991) Effect of local DNA sequence on topoisomerase I cleavage in the presence or absence of camptothecin. *J. Biol. Chem.* **266,** 20,418–20,423.
23. Svestrup, J. Q., Christiansen, K., Anderson, A. H., Lund, K., and Westergaard, O. (1990) Minimal DNA duplex requirements for topoisomerase I-mediated cleavage in vitro. *J. Biol. Chem.* **265,** 12,529–12,535.
24. Megonigal, M. D., Fertala, J., and Bjornsti, M.-A. (1997) Cell cycle arrest and lethality produced by alterations in the catalytic activity of yeast DNA topoisomerase I mutants. *J. Biol. Chem.* **272,** 12,801–12,808.

29

Studying DNA Topoisomerase I-Targeted Drugs In the Yeast

Saccharomyces cerevisiae

Michael H. Woo, John R. Vance, and Mary-Ann Bjornsti

1. Introduction

The budding yeast *Saccharomyces cerevisiae* has been a valuable model in establishing eukaryotic DNA topoisomerase I as the cellular target of specific antineoplastic agents, including camptothecin ***(1–3)***, aclacinomycin A ***(4)***, and R-3, a rebeccamycin analogue (Vance, J. R., Woo, M. H., Otero, A., Bailly, C., and Bjornsti, M. A., unpublished results). This genetically tractable eukaryote has also been useful in studying mechanisms of resistance to DNA topoisomerase I-targeted drugs and providing information about drug function, cell cycle specificity, mutations affecting drug sensitivity, and the cellular consequences of drug treatment (reviewed in **refs. *3,5***).

Yeast cells exhibit relatively high rates of homologous recombination, such that a given DNA sequence can be directed to a specific locus and integrated into the genome via one-step gene replacement ***(6)***. A wide range of vectors are available for the insertion, deletion, epitope-tagging and expression of genes (the reader is referred to **refs. *7–10***) for comprehensive reviews and yeast methods). These plasmids also contain selectable markers for the selection and maintenance of plasmid borne sequences in auxotrophic yeast strains. *ARS/CEN* vectors are maintained in yeast cells in low copy number, while episomal 2 µm based vectors exist in high copy number. Strong constitutive promoters (*pGPD*) or inducible promoters (*pGAL1* and *pGAL10*) enable the study of heterologous and/or mutant protein function ***(3,11)***.

From: *Methods in Molecular Biology, Vol. 95: DNA Topoisomerase Protocols, Part II: Enzymology and Drugs*
Edited by N. Osheroff and M.A. Bjornsti © Humana Press Inc., Totowa, NJ

Yeast cells can be maintained as haploid cells of opposite mating type (*MAT***a** or *MAT*α) or as **a**/α diploid cells. Under conditions of nutrient deprivation, diploid cells undergo meiosis to produce four haploid spores. The separation, germination and analysis of these meiotic products allow for the construction of isogenic yeast strains, such that downstream events of specific genetic alterations can be determined. Such studies are facilitated by annotated databases containing the entire sequence of the yeast genome (*Saccharomyces* Genome Database (http://genome-www.stanford.edu/Saccharomyces/) and information about yeast proteins (Yeast Proteome Database ***(12)*** ; MIPS [http://www.mips.biochem.mpg.de/proj/yeast/]). Each has links to other databases and websites detailing protein-protein interaction maps, promoter function, transposon tagging and cross referencing to mammalian phenotypes and human ESTs.

A drawback to examining mechanisms of drug action in yeast is that wild-type laboratory strains are relatively insensitive to most DNA topoisomerase-targeted drugs, possibly due to the impermeability of the cell wall/membrane or drug efflux ***(1,3,5,13)***. Several approaches circumvent such problems, including the development of drug permeable strains, deletion of specific ATP binding cassette (ABC) transporters and/or alterations in cellular levels of DNA topoisomerase activity. For example, sensitivity to camptothecin and mAMSA were increased in *ise1* and *ISE2* mutant strains ***(1)*** and further enhanced in cells defective in the recombinational repair of double-stranded DNA breaks, due to deletion of the *RAD52* gene ***(1,13)***. Mutation of the *ERG6* gene, which is required for ergosterol biosynthesis, also increases cell sensitivity to camptothecin and apparently contributes to the *ise1* phenotype ***(3)***. The transcription factors encoded by the *PDR1* and *PDR3* genes regulate the expression of ABC transporters and modulate cell sensitivity to a wide range of drugs ***(14,15)***. Deletion of *PDR1* or the downstream ABC transporter gene SNQ2 enhanced yeast cell sensitivity to camptothecin ***(16)***.

The analysis of DNA topoisomerase I poisons in yeast is aided by the fact that the gene encoding the enzyme, *TOP1*, is nonessential, i.e., *top1*Δ strains are viable ***(17)***. These cells are also resistant to camptothecin ***(1,13)***. However, if yeast or human *TOP1* is expressed from plasmid-borne sequences in permeable, *rad52*Δ cells, drug sensitivity is restored ***(2)***. Further, overexpression of *TOP1* from the galactose-inducible *GAL1* promoter is sufficient to confer camptothecin sensitivity to repair-proficient cells that lack permeability mutations ***(3,16,18,19)***. The phenotypic consequences of treating mammalian cells with camptothecin also appear to be faithfully reiterated in yeast ***(3,11,20)***. Thus, it is possible to establish DNA topoisomerase I as the cellular target of a cytotoxic agent by comparing the chemosensitivity of isogenic *top1*Δ and *TOP1* strains. The effects of DNA topoisomerase I mutations on enzyme func-

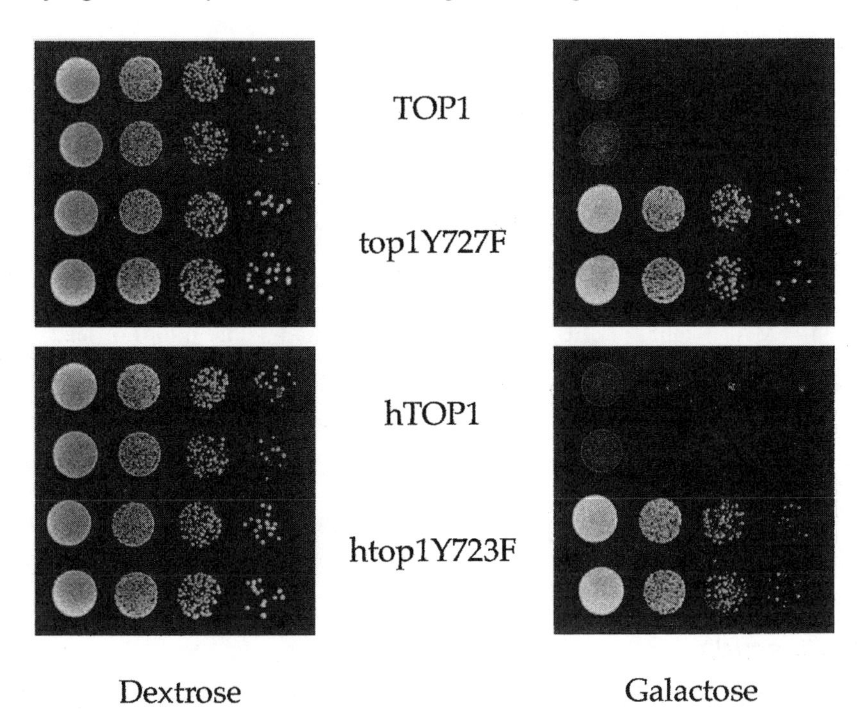

Fig. 1. Exponentially growing cultures of MBY3 (*MATα*, *ura3-52*, *his3Δ200*, *leu2Δ1*, *trp1Δ63*, *top1::TRP1*, *rad52::LEU2*) cells transformed with the indicated YCpGAL1-TOP1 (yeast TOP1, upper panel) or YCpGAL1-hTOP1 (human TOP1, lower panel) construct, were serially diluted and 5 μL aliquots spotted onto S.C. ura-HEPES plates containing 1 μg/mL camptothecin and either dextrose (left panel) or galactose (right panel). In yeast *top1Y727F* and human *top1Y723F*, the active site tyrosine residues were mutated to phenylalanine. Thus, these *top1* alleles encode catalytically inactive proteins. In each case, two independent transformants were examined.

tion and drug sensitivity can also be assessed by expressing *top1* mutants from the inducible GAL1 promoter in yeast *top1Δ* strains. For example, MBY3 cells (*top1Δ*, *rad52Δ*) expressing the catalytically inactive yeast *top1Y727F* or human *top1Y723F* mutant, where substitution of Phe for the active-site tyrosine in yeast or human DNA topoisomerase I, Tyr^{727} or Tyr^{723}, respectively, were resistant to camptothecin (**Fig. 1**). In contrast, galactose-induced expression of wild-type yeast or human DNA topoisomerase I restored *top1Δ* cell sensitivity to the drug.

The protocols described in this chapter focus on camptothecin and DNA topoisomerase I. However, the same approach may be used to establish the mechanism of action of other, putative DNA topoisomerase I poisons. Indeed, similar principles may also be applied to investigate drugs thought to target other gene products.

Table 1

Strain	Genotype
EKY3	MATα, ura3-52, his3Δ200, leu2Δ1, trp1Δ63, top1::TRP1
MBY3	MATα, ura3-52, his3Δ200, leu2Δ1, trp1Δ63, top1::TRP1, rad52::LEU2
RRY21	MATα, ura3-52, his3Δ200, leu2Δ1, trp1Δ63, top1::TRP1, pdr1Δ1::URA3
MWY3	MATα, ura3-52, his3Δ200, leu2Δ1, trp1Δ63, top1::TRP1, erg6::his5$^+$
MWY3-1	MATα, ura3-52, his3Δ200, leu2Δ1, trp1Δ63, top1::TRP1, pdr1Δ1::URA3, erg6::his5$^+$
JN2-134	MATa, rad52::LEU2, trp1, ade2-1, his7, ura3-52, ise1, top1-1, leu2

2. Materials

2.1. Yeast Strains

Yeast strains are described in Table 1 (*see* **Note 1**). In strains deleted for *ERG6* (*erg6::his5*$^+$), the *his5*$^+$ gene from the fission yeast *Schizosaccharomyces pombe* complements the histidine auxotrophy of *S. cerevisiae his3Δ* cells.

2.2. Plasmids

Yeast cells lacking DNA topoisomerase I, because of deletion of the *TOP1* gene (*top1Δ* strains), can be transformed with one of several expression vectors, to achieve inducible or constitutive expression of the desired *TOP1* allele from human, yeast or other eukaryotic cells. Representative examples are listed in **Table 2**. These shuttle vectors contains

1. ori and Amp^R sequences for plasmid DNA replication and selection in bacteria,
2. a yeast selectable marker, such as *URA*3 (*see* **Note 2**),
3. sequences necessary for DNA replication in yeast and (4) the galactose-inducible GAL1 promoter driving expression of yeast *TOP1* or human *TOP1* cDNA (*see* **Note 3**).

Low copy number, yeast centromeric plasmids, designated YCp , contain a yeast origin of replication (ARS) and centromere sequences derived from a yeast chromosome. In contrast, yeast episomal vectors (YEp) contain the 2 μm origin of replication and are present in high copy number (*see* **Note 4**).

2.3. Yeast Media

1. YPD media: 10 g yeast extract, 20 g bacto-peptone, 0.7 g adenine. Add dH_2O to a final volume of 900 mL. Autoclave, then add 100 mL of 20% dextrose.
2. 20% Raffinose: 20 g raffinose/100 mL dH_2O, filter-sterilized through a 0.45 μm filter.
3. 20% Galactose: 20 g galactose/100 mL dH_2O, filter-sterilized through a 0.45 μm filter.

Table 2

Plasmid	Characteristics
YCpGAL1-TOP1	ARS/CEN vector contains yeast TOP1 under the control of the galactose-inducible GAL1 promoter, URA3 selectable marker
YCpGAL1-hTOP1	ARS/CEN vector contains human TOP1 cDNA under the control of the galactose-inducible GAL1 promoter, URA3 selectable marker
YEpGAL1-TOP1	2μm episomal vector contains yeast TOP1 under the control of the GAL1 promoter, URA3 selectable marker
YEpGAL1-hTOP1	2μm episomal vector contains human TOP1 cDNA under the control of the GAL1 promoter, URA3 selectable marker

4. 20% Dextrose: 20 g dextrose/100 mL dH_2O, filter-sterilized through a 0.45 μm filter.
5. Synthetic complete media lacking uracil (S.C. ura-media): 1.7 g yeast nitrogen base without amino acids and ammonium sulfate (Difco Laboratories, Detroit, MI), 5 g ammonium sulfate, 0.72 g ura-dropout mix, 2 mL 1 N NaOH. Add dH_2O to 900 mL. Autoclave, then add 100 mL of the requisite sugar solution to give a final 2% (2 g/100 mL) (*see* **Note 5**).
6. Synthetic complete media lacking uracil with HEPES (S.C. ura-HEPES media): As for S.C. ura-media, except replace NaOH with 25 mL 1*M* HEPES pH 7.2 (*see* **Note 6**).
7. Ura-dropout mix: 2.4 g adenine sulfate, 2.4 g L-tryptophan, 1.2 g L-arginine, 6.0 g L-aspartic acid, 1.2 g L-histidine, 3.6 g L-leucine, 1.8 g L-lysine, 1.2 g L-methionine, 3.0 g L-phenylalanine, 12.0 g L-threonine, 1.8 g L-tyrosine. Shaking with acid washed glass beads will facilitate mixing.

2.4. Yeast Plates

1. YPD agar plates: 10 g yeast extract, 20 g bacto-peptone, 0.7 g adenine, 20 g agar. Add dH_2O to a final volume of 900 mL. Autoclave, cool to 55°C, and add 100 mL of 20% dextrose (sufficient for ~35 plates). Let set for 20–24 h at room temperature, then store in plastic bags at 4°C.
2. Synthetic complete media lacking uracil agar plates (S.C. ura-plates): 1.7 g yeast nitrogen base without amino acids and ammonium sulfate (Difco Laboratories, Detroit, MI), 5 g ammonium sulfate, 0.72 g ura-dropout mix, 2 mL 1 *N* NaOH. Add dH_2O to 450 mL. Filter-sterilize. Add 20 g agar to 450 mL dH_2O, autoclave, and cool to 55°C. Combine media and agar solutions, then add 100 mL of the requisite sugar solution to give a final concentration of 2% (2 g/100 mL) (sufficient for ~35 plates).
3. Synthetic complete media lacking uracil with HEPES agar plates (S.C. ura-HEPES plates): As for S.C. ura-media, except replace NaOH with 25 mL 1 *M* HEPES pH 7.2.

4. S.C. ura-HEPES, camptothecin plates: As for S.C. ura-HEPES plates, except add 1–5 μg/mL camptothecin in a final 0.125% Me_2SO just before pouring plates. Add Me_2SO alone for control plates (*see* **Note 7**).

2.5. Yeast Transformation

1. 10X LiOAc solution: 1 *M* LiOAc.
2. 10X TE buffer: 100 m*M* Tris-HCl, pH 7.5, 10 m*M* EDTA.
3. 50% PEG solution: 50 g PEG 3350 in 100 mL dH_2O. Filter-sterilize through a 0.45 μm filter (*see* **Note 8**).
4. 1X TE-LiOAc: 10 m*M* Tris-HCl, pH 7.5, 1 m*M* EDTA, 0.1*M* LiOAc. Make fresh from 10X stock solutions just prior to use (*see* **Note 9**).
5. 1X TE-LiOAc-PEG: 10 m*M* Tris-HCl, pH7.5, 1 m*M* EDTA, 0.1*M* LiOAc, 40% PEG. Make fresh from stock solutions just prior to use (*see* **Note 9**).
6. Salmon sperm DNA: 10 mg/mL salmon sperm DNA, sonicated and boiled (*see* **Note 10**).

2.6 Drug Solutions

Dissolve camptothecin (Sigma), or other putative DNA topoisomerase I poisons, in Me_2SO to a final concentration of 4 mg/mL. To use lower drug concentrations, serially dilute drug stocks in Me_2SO prior to medium addition. Store stock solutions at –20°C (*see* **Note 11**).

3. Methods

Since DNA topoisomerase I is not essential for yeast cell viability, the cytotoxic action of specific drugs can be assessed in yeast cells devoid of the enzyme, or expressing wild-type or mutant forms of Top1 from plasmid borne sequences. Using this approach, it is possible to determine if a drug specifically targets DNA topoisomerase I by interfering with enzyme catalysis. Assays of cell viability will further distinguish catalytic inhibitors from poisons, as drugs that suppress catalytic activity are unlikely to elicit a cytotoxic response (unless they trap the closed clamp form of the enzyme on the DNA). Further, the ability of specific mutations in Top1 to confer drug resistance may also be ascertained. However, in such cases, the specific activity of the mutant enzymes should also be evaluated in biochemical assays.

In these protocols, yeast strains defective in double-strand break repair and exhibiting enhanced permeability to a variety of agents are described. However, for each drug, the optimal combination of permeability mutants (such as *pdr1Δ*, *pdr3Δ*, *erg6Δ*, *pdr5Δ*, *snq2Δ*), DNA repair mutants (*rad52Δ*, *rad50Δ*), cell cycle checkpoint mutants (*rad9Δ*), growth conditions (such as pH) and/or overexpression strategies to observe drug-induced cell killing should be empirically determined.

3.1. Yeast Transformation

The topoisomerase I expression plasmid is transformed into the appropriate yeast cells using a modified LiOAc procedure *(21)*.

1. Grow *top1Δ* yeast strains in 40 mL YPD media to an OD_{595} of ~ 1.0 at 30°C (*see* **Notes 12** and **13**).
2. Centrifuge the cells at 4000*g* for 10 min, wash with 10 mL of freshly prepared 1X TE-LiOAc, and resuspend the cells in 600 μL 1X TE-LiOAc.
3. Mix 200 μL of the cell suspension with 150 μg salmon sperm DNA and 200–500 ng expression vector DNA in microcentrifuge tubes.
4. Add 700 μL 1X TE-LiOAc-PEG, mix thoroughly with a pipet, and incubate at 30°C for 30 min with gentle shaking.
5. Heat-shock for 15 min at 42°C. Centrifuge the cells at 15,000*g* for 30 s. Aspirate all but 100 μL of the supernatant. Resuspend the cells in the remaining 1X TE-LiOAc-PEG and plate on S.C. ura-plates. Individual transformants should be visible as distinct colonies following 2–3 d of incubation at 30°C (*see* **Note 14**). Subsequent growth in S.C. ura-medium will ensure the maintenance of *TOP1* plasmids.

3.2 Spot Tests

In this method, cell viability is scored by the formation of colonies either in the presence of drug, or following drug treatment. Spotting aliquots of serially diluted cultures onto agar plates provides a rapid and semi-quantitative measure of drug-induced cytotoxicity. With this approach, one -four log decreases in cell viability can easily be scored as a consequence of drug treatment. A more quantitative method is described below.

1. Transform *top1Δ* yeast strains with *TOP1* plasmid constructs by treatment LiOAc and select on S.C. ura-medium supplemented with 2% dextrose.
2. Pick individual transformants and inoculate 3 mL of S.C. ura-medium containing dextrose (*see* **Note 15**). Grow overnight at 30°C.
3. Serially ten-fold dilute overnight cultures in 1X TE buffer. This can easily be done in 96 well microtiter plates using a multichannel pipettor.
4. Spot 5 μL aliquots onto S.C. ura-HEPES, camptothecin plates containing 2% dextrose or galactose and varying concentrations of camptothecin in a final 0.125% Me_2SO. No drug control plates should also contain 0.125% Me_2SO. Score cell viability following incubation at 30°C for 2–4 d (*see* **Notes 16** and **17**).
5. Alternatively, dilute the overnight cultures 1 : 100 into S.C. ura-HEPES medium containing 2% raffinose and, at an OD_{595} = 0.3, add dextrose (to suppress *TOP1* expression) or galactose (to induce TOP1 expression) to a final concentration of 2%. Incubate the cultures with varying concentrations of camptothecin or Me_2SO alone (*see* **Note 18**). At various time points, serially ten-fold dilute the cells and spot 5 μL aliquots onto S.C. ura-plates containing 2% dextrose.

3.3 Cell Viability Assays

A more quantitative measure of cell viability is to count the number of viable cells that form colonies. This can be achieved by spreading aliquots of serially diluted cultures onto agar plates and counting the number of colonies that form. After correcting for dilution, these numbers can be plotted relative to the untreated control.

1. Inoculate 3–5 mL of S.C. ura-, dextrose medium with individual *top1Δ* transformants containing *TOP1* expression vectors or vector controls.
2. Grow cultures at 30°C to an OD_{595} of 1.0. and serially ten-fold dilute in 1X TE buffer.
3. Spread 100 µL aliquots of 10^{-1}, 10^{-2}, 10^{-3} and 10^{-4} onto S.C. ura-HEPES plates supplemented with 2% dextrose or galactose and varying concentrations of camptothecin (final Me_2SO concentration of 0.125%).
4. Count the number of cells forming colonies after incubation at 30°C. Correcting for dilution will give the number of viable cells forming colonies per mL. For example, if 100 µL of a 10^{-1} dilution gives rise to 150 colonies on a camptothecin plate, this corresponds to 150 per 0.1 mL of a 10^{-1} dilution, or 1.5×10^4 cells per mL of the original culture. The same dilutions plated on the Me_2SO control plates might yield 150 colonies per 100 µL of the 10^{-4} dilution, or 1.5×10^7 cells per mL. Thus, relative to the untreated control, cell viability drops to $(1.5 \times 10^4/1.5 \times 10^7)$ or 1×10^{-3} (*see* **Note 19**).
5. Alternatively, the cytotoxic action of drugs may also be assessed over time in liquid culture. As in **Subheading 3.2., step 5**, dilute overnight cultures 1:100 into S.C. ura-HEPES medium containing 2% raffinose. At an OD_{595} of 0.3, add dextrose or galactose to a final concentration of 2%. After one hour, treat with camptothecin or Me_2SO. At various time points, aliquots are serially ten-fold diluted and plated for colonies on S.C. ura-dextrose plates. Following incubation at 30°C for 2–3 d, the number of viable cells forming colonies are determined and plotted relative to that obtained at time zero (drug addition) (*see* **Note 19**).

4. Notes

1. With the exception of JN2-134 *(2)*, all yeast strains were derived from isogenic parent strains FY250 (*MATα*, *ura3-52*, *his3Δ200*, *leu2Δ1*, *trp1Δ63*) and FY251 (*MAT***a**, *ura3-52*, *his3Δ200*, *leu2Δ1*, *trp1Δ63*) by one-step gene replacement (as described in **ref.** ***6***) or using PCR amplified selectable markers (as detailed in **ref.** ***8***). The use of isogenic strains eliminates any variation in drug response owing to unknown differences in genetic background.
2. Although *URA3*-based expression vectors are denoted in this protocol, other vectors bearing *HIS3*, *LEU2*, *TRP1*, or *LYS2* are readily available or are easily constructed.
3. The yeast and human wild-type *TOP1* sequences in these vectors can easily be replaced with *TOP1* sequences derived from other eukaryotic cells or mutant

top1 alleles to assess species specific differences in drug action, or the effects of specific amino acid substitutions on DNA topoisomerase I sensitivity to drugs in vivo.

4. Although higher levels of *TOP1* expression may be achieved with YEpGAL1-TOP1 vectors, yeast strains defective in the repair of double stranded DNA breaks (for example, due to deletion of *RAD52, RAD51*, or *RAD50*) generally do not tolerate high intracellular concentrations of DNA topoisomerase I. In these strains, YCpGAL1-TOP1 vectors should be used.
5. Strains lacking the *URA3* gene product, i.e., *ura3-52* cells, are unable to synthesize uracil. In these uracil auxotrophs, growth in ura-media ensures the maintenance of plasmids bearing the *URA3* gene. Likewise, leu-, trp- or his- media can be used to maintain plasmid borne *LEU2*, *TRP1*, or *HIS3* sequences in *leu2Δ*, *trp1Δ*, or *his3Δ* strains.
6. Synthetic yeast media is typically adjusted to pH 5.8 for optimal cell growth. Buffering the media to pH 7.2 enhances the cytotoxic action of camptothecin, presumably because of alterations in membrane mediated drug influx/efflux. Nevertheless, cell growth will eventually cause acidification of the media to ~pH 5.8. With different classes of DNA topoisomerase poisons, the optimal pH for drug action has to be empirically determined.
7. Camptothecin is light sensitive, so cover plates with foil. Allow plates to set/dry at room temperature for 20–24 h. Plates may be stored at 4°C for ~1 wk.
8. Use disposable bottle top filters (0.45 μm) attached to a vacuum line.
9. Freshly prepared solutions will improve transformation efficiency.
10. Higher transformation efficiencies are typically obtained with single-stranded DNA.
11. Me_2SO is readily oxidized. Fresh aliquots of Me_2SO packaged under nitrogen (Aldrich) are essential.
12. Exponentially growing cultures are essential for efficient transformation with plasmid DNA. If saturated overnight cultures are used, dilute cells into fresh media and incubate cultures until cells are in the exponential phase of the growth curve (typically ~2 h).
13. Many yeast strains spontaneously acquire mutations affecting the induction of the *GAL1/10* promoters. To obviate any effects on *GAL1*-promoted expression of plasmid borne *TOP1* alleles, the cells may be grown overnight in YPG media (YPD media containing a final 2% galactose in place of dextrose) and diluted into fresh YPD media to achieve exponential growth.
14. To ensure optimal expression of *TOP1*, fresh transformants should be used (within 4–7 d).
15. This should be done in duplicate or triplicate to avoid picking spontaneous mutants that affect enzyme function or drug action.
16. Ensure the plates are sufficiently dry or the samples will spread over the surface of the plate. Leaving the plates at room temperature for 20–24 h to allow the agar to set is usually sufficient. If stored at 4°C, warm to room temperature before plating.
17. Dextrose is a better carbon source than galactose. Score yeast cell viability on dextrose containing plates after 2 d, and cell growth on galactose plates at 3–4 d.

Avoid prolonged incubation (more than 5 d) as spontaneous, drug resistant colonies will appear.

18. Camptothecin concentrations in excess of 50–100 μM (final 1% Me_2SO) will precipitate out of solution.
19. Correcting for the untreated control (or t = 0 time point), allows a direct comparison and statistical evaluation of values obtained from multiple experiments.

Acknowledgments

The Authors thank past and present members of the lab and Piero Benedetti for help in developing these assays. This work was supported by NIH grant CA70406 to M-A.B., training grant CA09346 (to J.R.V.), CA21675 Cancer Center Grant and the American Lebanese Syrian Associated Charities (ALSAC).

References

1. Nitiss, J. and Wang, J. C. (1988) DNA topoisomerase-targeting antitumor drugs can be studied in yeast. *Proc. Natl. Acad. Sci. USA* **85,** 7501–7505.
2. Bjornsti, M.-A., Benedetti, P., Viglianti, G. A., and Wang, J. C. (1989) Expression of human DNA topoisomerase I in yeast cells lacking yeast DNA topoisomerase I: restoration of sensitivity of the cells to the antitumor drug camptothecin. *Cancer Res.* **49,** 6318–6323.
3. Reid, R. J. D., Benedetti, P., and Bjornsti, M.-A. (1998) Yeast as a model organism for studying the actions of DNA topoisomerase-targeted drugs. *BBA* **1400,** 289–300.
4. Nitiss, J. L., Pourquier, P., and Pommier, Y. (1997) Aclacinomycin A stabilizes topoisomerase I covalent complexes. *Cancer Res* **57,** 4564–4569.
5. Benedetti, P., Benchokroun, Y., Houghton, P. J., and Bjornsti, M.-A. (1998) Analysis of camptothecin resistance in yeast: relevance to cancer. *Drug Res. Updates* **1,** 176–183.
6. Rothstein, R. (1991) Targeting, Disruption, Replacement, and Allele Rescue: Integrative DNA Transformation in Yeast, in *Methods in Enzymology*, vol. 194 (Abelson, J. N., and M. I. Simon, ed.), Academic Press, San Diego, pp. 281–301.
7. Adams, A., Gottschling, D. E., Kaiser, C. A., and Stearns, T. (1997) *Methods in yeast genetics*, Cold Spring Harbor Laboratory Press, Cold Spring Harbor.
8. Longtine, M. S., McKenzie, A., 3rd, Demarini, D. J., Shah, N. G., Wach, A., Brachat, A., Philippsen, P., and Pringle, J. R. (1998) Additional modules for versatile and economical PCR-based gene deletion and modification in Saccharomyces cerevisiae. *Yeast* **14,** 953–961.
9. Abelson, J. N. and M.I. Simon (1991) in *Guide to Yeast Genetics and Molecular Biology*, vol. 194 (Guthrie, C. and G. R. Fink, eds.), Academic Press, San Diego.
10. Sherman, F. (1997) Yeast genetics, in *The Encyclopedia of Molecular Biology and Molecular Medicine*, vol. 6 (Meyers, R. A., ed.), VCH Pub.,Weinheim, Germany, pp. 302–325.

11. Kauh, E. A. and Bjornsti, M.-A. (1995) SCT1 mutants suppress the camptothecin sensitivity of yeast cells expressing wild-type DNA topoisomerase I. *Proc. Natl. Acad. Sci. USA* **92,** 6299–6303.
12. Costanzo, M. C., Hogan, J. D., Cusick, M. E., Davis, B. P., Fancher, A. M., Hodges, P. E., Kondu, P., Lengieza, C., Lew-Smith, J. E., Lingner, C., Roberg-Perez, K. J., Tillberg, M., Brooks, J. E., and Garrels, J. I. (2000) The yeast proteome database (YPD) and caenorhabditis elegans proteome database (WormPD): comprehensive resources for the organization and comparison of model organism protein information (In Process Citation). *Nucleic Acids Res* **28,** 73–76.
13. Eng, W.-K., Faucette, L., Johnson, R. K., and Sternglanz, R. (1988) Evidence that DNA topoisomerase I is necessary for the cytotoxic effects of camptothecin. *Mol. Pharmacol.* **34,** 755–760.
14. Bauer, B. E., Wolfger, H., and Kuchler, K. (1999) Inventory and function of yeast ABC proteins: about sex, stress, pleiotropic drug and heavy metal resistance. *BBA* **1461,** 217–236.
15. Taglicht, D. and Michaelis, S. (1998) Saccharomyces cerevisiae ABC proteins and their relevance to human health and disease. *Methods Enzymol* **292,** 130–162.
16. Reid, R. J. D., Kauh, E. A., and Bjornsti, M.-A. (1997) Camptothecin sensitivity is mediated by the pleiotropic drug resistance network in yeast. *J. Biol. Chem.* **272,** 12,091–12,099.
17. Goto, T. and Wang, J. C. (1985) Cloning of yeast TOP1, the gene encoding DNA topoisomerase I, and construction of mutants defective in both DNA topoisomerase I and DNA topoisomerase II. *Proc. Natl. Acad. Sci. USA* **82,** 7178–7182.
18. Knab, A. M., Fertala, J., and Bjornsti, M.-A. (1993) Mechanisms of camptothecin resistance in yeast DNA topoisomerase I mutants. *J. Biol. Chem.* **268,** 22,322–22,330.
19. Knab, A. M., Fertala, J., and Bjornsti, M.-A. (1995) A Camptothecin-resistant DNA Toposiomerase I Mutant Exhibits Altered Sensitivities to Other DNA Topoisomerase Poisons. *J. Biol. Chem.* **270,** 6141–6148.
20. Megonigal, M. D., Fertala, J., and Bjornsti, M.-A. (1997) Cell cycle arrest and lethality produced by alterations in the catalytic activity of yeast DNA topoisomerase I mutants. *J. Biol. Chem.* **272,** 12,801–12,808.
21. Kaiser, C., Michaelis, S., and Mitchell, A. (1994) *Methods in Yeast Genetics*, Cold Spring Harbor Laboratory Press.

30

Yeast Systems for Demonstrating the Targets of Anti-Topoisomerase II Agents

John L. Nitiss and Karin C. Nitiss

1. Introduction

A key aspect of anti-topoisomerase drug action is that most anti-topoisomerase drugs act by stabilizing an intermediate of the topoisomerase reaction *(1)*. This intermediate consists of the enzyme covalently bound to DNA by a phosphotyrosine linkage, where the DNA strand scission has occurred, and is referred to as a covalent or cleavage complex. Complex-stabilizing topoisomerase II agents kill cells mainly because the topoisomerase:DNA covalent complex is DNA damage that interferes with DNA metabolic processes, and in mammalian cells, commits cells to apoptotic cell death *(2)* A substantial body of evidence has shown that DNA cleavage, rather than inhibition of enzyme activity is responsible for cell killing. Hence, complex-stabilizing anti-topoisomerase agents are referred to as topoisomerase poisons *(1)*.

We have developed a system using *Saccharomyces cerevisiae* to study the mechanisms of action of anti-topoisomerase agents *(3–5)*. Yeast is an ideal system for studying drug mechanisms in eukaryotic cells, because of the ability to make targeted changes in the yeast genome and to easily assess the consequences of the changes on drug sensitivity. The mechanisms of action of topoisomerase poisons and other topoisomerase inhibitors predict that changes in the levels of topoisomerases will alter cellular sensitivity to these agents. These changes in drug sensitivity can be used to test the mechanisms of cell killing of novel therapeutic agents.

The systems described here for examining the targets of putative anti-topoisomerase II drugs are based on the mechanism of action of topoisomerase poisons. Increased levels of topoisomerase II will lead to increased levels of

From: *Methods in Molecular Biology, Vol. 95: DNA Topoisomerase Protocols, Part II: Enzymology and Drugs*
Edited by N. Osheroff and M.A. Bjornsti © Humana Press Inc., Totowa, NJ

DNA damage, and therefore greater sensitivity to cell killing, while decreases in topoisomerase II level or activity will lead to drug resistance due to reduced levels of DNA damage. Since yeast cells are the only genetically tractable eukaryotic system that can readily tolerate altered levels of expression of topoisomerases *(6)*, they are ideal for the approach described here. In addition, a large number of mutations leading to resistance to topoisomerase II poisons have been described in yeast *(7)*. Coupled with the ability to carry out gene replacement by homologous recombination, the mutations conferring resistance to topoisomerase II poisons also represent a unique set of reagents for assessing the ability of agents to target topoisomerase II in eukaryotic cells.

While most clinically used topoisomerase inhibitors stabilize covalent complexes, other cytotoxic agents have been described that are catalytic inhibitors of topoisomerase II. Examples include the bisdioxopiperazines ICRF-187 and ICRF-193, merbarone, and the anthracyline aclarubicin ***(8–10)***. Cell killing by such agents should depend on the inhibition of topoisomerase II, an essential enzyme, and the mechanism of cell killing in yeast by catalytic inhibitors would be expected to parallel the effects of temperature sensitive topoisomerase II mutations ***(11)***. The pattern of sensitivity for such agents in topoisomerase mutations follows the same pattern as classical enzyme inhibitors, and is opposite to the pattern for topoisomerase II poisons described above. Overexpression of topoisomerase II results in resistance to these compounds, while reduced enzyme activity leads to drug hypersensitivity.

2. Materials

2.1. Required Laboratory Equipment and Supplies

1. Shaking water bath that can stably maintain temperatures from 25°C to 36°C.
2. Incubators suitable for incubating petri dishes.
3. Microscope (preferably phase contrast, with a 10× objective).
4. Hemacytometer.
5. Automatic pipetors.
6. Sterile pipet tips.
7. Plating wheel (Optional).
8. Bunsen burner.
9. Disposable 50-mm petri dishes
10. Small disposable flasks, or small Erlenmeyer flasks (*see* **Note 1**).

2.2. Yeast Strains

A key consideration in using yeast to examine sensitivity to topoisomerase targeting drugs is the use of yeast strains that are sensitive to agents under examination. Traditionally, it has been thought that yeast strains are impermeable to large molecular weight compounds (i.e., ranging from 250–1000 MW), however recent work has suggested that yeast express many proteins that can confer pleiotropic drug resistance ***(12)***. Regardless of the precise mechanism, wild type yeast strains are relatively insensitive to many drugs targeting topoisomerase II, including drugs such as etoposide, doxorubicin, and mitoxantrone. We and others have developed yeast strains carrying mutations that confer sensitivity to a many of these test agents ***(3,5,13)***. Most of the published work has used mutations in one of two genes. Mutations that disrupt the ERG6 have been widely used for assessing drug action in yeast strains. ERG6 encodes S-adenosylmethionine:delta 24-sterol- C-methyltransferase, and is required for biosynthesis of ergosterol, the normal membrane sterol of yeast ***(14)***. Loss of ERG6 function presumably alters membrane function in such a way that drug exporters fail to function, resulting in enhanced drug accumulation. Using an allele of ERG6 termed ise1, we showed that erg6- cells were hypersensitive to the topoisomerase II poisons mAMSA, etoposide and doxorubicin ***(5)***, as well as the topoisomerase I poison camptothecin ***(13,15)***. A second mutation, ise2, with altered drug sensitivity shows enhanced sensitivity to a broad range of topoisomerase II poisons including etopoisde, mAMSA, and fluoroquinolones. The biochemical defect in ise2 strains has not been determined. Finally, many other genes have been identified that encode putative drug transporters, or proteins that regulate their expression. Mutants defective in these genes will likely have enhanced drug sensitivity. For example, yeast strains defective in PDR1 and PDR3, two transcriptional regulators that affect the expression of several yeast drug transporters have been applied to examine sensitivity to a broad range of anti-cancer drugs ***(16–18)***.

In many cases, the drug sensitivity conferred by mutations increasing drug accumulation is insufficient to result in cell killing. Cell killing can be enhanced by mutations affecting DNA repair processes. Cells defective in recombinational repair are extremely sensitive to both topoisomerase I and topoisomerase II poisons ***(5,13,15)***. Therefore, including a mutation that reduces recombinational repair, such as in the *RAD52* gene, greatly increases the sensitivity of yeast based assays.

A third requirement is to alter the level of drug sensitive topoisomerase. For studying topoisomerase I-targeting agents, this can be readily accomplished in yeast. *TOP1* is not essential for cell viability, so all drug sensitive activity can be abolished by using strains deleted for the *TOP1* gene. Since *TOP2* is essential for viability, this approach requires modification when studying topoisomerase II targeting drugs. This can be accomplished in three ways. First, the levels of active enzymes can be enhanced, by overexpression of the wild type enzyme. Second, enzyme levels can be reduced using alleles of *TOP2* with reduced enzymatic activity. Finally, the wild type *TOP2* gene can be replaced with a drug resistant allele. Although this third possibility is the easi-

Table 1
Yeast Strains

Strain	Genotype
JN394	a ise2 ura3-52 leu2 trp1 ade2 his7 rad52: :LEU2
JN394t2-1	As JN394 but top2-1
JN394t2-5	As JN394 but top2-5
JN394t1	As JN394 but top1: :LEU2 rad52: :TRP1
JN394t2-4 pDED1TOP2	As JN394 but top2-4, and carrying the plasmid pDED1TOP2
JN394t2-4 pMJ1	As JN394 but top2-4, and carrying the plasmid pMJ1 (human TOP2 α expression plasmid)

est one to carry out in practice, there is the potential problem that an allele of *TOP2* that is resistant to one class of drugs, may be fully sensitive to other drug classes. In the experimental system described here, the *top2-5* allele is used. This mutation confers resistance to epipodophyllotoxins, intercalating topoisomerase II poisons, and fluoroquinolones ***(19,20)***. However, if a strain carrying this mutation is still drug sensitive, it does not exclude the possibility that the agent targets topoisomerase II.

A final consideration in the choice of strains is integrating all of the variables described above into one isogenic set of strains. Use of isogenic strains constructed using gene replacement techniques is needed to exclude the possibility that other genetic changes contribute to alterations in drug sensitivity. The experiments described below use the set of isogenic strains described in **Table 1**.

2.3. Plasmids

As described previously, yeast strains overexpressing topoisomerase II are hypersensitive to drugs targeting this enzyme. To overexpress yeast topoisomerase II, the plasmid pDED1TOP2 is used. This plasmid carries the yeast TOP2 open reading frame under the control of the yeast DED1 promoter ***(21)***. The plasmid carries the yeast URA3 in order to select for the plasmid. It also carries an origin of replication and a yeast centromere, and is stably maintained as a single copy plasmid.

As described below, it is frequently of interest to determine whether a drug is active against human topoisomerases. Expression of human topoisomerase II α or topoisomerase II β will complement yeast cells carrying yeast TOP2 mutations ***(22–24)***. For example, yeast cells carrying a temperature sensitive topoisomerase II mutation (such as top2-4) are unable to grow at 36°C because of a thermolabile topoisomerase II protein. Expression of either human isozyme allows the cells to grow at that temperature. At 36°C, the only active topoisomerase will be the human enzyme, therefore, all drug sensitivity due to topoisomerase II will be solely because of the human enzyme. A simple vector carrying human TOP2 α under the control of the yeast TOP1 promoter results in a level of expression of human TOP2 that confers sensitivity to etoposide, mAMSA, and other TOP2 targeting drugs *(22)*. This plasmid, termed pMJ1, also carries the yeast URA3 in order to select for the plasmid in yeast, an origin of replication and a yeast centromere.

2.4. Yeast Media

2.4.1. YPDA Broth

20 g Bacto-peptone
20 g Dextrose
10 g Yeast extract
2 mL Adenine sulfate solution
Water to 1 liter

Dissolve the media components, dispense into bottles, and autoclave at 121°C for 20 min.

Adenine sulfate solution is prepared by dissolving 0.5 g adenine sulfate in water, and then sterilizing the solution by filtration. The adenine sulfate solution should be stored at room temperature.

2.4.2. YPDA Agar Plates

Prepare 1 liter of YPDA broth. Prior to autoclaving, add 17.5 g agar. Autoclave at 121°C for 20 min. After removing the medium from the autoclave, gently swirl to make sure the agar is completely dissolved. Allow to cool to 50°C, then pour approximately 25 mL of medium per petri dish.

2.4.3. Ura⁻ Dropout Powder

The recipe listed below is for a mixture of amino acids and adenine used to prepare synthetic complete medium lacking uracil (SC-ura) in sufficient quantity to prepare 100 L of media. The mixture of amino acids and adenine is stable indefinitely if kept protected from moisture. The recipe can be scaled up or down as required.

Adenine sulfate	1.0 g
L-arginine	5.0 g
L-aspartic acid HCl	7.5 g
L-glutamic acid, monosodium salt	10. g
L-histidine HCl	2.0 g
L-isoleucine	5.0 g

L-leucine	10. g
L-lysine HCl	5.0 g
L-methionine	2.0 g
L-phenylalanine	5.0 g
L-serine	37.5 g
L-threonine	10. g
L-tryptophan	5.0 g
L-tyrosine	5.0 g
L-valine	15. g

It is convenient to weigh out the components into a 250 mL plastic bottle. After adding all the components, mix thoroughly. An air shaker can be conveniently used, in which case, the bottle containing the powders should be shaken for 2–3 days.

2.4.4. Ura⁻ Broth

20 g dextrose
1.7 g yeast nitrogen base without amino acids and without ammonium sulfate
5.0 g ammonium sulfate
1.3 g Ura⁻ dropout powder (*see* **Subheading 2.4.3**)
Water to 1 L

Add all the components, and stir to dissolve. After all components have dissolved, adjust pH of the medium to 6.0 using 1 N NaOH. About 2.5 mL will be required for 1 L of medium. After adjusting pH, dispense into bottles, and autoclave at 121°C for 20 min.

2.4.5. Ura⁻ Agar Plates

Prepare ura⁻ broth as described in **Subheading 2.4.4.** Prior to autoclaving, add 17.5 g agar. Autoclave at 121°C for 20 min. After removing the medium from the autoclave, gently swirl to make sure the agar is completely dissolved. Allow to cool to 50°C, then pour approximately 25 mL of medium per petri dish.

3. Methods

3.1. Strategy for Determining Targeting of Putative Topoisomerase II Agents

There are three stages to the experiments described below. The first stage tests whether the yeast strain is sensitive to the test agent. These experiments test drug sensitivity in a strain with a wild type topoisomerase complement. The second stage tests the mechanism of action of the agent to determine whether it is a topoisomerase II poison, a topoisomerase II catalytic inhibitor, or a topoisomerase I poison. These experiments test drug sensitivity in yeast strains overexpressing topoisomerase II or in cells lacking topoisomerase I. The final set of experiments tests the specificity of topoisomerase II poisons. Two experimental approaches are described; one using a mutation in yeast topoisomerase II that is resistant to many different topoisomerase II poisons. Since that mutant allele, *top2-5*, may not be resistant to all topoisomerase II poisons, an alternate approach is also described that is more technically demanding, but is applicable to all topoisomerase II poisons.

3.2. Determination of Drug Sensitivity in Strains with Wild Type Topoisomerase Levels

1. Inoculate a 10 mL overnight culture of JN394 or other suitable yeast strain in YPDA broth. Incubate at 30°C in a shaking water bath.
2. The following day, determine the number of yeast cells per mL using a hemacytometer. Dilute the cells with YPDA to a titer of 2×10^6 cells/mL. For each condition, 3 mL of diluted culture is required. Include one sample to which drug solvent only is added. Aliquot the diluted cells to small flasks. Add the required volume of the test agents, or the solvent the agent is dissolved in (*see* **Note 2**). Place the flasks in a shaking water bath at 30°C. Prior to putting the flasks in the water bath, remove 100 µL from the solvent control flask to determine the viable titer at time = 0. Add the 100 µL aliquot of cells to 900 µL sterile distilled water. Carry out two more serial 1 : 10 dilutions, and plate 200 µL of the last dilution to each of two YPDA plates. All plates are incubated for three days at 30°C, and then counted for total number of colonies. If the starting titer is 2×10^6 cells/ml, this will result in about 300 colonies/plate at t = 0.
3. Incubate the cells with shaking at 30°C for 24 h. After 24 h remove 100 µL aliquots from each flask, carry out serial dilutions, and plate the dilutions to duplicate YPDA plates. For strain JN394, a suitable dilution in the absence of drug is 10^{-5} (i.e., five serial 1 : 10 dilutions). Cultures containing drug will require plating of multiple dilutions. For an unknown agent, plating dilutions from 10^{-3}, 10^{-4} and 10^{-5} will usually be sufficient to get an appropriate number of colonies. An ideal dilution will result in about 250 colonies/plate.
4. Determine the viable titer relative to the viable titer at t = 0. For example, if the t = 0 no drug sample results in an average of 200 colonies/plate at a dilution of 10^{-3}, and the no drug sample has an average of 100 colonies at 24 h from the 10^{-5} dilution, the relative survival will be $100/10^{-5}$ divided by $200/10^{-3}$ =5000. A drug that is cytostatic will give the same viable titer at 24 h as a 0 h. Lower viable titers at 24 h compared to t = 0 indicates that the drug is cytotoxic.
5. In order to proceed with a determination of whether topoisomerases are drug targets, the drug has to have an observable affect on the strain having a wild type topoisomerase complement. If no effect on growth is seen in this experiment, yeast cells are insensitive to the test agent, or yeast topoisomerase II is insensitive to the test agent. Strains expressing human topoisomerases can be used to test this latter possibility (*see* **Subheading 3.5.**).

3.3. Determination of Drug Sensitivity in Strains with an Altered Topoisomerase Complement

The protocol described in **Subheading 3.2.** can be adapted to all drug sensitivity measurements. This section describes the alterations needed to determine drug sensitivity in strains with altered DNA topoisomerases. This section covers both stage II experiments, which test drug mechanism, and stage III experiments that test drug specificity.

3.3.1. Determination of Drug Sensitivity in Strains Overexpressing Topoisomerase II

Strain JN394 pDED1TOP2 carries a plasmid that expresses TOP2 from the yeast DED1 promoter. The DED1 promoter is constitutively expressed, so the only consideration is to selectively examine cells that carry the plasmid. This is accomplished by growing the cells, in Ura^- broth (**Subheading 3.2., steps 1** and **2**), and determining viable titer by plating to Ura^- plates (**Subheading 3.2., steps 3** and **4**). Since cell growth is determined using Ura^- media, the results obtained under these conditions are not directly comparable to the results obtained with JN394 in YPDA. Therefore, a yeast strain carrying an empty vector (e.g., JN394 carrying yCP50) should also be tested under the same conditions as JN394 pDED1TOP2.

It should also be noted that the plating efficiency of this strain (the number of colonies arising following growth in drug free medium) will be lower than for JN394 cells grown in YPDA. Therefore, the dilutions used after 24 h of growth in the absence of drug should also include both 10^{-4} and 10^{-5} dilutions. Since drug sensitivity is determined by relative survival (**Subheading 3.2.**, **step 4**), the reduced growth in drug free medium is accounted for.

3.3.2. Determination of Drug Sensitivity in Strains Lacking Topoisomerase I

The protocol described in **Subheading 3.2.** can be applied directly to strains lacking topoisomerase I, e.g., JN394t1. Since the topoisomerase I gene is deleted, and the strain carries no plasmid, this experiment can be carried out in YPDA medium, and can be compared directly to results obtained with JN394. The recommended dilutions in **Subheading 3.2.** can be applied to strain JN394t1, except that the strain will be more sensitive to topoisomerase II poisons than JN394, requiring plating more cells (i.e., a less diluted culture) at high drug concentrations.

3.3.3. Determination of Drug Sensitivity in Strains Expressing Drug Resistant Topoisomerase II Activity

The protocol described in **Subheading 3.2.** can be applied to strains carrying the drug resistant allele *top2-5* with one critical modification. The *top2-5* allele is drug resistant at its permissive temperature, 25°C. The strain grows very poorly at 30°C, and does not grow above this temperature. Therefore the drug sensitivity must be determined at 25°C. The culture should be inoculated at 25°C (**Subheading 3.2., step 1**), the drug exposure should be carried out at 25°C, and the plates used to determine viable titer should be incubated at 25°C. Note that cell growth is somewhat slower at 25°C, therefore the plates may need an additional days incubation prior to counting. Because the strain JN394t2-5 carries no plasmid, this experiment can be carried out in YPDA medium. To ensure that drug sensitivity is not affected by temperature, this same protocol should be applied to strain JN394 to serve as a control for the JN394t2-5 results.

3.3.4. Determination of Drug Sensitivity Under Conditions of Low Topoisomerase II Activity

The *top2-5* allele confers resistance to topoisomerase II targeting agents at its permissive temperature. Although it confers resistance to many different topoisomerase II poisons, it remains possible that the *top2-5* allele does not confer resistance to all classes of topoisomerase II poisons. An alternate approach is to reduce the level of active topoisomerase II, thereby generating resistance to all topoisomerase II poisons. This can be accomplished by using a temperature sensitive allele, provided that the enzyme activity is (inversely) proportional to temperature. The *top2-1* allele fits this criterion ***(25,26)***.

1. Inoculate a 10 mL overnight culture of JN394t2-1 or other suitable yeast strain in YPDA broth. Incubate at 25°C in a shaking water bath.
2. The following day, determine the number of yeast cells per ml using a hemacytometer. Dilute the cells with YPDA to a titer of 2×10^6 cells/mL. For each condition, 3 mL of diluted culture is required. Include one sample to which drug solvent only is added. Aliquot the diluted cells to small flasks. Add the required volume of the test agents, or the solvent the agent is dissolved in.. Place the flasks in a shaking water bath at 30°C. Prior to putting the flasks in the water bath, remove 100 µL from the solvent control flask to determine the viable titer at time = 0. Add the 100 µL aliquot of cells to 900 µL sterile distilled water. Carry out two more serial 1:10 dilutions, and plate 200 µL of the last dilution to each of two YPDA plates. All plates are incubated for three to four days at 25°C, and then counted for total number of colonies. If the starting titer is 2×10^6 cells/ml, this will result in about 300 colonies/plate at t = 0.

Table 2
Expected Patterns of Sensitivity of Topoisomerase Inhibitors to Strains Carrying Alterations in Topoisomerase Genes

	Strain genotype					
Drug class	WT	top1⁻	top2-1 25°	top2-1 30°	top2-5 25°	TOP2 elevated
TOP1 poison	+	+	+	++	+	+
TOP2 poison	+	++	+	–	–?	++
TOP2 catalytic	+	++	+	++	+?	+/–
Not a top Inhibitor	+	+	+	+	+	+

– resistant
+/–reduced sensitivity
+ wild type sensitivity
++ hypersensitive
? possibly indeterminate

3. Incubate the cells with shaking for 24 h. After 24 h remove 100 µL aliquots from each flask, carry out serial dilutions, and plate the dilutions to duplicate YPDA plates. For strain JN394t2-1, a suitable dilution in the absence of drug is 10^{-3}. Cultures containing drug will require plating of multiple dilutions. For an unknown agent, plating dilutions from 10^{-3}, 10^{-2} and 10^{-1}will usually be sufficient to get an appropriate number of colonies.
4. Determine the viable titer relative to the viable titer at t = 0.

3.4. Interpretation of Results

Table 2 shows the predicted results for topoisomerase I or II poisons or topoisomerase II catalytic inhibitors. As can readily be seen, each of these three classes of agents gives a unique pattern of drug sensitivity. As can also be seen, the pattern is distinct from agents that do not act against topoisomerases. There is one significant alternate case not covered in **Table 2**. This is the possibility that an agent may act against both topoisomerase I and topoisomerase II. In the case of an agent that acts as a poison against both topoisomerase I and topoisomerase II, the *top2-5* mutation would be distinctly useful, since the *top2-5* mutation would probably confer (partial) resistance to the agent, as would the *top1* deletion. A similar case arises for an agent that is a catalytic inhibitor of topoisomerase II and a topoisomerase I poison. In this case, the readout will be similar to a topoisomerase II catalytic inhibitor, as described in **Table 2**, except the top1 deleted strain will also show some resistance. An example of this latter class of drug is the anthracycline aclarubicin *(27)*.

3.5. Application to Human Topoisomerases

Anticancer drugs target human not yeast topoisomerases. Although human topoisomerase II α and β are very similar to yeast topoisomerase II in their amino acid sequence and biochemical properties, the human and yeast enzymes have some differences in their sensitivities to topoisomerase II targeting drugs. The experimental approach described in **Subheading 3.3.1.** can be modified to determine the sensitivity of yeast cells that depend on human topoisomerase II for viability. The plasmid pMJ1 carries human topoisomerase II α under the control of the yeast TOP1 promoter. The plasmid also carries URA3 as a selectable marker. When transformed into JN394t2-4, this plasmid confers the ability to grow at 34°C, a temperature that the untransformed strain cannot grow at. It is important to determine the drug sensitivity at 34°C, so that the measured drug sensitivity is not contributed to by the yeast enzyme. However, once the cells are plated to Ura- medium to determine viable titer, the plates can be incubated at 30°C (*see* **Note 3**).

Notes

1. Determination of drug sensitivity requires the use of a large number of disposable flasks. We have found that 25 cm^2 tissue culture flasks are convenient for growing cultures for testing drug sensitivity. Cultures should be grown in flasks with vented filter caps (e.g., Costar 3056). Alternately, if flasks without vented filter caps are used, the cap should not be tightened, to allow gas exchange.
2. The DMSO concentration should not exceed 2% (v/v).
3. All of the strains and plasmids described in this paper are available from the author.

Acknowledgments

This work was supported by CA52814 and core grant CA21765 from the National Cancer Institute, and the American Lebanese Syrian Associated Charities (ALSAC).

References

1. Froelich-Ammon, S. J. and Osheroff, N. (1995) Topoisomerase poisons: harnessing the dark side of enzyme mechanism. *J. Biol. Chem.* **270,** 21,429–21,432.
2. Nitiss, J. L. and Beck, W. T. (1996) Antitopoisomerase drug action and resistance. *Eur. J. Cancer* **32A,** 958–966.
3. Nitiss, J. L. (1994) Using yeast to study resistance to topoisomerase II-targeting drugs. *Cancer Chemother. Pharmacol.* **34(Suppl),** S6–S13.
4. Nitiss, J. L. (1994) Yeast as a genetic model system for studying topoisomerase inhibitors. *Advances in Pharmacology* **29B,** 201–226.
5. Nitiss, J. L. and Wang, J. C. (1991) Yeast as a genetic system in the dissection of the mechanism of cell killing by topoisomerase-targeting anti-cancer drugs in

DNA Topoisomerases and Cancer (Potmesil, M. and Kohn, K., eds.), Oxford University Press, London, pp. 77–91.

6. Nitiss, J. L. (1998) Investigating the biological functions of DNA topoisomerases in eukaryotic cells. *Biochimica Et Biophysica Acta* **1400,** 63–81.
7. Nitiss, J. L. (1996) Mutational analysis of topoisomerase II drug action: the yeast test tube. *Anticancer Drugs* **7(Suppl 3),** 27–34.
8. Drake, F. H., Hofmann, G. A., Mong, S. M., Bartus, J. O., Hertzberg, R. P., Johnson, R. K., Mattern, M. R., and Mirabelli, C. K. (1989) In vitro and intracellular inhibition of topoisomerase II by the antitumor agent merbarone. *Cancer Res.* **49,** 2578–2583.
9. Ishida, R., Miki, T., Narita, T., Yui, R., Sato, M., Utsumi, K. R., Tanabe, K., and Andoh, T. (1991) Inhibition of intracellular topoisomerase II by antitumor bis(2,6-dioxopiperazine) derivatives: mode of cell growth inhibition distinct from that of cleavable complex-forming type inhibitors. *Cancer Res.* **51,** 4909–4916.
10. Jensen, P. B., Sorensen, B. S., Demant, E. J., Sehested, M., Jensen, P. S., Vindelov, L., Hansen, H. H., Srensen, B. S., and Vindelv, L. (1990) Antagonistic effect of aclarubicin on the cytotoxicity of etoposide and 4'-(9-acridinylamino) methanesulfon-m-anisidide in human small cell lung cancer cell lines and on topoisomerase II-mediated DNA cleavage. *Cancer Res.* **50,** 3311–3316.
11. Ishida, R., Hamatake, M., Wasserman, R. A., Nitiss, J. L., Wang, J. C., and Andoh, T. (1995) DNA topoisomerase II is the molecular target of bisdioxopiperazine derivatives ICRF-159 and ICRF-193 in Saccharomyces cerevisiae. *Cancer Res.* **55,** 2299–2303.
12. Balzi, E. and Goffeau, A. Yeast multidrug resistance: the PDR network. *J. Bioenerg. Biomembr.* **27,** 71–76.
13. Nitiss, J. and Wang, J. C. (1988) DNA topoisomerase-targeting antitumor drugs can be studied in yeast. *Proc. Natl. Acad. Sci. USA* **85,** 7501–7505.
14. Gaber, R. F., Copple, D. M., Kennedy, B. K., Vidal, M., and Bard, M. (1989) The yeast gene ERG6 is required for normal membrane function but is not essential for biosynthesis of the cell-cycle-sparking sterol. *Mol. Cell Biol.* **9,** 3447–3456.
15. Eng, W. K., Faucette, L., Johnson, R. K., and Sternglanz, R. (1988) Evidence that DNA topoisomerase I is necessary for the cytotoxic effects of camptothecin. *Mol. Pharmacol.* **34,** 755–760.
16. Katzmann, D. J., Hallstrom, T. C., Mahe, Y., and Moye-Rowley, W. S. Multiple Pdr1p/Pdr3p binding sites are essential for normal expression of the ATP binding cassette transporter protein-encoding gene PDR5. *J. Biol. Chem.* **271,** 23,049–23,054.
17. Katzmann, D. J., Burnett, P. E., Golin, J., Mahe, Y., and Moye-Rowley, W. S. Transcriptional control of the yeast PDR5 gene by the PDR3 gene product. *Mol. Cell Biol.* **14,** 4653–4661.
18. Wolfger, H., Mahe, Y., Parle-McDermott, A., Delahodde, A., and Kuchler, K. The yeast ATP binding cassette (ABC) protein genes PDR10 and PDR15 are novel targets for the Pdr1 and Pdr3 transcriptional regulators. *FEBS Lett.* **418,** 269–274.

19. Jannatipour, M., Liu, Y. X., and Nitiss, J. L. (1993) The top2-5 mutant of yeast topoisomerase II encodes an enzyme resistant to etoposide and amsacrine. *J. Biol. Chem.* **268,** 18,586–18,592.
20. Nitiss, J. L., Zhou, J., Rose, A., Hsiung, Y., Gale, K. C., and Osheroff, N. (1998) The bis(naphthalimide) DMP-840 causes cytotoxicity by its action against eukaryotic topoisomerase II. *Biochemistry* **37,** 3078–3085.
21. Nitiss, J. L., Liu, Y. X., Harbury, P., Jannatipour, M., Wasserman, R., and Wang, J. C. (1992) Amsacrine and etoposide hypersensitivity of yeast cells overexpressing DNA topoisomerase II. *Cancer Res.* **52,** 4467–4472.
22. Hsiung, Y., Jannatipour, M., Rose, A., McMahon, J., Duncan, D., and Nitiss, J. L. (1996) Functional expression of human topoisomerase II alpha in yeast: mutations at amino acids 450 or 803 of topoisomerase II alpha result in enzymes that can confer resistance to anti-topoisomerase II agents. *Cancer Res.* **56,** 91–99.
23. Wasserman, R. A., Austin, C. A., Fisher, L. M., and Wang, J. C. (1993) Use of yeast in the study of anticancer drugs targeting DNA topoisomerases: expression of a functional recombinant human DNA topoisomerase II alpha in yeast. *Cancer Res.* **53,** 3591–3596.
24. Austin, C. A., Marsh, K. L., Wasserman, R. A., Willmore, E., Sayer, P. J., Wang, J. C., and Fisher, L. M. (1995) Expression, domain structure, and enzymatic properties of an active recombinant human DNA topoisomerase II beta. *J. Biol. Chem.* **270,** 15,739–15,746.
25. Nitiss, J. L., Liu, Y. X., and Hsiung, Y. (1993) A temperature sensitive topoisomerase II allele confers temperature dependent drug resistance on amsacrine and etoposide: a genetic system for determining the targets of topoisomerase II inhibitors. *Cancer Res.* **53,** 89–93.
26. Elsea, S. H., Osheroff, N., and Nitiss, J. L. (1992) Cytotoxicity of quinolones toward eukaryotic cells. Identification of topoisomerase II as the primary cellular target for the quinolone CP- 115,953 in yeast. *J. Biol. Chem.* **267,** 13,150–13,153.
27. Nitiss, J. L., Pourquier, P., and Pommier, Y. (1997) Aclacinomycin A stabilizes topoisomerase I covalent complexes. *Cancer Res.* **57,** 4564–4569.

Index

A

Agarose gel electrophoresis
  DNA degradation during apoptosis, 246, 248
  DNA unwinding test, 151–152
  plasmid DNA supercoiling assay, 30
  topoisomerase I-catalyzed DNA relaxation, 6
  topoisomerase I-mediated DNA nicking, 83
  topoisomerase II-catalyzed DNA decatenation, 16–17
  topoisomerase II-catalyzed plasmid DNA catenation, 278
  topoisomerase II-catalyzed plasmid DNA relaxation, 277
Apoptosis, 241–253
  cell morphology, 242–244, 247–248
  detection of apoptotic cells, 244–249
  DNA degradation, 244, 248
  DNA strand-break labeling, 47, 249
  factional DNA content, 244–249
Archebacterial reverse gyrase activity, 35–48
  ATPase activity, 43
  ATP-dependent DNA relaxation, 39–40
  ATP-dependent positive DNA supercoiling, 40
    stimulation by DNA condensing agents, 40–43
  DNA cleavage activity, 44
  DNA unwinding activity, 44–45
  purification, 39
ATPase activity
  reverse gyrase, reverse gyrase, 43–44
  real-time coupled assay for topoisomerase II-catalyzed ATP hydrolysis, 57–64
    advantages and disadvantages, 58–60
    stimulation by DNA, 61
  thin-layer chromatography method for topoisomerase II-catalyzed AT hydrolysis, 51–55
  advantages and disadvantages, 51–52
  stimulation by DNA, 55

B

Bacterial DNA gyrase, 25,267
  GyrA subunit, 25
  GyrB subunit, 25
  plasmid DNA supercoiling (*see* plasmid DNA supercoiling assay)
  purification, 173
  quinolone interactions, 171–182
Bactericidal assays for fluoroquinolones, 185–193
  killing curve test, 191–192
  minimal bactericidal concentration (MBC) determination, 191
  minimal inhibitory concentration (MIC) test, 185–188, 190–191
5'-Bridging phosphorothioate-containing DNA suicide substrates for topoisomerase I, 119–127
  advantages and disadvantages, 120–121
  solid phase synthesis of oligonucleotides, 125–127

solid phase synthesis of
phosphorothioate, 121–124

C

Camptothecin, 195–198, 217, 291–292, 308
lactone-carboxylate equilibrium, 217–218
Cell-cycle analysis 229–239
DNA content measurements, 30, 235–236
DNA vs cyclin measurements, 230–232, 236
DNA vs DNA synthesis measurements, 233, 236–237
Cesium chloride gradients, 139–143
Coumarin antibacterials, 25
coumermycin, 25
novobiocin, 25
Crithidia fasciculata, 13

D

Daunomycin, 220
DNA binding by topoisomerases
electrophoretic mobility shift assay (EMSA), 63–73
DNA ligand, 66–67
labeling the DNA probe, 69–70
native polyacrylamide gel, 70
filter binding assay 75–80
DNA catenation activity of topoisomerase II, 275–279
agarose gel electrophoresis, 278
DNA cleavage activity of topoisomerases
drug-induced stabilization of covalent DNA-topoisomerase I complexes, 291–300
DNA sequencing, 296–297
DNA substrate, 292–293
polyacrylamide gel electrophoresis, 297–298
reverse gyrase, 44
topoisomerase I-mediated DNA nicking, 81–86
agarose gel electrophoresis, 83
data analysis, 85
topoisomerase II-mediated DNA cleavage, 283–287
agarose gel electrophoresis, 285
uncoupling from DNA religation, 101–117
DNA cleavage complexes
analysis using reactive inhibitor derivatives, 89–97
topoisomerase I-DNA cleavage complexes, 96–97
topoisomerase II-DNA cleavage complexes, 95–96
ICE bioassay, 137–149
cesium chloride gradients, 139–143
immunoblotting, 141, 143–144
isolation of covalent complexes by K^+/SDS method, 129–135
in vitro, 131–132
in cultured cells, 132–135, 212
topoisomerase I-DNA covalent intermediates, 291–292
topoisomerase II-DNA cleavage complexes, 283
trapping using DNA suicide substrates, 101–117
migration in polyacrylamide gels, 105, 110
DNA decatenation activity of topoisomerase II, 13–22
agarose gel electrophoresis, 16–17
centrifugation assay, 17–18
isolation of unlabeled and [^{3}H]kDNA, 15–16
DNA relaxation activity
topoisomerase I, 1–9
agarose gel electrophoresis, 6, 277

distributive vs processive activity, 7–8
relaxed DNA topoisomer distribution, 8
serial dilution method, 3–4
time course assay, 4–5
topoisomerase II, 275–277
DNA religation activity, 101–117
DNA sequencing, 269, 296–297
DNA suicide substrates, 101–117
DNA supercoiling activity
DNA gyrase, 25–32
reverse gyrase, 40
DNA topoisomerase I, 1
DNA relaxation activity (*see* DNA relaxation activity)
human, 195–198
single-strand conformational analysis, 255–263
suicide substrates, 101–117, 119–127
uncoupling DNA cleavage and religation, 101–117
use in DNA unwinding test, 149–159
DNA topoisomerase II, 13
cleavage of plasmid DNA , 283–287
DNA decatenation activity (*see* DNA decatenation activity of topoisomerase II)
Drosophila topoisomerase II (*see* *Drosophila* topoisomerase II)
single-strand conformational analysis, 255–263
suicide substrates, 101–117
uncoupling DNA cleavage and religation, 101–117
DNA unwinding test, 149-159
agarose gel electrophoresis, 151–152
DNA unwinding by quinolones and quinobenoxazines, 154–157
unwinding reactions, 11–152
Drosophila topoisomerase II
ATPase activity, 51–55
DNA cleavage activity, 283
Drug-DNA interactions, 161–168
absorbance titration, 164–165
DNA preparation, 162–163
equilibrium binding parameters, 167–168
equilibrium dialysis, 166–167
ethidium displacement assay, 165–166
fluorescence titration, 165
quinolone-DNA interactions (*see* quinolone antibacterials)
spectral properties of DNA binding ligands, 163–164
Drug-induced cytotoxicity in tissue culture, 205–212
growth curve assay, 208–209
colony formation assay, 209–212
MTT assay, 208
Drug toxicity in *E. coli* cells
fluoroquinolones (*see* bactericidal assays for fluoroquinolones)
drug toxicity in cells expressing human topoisomerase I, 195–202
plasmid pMStopI, 196
agar assay, 20–202
liquid assay, 199–200
Drug uptake-low cytometry, 215–225
drug accumulation in multidrug resistant cell lines, 221–222
effect of human serum albumin on camptothecin uptake, 218
uptake of lactone vs carboxylate form of camptothecin, 218–220

E

Electrophoretic mobility shift assay (EMSA), 65–73
Equilibrium binding, 167–168
Escherichia coli, 185–188, 195–196
Etoposide, 285

F

Flow cytometry (*see* drug uptake-flow cytometry; cell-cycle analysis)
Fluoroquinolones (*see* quinolone antibacterials)

H

HL-60 cells

I

ICE bioassay (isolating in vivo complexes of enzyme to DNA) 137–146

K

Kinetoplast DNA (kDNA), 13
  minicircles and maxicircles, 13

L

Lactate dehydrogenase, 57

M

MBC (*see* bactericidal assays for fluoroquinolones)
Membrane ultrafiltration, 175
MIC (*see* bactericidal assays for fluoroquinolones)
Microfiber glass filters, 76

P

PCR-based cloning of topoisomerase genes, 265–272
Pentapyrimidine sequence, 67
Plasmid pBR322
  isolation of supercoiled pBR322 DNA, 27–28
  preparation of relaxed pBR322 DNA, 29
Plasmid pMStopI, 196–197
Polyacrylamide gel electrophoresis
  resolution of topoisomerase I-DNA cleavage products, 297–298
  single-strand conformational polymorphism analysis (SSCP), 262
  topoisomerase-DNA binding assay, 70
Polyethyleneimine-impregnated cellulose TLC plates, 52
Pyruvate kinase, 57

Q

Quinolone antibacterials, 25, 154, 171, 185
  bactericidal assays (*see* bactericidal assays for fluoroquinolones)DNA unwinding by, 154–155
  ciprofloxacin, 25
  interactions with DNA and gyrase, 171–182
    data analysis, 176–181
    membrane ultrafiltration procedure, 175
    spin-column binding method, 175, 176
      nalidixic acid, 200
      norfloxacin, 25

R

Reactive inhibitor derivatives
  alkylating camptothecin derivative, 91–92
  potoactivatible *m*-AMSA derivative, 89–90
Red blood cells, 218–220
  preparation, 223

S

Saccharomyces cerevisiae (*see* yeast)

Single-strand conformational polymorphism analysis (SSCP), 255-263
advantages, 256–258
efficiency, 258–259
interpretation of results, 259–260
Spin-column, 175–176

T

Type I topoisomerases, 1
archebacterial reverse gyrase, 1, 35
activity, 35–48
helicase-like domain, 35–36
archebacterial topoisomerase III, 1
archebacterial topoisomerase V, 1
eukaryotic topoisomerase I (*see* DNA topoisomerase I), 1, 265
prokaryotic topoisomerase I, 1, 265
prokaryotic topoisomerase III, 1
type I-3', 1
type I-5', 1

U

Uncoupling topoisomerase-mediated DNA cleavage and religation, 101–117
DNA suicide substrates, 102–103, 106–109, 111
topoisomerase I, 104-113
topoisomerase II, 106–111, 114–115

V

Vaccinia topoisomerase, 67

Y

Yeast (*Saccharomyces cerevisiae*)
studying topoisomerase I-targeted drugs in yeast, 303–312
cell viability assays, 310
spot tests, 309
TOP1 expression plasmids, 307
transformation, 309
yeast strains, 306
studying anti-topoisomerase II-targeted, 315–325
complex-stabilizing anti-topoisomerase agents, 315
covalent complex, 315–316
plasmids, 318
Saccharomyces cerevisiae, 315
yeast strains, 316–318

topoisomerase II ATPase activity, 57–64